Manfred von Richthofen's
Royal Prussian Jagdstaffel 11

Volume 1

October 1916 – March 1917

Its History
Its Pilots
Its Aircraft Markings and Colors

Bruno Schmäling

with contributions by Lance J. Bronnenkant, Ph.D.,
Greg VanWyngarden, and Jörn Leckscheid
War diary elaboration and compilation by Winfried Bock

Translated by Lance J. Bronnenkant, Ph.D.

Manfred von Richthofen's
Royal Prussian Jagdstaffel 11

Volume 1

October 1916 – March 1917

Bruno Schmäling

Design and layout: Jack Herris
Cover design: Aaron Weaver
Front cover painting: Russell Smith
Color profiles: Alexsandr Kasakov
Digital photo editing: Jack Herris

Aeronaut Books

Books for Enthusiasts by Enthusiasts
www.aeronautbooks.com

Publisher's Cataloging-in-Publication data

Schmäling, Bruno
Royal Prussian Jagdstaffel 11 / by Bruno Schmäling, translated by Lance J. Bronnenkant, Ph.D.
p. cm.
ISBN 978-1-964637-05-1
1. World War, 1914–1918 -- Aerial operations, German. 2. Fighter pilots -- Germany. 3. Aeronautics, Military -- Germany -- History. II. Title.

ND237 .S6322 2011
759.13 --dc22 2011904920

This book is dedicated to the members of Jagdstaffel 11, especially: Carl Allmenröder, Eduard Lübbert, Karl-Emil Schäfer, and Kurt Wolff, whose notes, letters, and diaries made key portions of this book possible.

Above: Lt. Carl Allmenröder

Above: Lt. Eddy Lübbert (L. Bronnenkant)

Above: Lt. Emil Schäfer

Above: Oblt. Kurt Wolff

The Author

Bruno Schmäling has dealt intensively with the history of the German air service in the 1st World War for many years. In the 1970s he interviewed 70 former members of the German Jagdstaffelen of the 1st World War. He is the primary author of *Royal Prussian Jagdstaffel 30, Royal Bavarian Jagdstaffel 23, Jasta Colors* (Volumes I & II), and *Royal Bavarian Jagdstaffel* 76.

Collaborators

A book such as this is never the result of the authors' efforts alone, but rather always represents the teamwork of a number of individuals. For this reason, our thanks are due to the generous and dedicated collaboration of the following:

The late Alex Imrie & the late Dr. Gustav Bock, and
Rainer Absmeier
Winfried Bock
Lance Bronnenkant PhD
Colin Huston
Reinhard Kastner
Peter Kilduff
Colin Owers
The late Les Rogers
Michael Schmeelke
Dr. Hannes Täger
Oskar Joachim Tinnerholm
Greg VanWyngarden
Aaron Weaver
Tobias Weber
Reinhard Zankl

Further Acknowledgements

The author owes a debt of gratitude to those former pilots of the First World War who shared their knowledge and experiences through interviews and correspondence. They include:

Karl-August von Schoenebeck, Jagdstaffel 11, 59, Staffelführer Jagdstaffel 33.
Hans-Georg von der Osten, Jagdstaffel 11 and 4.
Hermann Lohmeyer, mechanic of Jagdstaffel 11
Wilhelm-Gisbert Groos, Jagdstaffel 11
Xaver Leinemüller, mechanic of Jagdstaffel 10
Alois Heldmann, Jagdstaffel 10
Kurt Delang, Jagdstaffel 10 and 54
Franz Hemer, Jagdstaffel 6
Carl Bodenschatz, Adjutant Jagdgeschwader I
Otto Fuchs, Jagdstaffel 11, 30, Staffelführer Jagdstaffel 77 and Jagdstaffel 35.
Walter Böning, Jagdstaffel 19, Staffelführer Jagdstaffel 76
As well as:
Georg Simon, Jagdstaffel 11, interviewed by Alex Imrie.

A number of individuals closely related to the personalities who are the subject of this book also deserve my grateful acknowledgment of their contributions. These would include the late Frau Helene Allmenröder (fiancée of Carl Allmendröder and wife of Wilhelm Allmenröder), Frau Elisabeth Papendiek (sister of Otto Brauneck), Eddy Lübbert (son of Friedrich-Wilhelm Lübbert, nephew of Eddy Lübbert, and cousin of Carlos Mayer-Baldo), the Döhle family (relatives of Karl-Emil Schäfer), Freifrau von Döring (widow of Kurt-Bertram von Döring), Kai Frhr. von Döring (son of Kurt-Bertram von Döring). I am also thankful to Oberstlt. Reinhard Schröder and OStFw. a.D. Falk Hallensleben of "Taktisches Luftwaffengeschwader 71 "Richthofen".

A very special thanks goes as well as to Trevor Henshaw who has been very helpful tracking down British Combat Reports, Casualty Reports, and Repatriated Prisoner Statements from the AIR1 Files at The National Archives in London. It should be noted that the identification of British pilots shot down or forced to land are based on his book *The Sky their Battlefield II* (2nd edition).

I owe a very special thanks to Frau Dr. Haggenmüller, the director of Department 4 of the Bayerisches Hauptstaatsarchiv (Bavarian Main State Archives) in Munich, and her coworkers for their dedicated support during the research for this book.

A special thanks also goes to German aviation historians of the first generation: Erich Tornuss and Ludwig Zacharias.

I would also like to thank my American friends and colleagues Lance Bronnenkant and Greg VanWyngarden who generously made their knowledge available to me and Lance for his congenial translation. Another special thanks go to Russell Smith for the great cover painting, Alexandr Kasakov for the congenital profiles, and Tom la Padula for the interior painting on page 154.

Last but very far from least my heartful thanks goes to my wife Kerstin for all her understanding and support for my passion. She is an author of historic novels, and I learned a lot from her about how to write books.

Unless otherwise noted, all photographs originate from the author's collection.

Bruno Schmäling

Front Cover: ***Flying Colors*** **by Russ Smith showing a morning patrol of Jagdstaffel 11 over the Scarpe River; commissioned specifically for this volume. The Albatros D.III fighters of Lt. Eddy Lübbert (foreground), Manfred von Ricthtofen (red Albatros in middle), and Oblt. Kurt Wolff in background on patrol over the front.**

Back Cover: **Portrait of Manfred von Richthofen owned by Taktisches Luftwaffengeschwader 71 "Richthofen".**

Table of Contents

Foreword

This is another book in my series about the German fighter squadrons of World War I. However, the idea of writing a book about the history of Jagdstaffel 11 originated many years ago. In the 1970s, Dr. Gustav Bock and I searched intensively for documents concerning Jagdstaffel 11 in the Bavarian Main State Archives and discovered a whole series of previously unpublished records. Dr. Bock was my mentor during this collaboration and showed me how such records should be compiled and expertly evaluated – work that my friend and colleague Winfried Bock continued.

The letters and notes of pilots Eddy Lübbert, Carl Allmenröder, and Karl-Emil Schäfer, as well as the personal diary of Kurt Wolff and the letters he wrote to his family members were particularly valuable for compiling this work. The wealth of information and photographs that became available made me realize that there was too much for a single volume. So I decided to divide the history of Jagdstaffel 11 into multiple volumes to avoid any omissions. For his help in this effort, my sincere thanks go to my friend and publisher, Jack Herris.

As in my previous Jagdstaffel histories, copies of Jagdstaffel 11's war diary that were made by Erich Tornuss, prepared by Dr. Gustav Bock, and revised and supplemented by Winfried Bock have served as the basis for this work. German 6th Army documents generously made available to us by the Bayerisches Hauptstaatsarchiv were also of great importance, as were the other official and semi-official documents included in this work. Also like my prior histories, special attention has been paid to the colors and markings of the unit's aircraft. Those that are presented herein are primarily based on the oral and written testimony of the unit's former members. It is expressly noted wherever this is not the case and extrapolations or interpretations are offered instead.

Historical research is always in a state of flux and as such, this work does not claim to be the final words on the subject. Accordingly, the author welcomes any new information and corrections.

A few readers have criticized the quality of some of the photos in previous publications. In this context, I would like to point out that the vast majority of the photos come from private photo collections. The quality of these photos varies greatly. Some photos are only 6 x 4.5 cm done with low quality cameras. As the author, I have decided to include even photos of poor quality in order to present as comprehensive a picture of the events as possible.

Bruno Schmäling
June 2024

Establishment and First Frontline Flights of Jagdstaffel 11, October 1916

Above: Oblt. Boelcke visiting Jagdstaffel 11 in October 1916. From left: Hptm. Oswald Boelcke, (unknown), Lt. Albert Oesterreicher. Two Fokker E IIIs of the Kampf-Einsitzer-Kommando III are still in the aircraft shed.

During the first days of October 1916, the usual hustle and bustle prevailed at La Brayelle airdrome, just west of Douai in the German 6th Army's sector. Two-seat Feldflieger-Abteilungen (field aviation units) aircraft stationed there took off for and returned from their missions, and machines were pushed into their aircraft hangars and worked on by mechanics. The same was true for the single-seat Fokker monoplane fighters stationed there as well, yet these were days that were somewhat special for the pilots of Kampfeinsitzer-Kommando (Single-Seat Fighter Command) III. Whenever they got together, they usually spoke of two things: the visit by Pour le Mérite airman Hptm. Oswald Boelcke, Staffelführer of Jagdstaffel 2, stationed at the German 1st Army located south of the 6th Army, and the deployment of the new Jagdstaffeln in the German 6th Army region.

The air combat situation at the front, and particularly the Somme Offensive which had been initiated by the Allies on 1 July 1916, showed that the Fokker pilots concentrated into Kampfeinsitzer-Kommandos were fighting a losing battle against the new types of fighter aircraft fielded by the Allies, which were flying and fighting in close formation.[1] The air superiority that the Allies enjoyed at the beginning of the Somme offensive had been wrested from them only by the hasty assembly and deployment of the first German Jagdstaffeln. Equipped with their new Halberstadt D-type fighters, which were closely followed by the Fokker biplane fighters and the new Albatros D-type fighters in August 1916, these new units demonstrated the military value of such "Spezialeinheiten" (special units). This led to the conclusion that more Jagdstaffeln needed to be established as soon as the supply of the new fighter aircraft types allowed their formation.

The Chief of the General Staff of the German 6th Army issued an order on 21 September 1916 that Jagdstaffel "Linck" – named after their Staffelführer, Oblt. Ludwig Linck - were to be established from the Prussian Fokker-pilots of the 6th Army.[2] It was subsequently re-designated Jasta 10 on 6 October when it was still based at Phalempin, until it was reassigned to the 5th Army on 26 October and

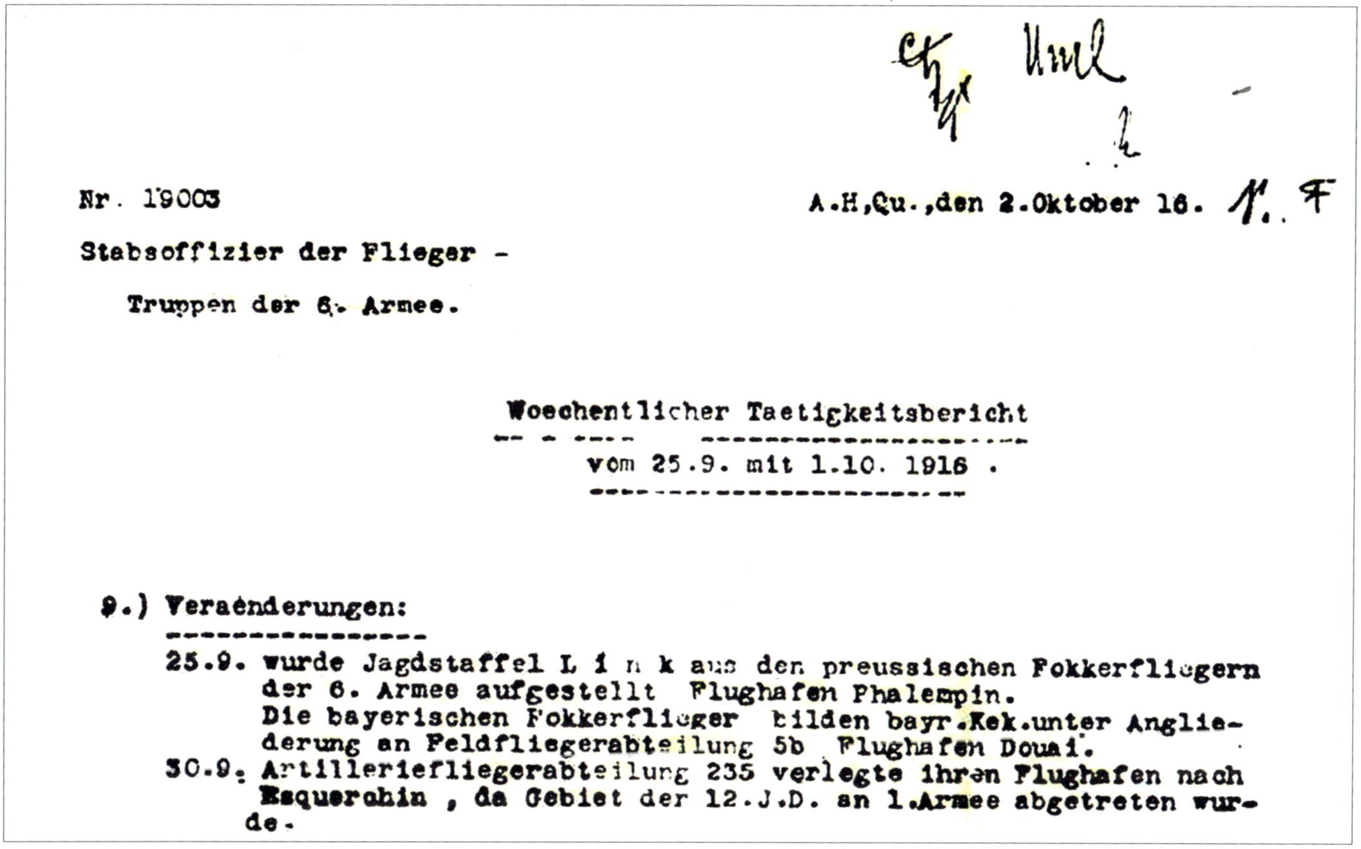

Nr. 19003 A.H.Qu.,den 2.Oktober 16.

Stabsoffizier der Flieger -

Truppen der 6. Armee.

Woechentlicher Taetigkeitsbericht

vom 25.9. mit 1.10. 1916 .

9.) Veraenderungen:

25.9. wurde Jagdstaffel L i n k aus den preussischen Fokkerfliegern der 6. Armee aufgestellt Flughafen Phalempin.
Die bayerischen Fokkerflieger bilden bayr.Kek.unter Angliederung an Feldfliegerabteilung 5b Flughafen Douai.

30.9. Artilleriefliegerabteilung 235 verlegte ihren Flughafen nach Esquerchin , da Gebiet der 12.J.D. an 1.Armee abgetreten wurde.

Above: The 6th Army Aviation Staff Officer's report of activities from 2 October 1916 with the Order: *"25.9. Jagdstaffel Link was formed from the Prussian Fokkerflieger of the 6th Army, Phalempin airfield. The Bavarian Fokkerflieger form Bavarian a Kampfeinsitzer-Kommando (KEK) under affiliation with Feldfliegerabteilung 5b, Douai airfield."* (Bayerisches Hauptstaatsarchiv, Abt. 4, München)

moved to Jametz, also in the area of the 5th Army, just two days later.[3]

On 25 September 1916, the order was issued that the Bavarian Fokker-pilots of the German 6th Army were to form a Bavarian "Kampfeinsitzer-Kommando" affiliated to Bavarian Feldflieger-Abteilung 5 at the airfield of Douai.[4]

Just three days later, on 28 September, yet another newly issued directive ordered the immediate establishment of another fighter unit, to be designated Jagdstaffel 11, at La Brayelle airfield. The nucleus of this new Jagdstaffel was to be provided by the Bavarian Fokker-pilots of the Kampfeinsitzer-Kommando III affiliated to Feldflieger-Abteilung 5.[5] One week later, on 5 October 1916, Jasta 11 was officially established in accordance with the A.O.K. (Army Command) 6's order of the day.[6]

At first, Jagdstaffel 11 primarily had only old Fokker E.III and E.IV monoplanes at its disposal to fulfill its flying duties at the front. The first day of mobilization of the new unit was 11 October 1916. On this date, the remaining flying members of KEK III, along with their groundcrew, were transferred to Jagdstaffel 11.[7] Also on this very day, the unit received an example of one of the new biplane fighters, when Albatros D.I D.425/16 was taken on charge. As it would turn out later, it was to remain the only example of this type of aircraft to fly with Jagdstaffel 11.[8]

The Kampfeinsitzer-Kommando III pilots who had formerly reported to Royal Bavarian Feldflieger-Abteilung 5 and were now members of Royal Prussian Jagdstaffel 11 were Lt. Albert Oesterreicher, Gefr. Sebastian Festner, Lt. Konstantin Krefft and Lt. Karl Kleinhenz. The latter would also serve as the unit's Offizier zur besonderen Verwendung (Officer for Special Duties), who was responsible for all administrative matters, including the supervision of its ground personnel. On the same day, Uffz. Erwin Howe joined them, arriving from Prussian Feldflieger-Abteilung 24. Vzfw. Alfred Prehn from Jagdstaffel 10 also joined the unit.

Albert Oesterreicher was born at Mühlen am Neckar, Oberamt (head office) Horb, in Württemberg. He earned Pilot's License No. 515 on 15 September 1913 as a civilian at Munich-Oberwiesenfeld, and then joined the Royal Bavarian Kraftfahr- und Fliegerbataillon (Motor Vehicle and Aviation Battalion) at Schleissheim as a one-year volunteer on 8 January 1914. On 1 June 1914, he was assigned to its Flieger-Kompanie (Aviation Company) and became an Unteroffizier on 1 August. Following his training as a military pilot, he was assigned to Bavarian Feldflieger-Abteilung 5 on 15 October 1914 and was promoted to Vizefeldwebel on 7 January 1915, and again to Offizier-Stellvertreter on 28 January.

Nr. 19228.　　　　　　　　　　　　　　　　A.H.Qu.,den 9.10.16.

Stabsoffizier der Flieger -

Truppen der 6. Armee.

Woechentlicher Taetigkeitsbericht

vom 2.10 mit 8.10.16.

1.) Feindl. Fliegertaetigkeit:

Wegen andauernd unguenstigen Wetters aeusserst schwach . 3 Geschwader mit 18 Flugzeugen (Vorwoche 55 Geschwader mit 328 Flugzeugen) Einige Flieger haben den Eindruck,dass gegnerische Flieger weggezogen worden sind.

Bomben wurden abgeworfen von:
1 Geschwader zu 5 Flugzeugen.......................................18
10 Einzelflieger...18
36

(Vorwoche 395)

Fliegerschiessen:
hat der Gegner durchgefuehrt am 5.10.16 gegen Fort Premesques bei Lille.

2.) Veraenderungen:

Am 2.10.16 wurden auf Befehl der Heeresgruppe von Bayern Ic Nr.662 vom 1.10.16 4 Artillerieflugzeuge mit Besatzung und Monteuren zur 2.Armee abgegeben:
2 Flugzeuge von Artilleriefliegerabteilung 201
1 Flugzeug " " " " 227
1 " " " " " 235

Am 5.10.16 wurde auf Befehl des A.O.K. 6 Nr. 70267 v.5.10.16 die Feldfliegerabteilung 13 mit Parkzug zur 1.Armee abgegeben.
Die Jagdstaffel 10 in Phalempin wurde etatsmaessig.
Die Aufstellung der Jagdstaffel 11 in Douai wurde befohlen.
Am 6.10.16 wurde auf Befehl der Heeresgruppe die Feldfl.Abtlg.24 und Artilleriefliegerabteilung 201 zur 1.Armee abgegeben.

Above: The 6th Army Aviation Staff Officer's report of activities from 10 October 1916 with the order for the establishment of Jagdstaffel 11 under point 2: *"The deployment of Jagdstaffel 11 at Douai has been ordered."* (Bayerisches Hauptstaatsarchiv, Abt. 4, München)

He trained on a 'Parasol' high-wing monoplane (likely a product of the "Pfalz" company) at the Flugzeug-Hilfsdepot (Aircraft Auxiliary Depot) at Speyer from 11 February to 17 April 1915, after which he was elevated to Leutnant der Reserve on 28 September 1915. He was a member of Feldflieger-Abteilung 62's Fokker contingent in November 1915 and later served in Kampfeinsitzer-Kommando III, where he participated in bringing down an airplane near Dury on 21 July 1916, followed by a victory of his own on 9 August 1916.(9) When he joined Jagdstaffel 11, he was an experienced aviator and combat pilot and had already earned the Iron Cross, 2nd Class and Bavaria's Military Merit Cross, 2nd Class with Swords.(10)

Sebastian Festner was born on 30 June 1894 in Holzkirchen, Lower Bavaria. The family moved later to Chiemsee Lake, where Sebastian worked alongside his father as a gardener at Schloss Ising, owned by the wealthy Czermak family.(11) The head of the family, Leo Czermak, owned several automobile agencies in Bavaria and gave young Sebastian, who had a strong interest in technology, the opportunity to train as an auto mechanic. He was working in their motor vehicle business when the war broke out and as a result was drafted into the Bavarian Fliegertruppe (Aviation Troops) as a mechanic.

On 5 October 1914, he was sent to Bavarian Flieger-Ersatz-Abteilung 1 at Schleißheim to train as an aircraft

Above: Formal portrait of Lt. Albert Oesterreicher. (L. Bronnenkant)

Above: Lt. Albert Oesterreicher recovering in the fall sunshine between frontline sorties in 1916.

attendant. He went to Bavarian Feldflieger-Abteilung 7 in that capacity on 15 January 1915, where he was promoted to Gefreiter. On 11 February 1916, he was sent to Feldflieger-Abteilung 19 as the personal mechanic of Vzfw. Diekmannshemke. He applied for pilot training and was transferred back to Bavarian Flieger-Ersatz-Abteilung 1 for this purpose, arriving there on 24 April 1916. Following several months of training, he went to Armeeflugpark (Army Aviation Supply Depot) 6 to 30 June, and from there to Bavarian Feldflieger-Abteilung 2 on 5 July 1916. He was sent to Feldflieger-Abteilung 18 as a Kampfeinsitzerflieger (single-seat fighter pilot) on 10 September 1916 and then to Bavarian Feldflieger-Abteilung 5's Kampfeinsitzer-Kommando III on 25 September.(12)

Konstantin Krefft was born in Ölde, Westphalia on 18 January 1892, but the family had moved to Nürnberg by the time the war broke out. Konstantin Krefft volunteered for the Bavarian Army, and as an engineering student before the war, applied for service with the Fliegertruppe in September 1914. His request was granted on 8 October 1914, and he was ordered to Bavarian Flieger-Ersatz-Abteilung 1 at

Above: Vzfw. Sebastian Festner as member of Jagdstaffel 11

Above: Lt. Konstantin Krefft (L. Bronnenkant)

Above: Konstantin Krefft as a pilot of the Kampfeinsitzer-Kommando III in Douai in front of his Fokker in September 1916.

Schleißheim for pilot training. Following his promotion to Unteroffizier on 6 February 1915, he was assigned to Bavarian Feldflieger-Abteilung 5 on 26 February 1915. He was awarded the Iron Cross, 2nd Class on 12 April 1915 as a result of several successful flights at the front and was then promoted to Vizefeldwebel on 8 May.(13) On 17 June, he was granted Bavaria's Military Merit Cross, 2nd Class with Swords. Following his promotion to Offizier-Stellvertreter on 28 September 1915, he was raised to officer status as a Leutnant der Reserve on 28 November 1915. On 5 February 1916 he was commanded to Feldflieger-Abteilung 18 and two days later to the Kampfeinsitzer-Kommando Krug.(14) His stay was short, as on 21 February 1916 he was ordered to Bavarian Flieger-Ersatz-Abteilung 1 in Schleißheim as a flight instructor. He was back at the front on 12 April 1916 with Bavarian Feldflieger-Abteilung 5's Kampfeinsitzer-Kommando III and on 25 May 1916 he was posted to Bavarian Feldflieger-Abteilung 2. On 25 August 1916 he was a back at Kampfeinsitzer-Kommando III. After a short intermezzo with Feldflieger-Abteilung 18, he was with Bavarian Feldflieger-Abteilung 5's Kampfeinsitzer-Kommando III again on 19 September and then at Jasta 11 on 11 October.(15)

The many transfers of Lt. Konstantin Krefft are a good example of how the few single-seater fighter pilots were shifted back and forth between the individual Feldflieger-Abteilungen, depending on the military situation.

Carl-August von Schoenebeck, who served with Jagdstaffel 11 from July 1917 to January 1918, described Krefft as a "typical Westphalian:" a bit gruff but always dependable, straightforward, honest, open, and clear in his statements and decisions. He was definitely somebody one you could rely upon. Manfred von Richthofen appreciated his calm, dependable, and civil nature even though he was an excellent flyer, but he was not a good fighter pilot. It was his technical expertise and talent that were of great value to the Jagdstaffel and later to the Jagdgeschwader, as emphasized in von Richthofen's assessment of him dated 16 January 1918:

"Lt.d.R. Krefft is a tireless pilot who controls his airplane brilliantly. He is excellently trained in technical matters and is a highly valued worker as a technical officer. Lt. Krefft is socially adept and well-educated."(16)

In addition, Lt. Konstantin Krefft had a respectful relationship with the Staffel's mechanics. One of his pieces of advice for new pilots was:

"Get on well with your mechanics, because your life depends on their conscientious work."(17)

It was also said that Lt. Krefft despised arrogance and would call out every pilot who set himself up as someone "better" than his mechanics.(18)

Karl Kleinhenz came from Ludwigshafen in the Palatinate, where he was born on 28 August 1892. At the start of the war, he went into the field with Königlich Bayerisches 8. Infanterie-Regiment 'Großherzog Friedrich II. von Baden'. He joined the Fliegertruppe by the end of 1915 and was sent to Bavarian Flieger-Ersatz-Abteilung 1 at Schleißheim for pilot training on 8 January 1916. On 6 May 1916, he was sent to Bavarian Feldflieger-Abteilung 8 via Armeeflugpark Gaede. After a transfer to Bavarian Flieger-Ersatz-Abteilung 1, he came to Bavarian Feldflieger-Abteilung 2 and was assigned to its Kampfeinsitzer-Kommando II on 7 August 1916. A transfer to Bavarian Feldflieger-Abteilung 5`s Kampfeinsitzer-Kommando III at Douai followed on 24 September 1916.[19]

Erwin Howe was born on 23 March 1894 in Riesenburg, Kreis Rosenberg, in West Prussia, but was living in Berlin by the time the war broke out. On 8 December 1914, he reported to Flieger-Ersatz-Abteilung 1 in Döberitz as a Kriegsfreiwilliger (war volunteer). He was subsequently sent via Armeeflugpark 6 to Artillerieflieger-Abteilung (Artillery Aviation Unit) 201 on 17 September 1915. He returned to Armeeflugpark 6 on 2 December 1915 to train as a Fokker pilot. He was ordered in this capacity to Feldflieger-Abteilung 5 on 22 January 1916 and then to Feldflieger-Abteilung 24 the next day. Thus, he had eight months of frontline experience before he joined Jagdstaffel 11.[20]

Finally, Vzfw. Alfred Prehn arrived from Jasta 10. He was born on 20 April 1895 at Pritzwalk, in the Prussian province of Brandenburg. By 10 August 1914, he was a ground crewman with Flieger-Ersatz-Abteilung 2 at Adlershof and was posted as ground crew man to Feldflieger-Abteilung 33 on 10 October 1914. He volunteered to be trained as a pilot and on 15 May 1915 he was transferred to the Flieger-Ersatz-Abteilung 4 at Posen. He continued his pilot training at Flieger-Ersatz-Abteilung 8 at Graudenz from 21 August 1915 to 30 October 1915. Soon after, on 5 November 1915, he was assigned to Feldflieger-Abteilung 62 and then to the unit's Kampfeinsitzer-Kommando III on 5 May 1916. He moved to Feldflieger-Abteilung 24 as a single-seat fighter pilot on 12 August 1916 and was then transferred to Jasta "Linck" on 25 September before its designation was changed to Jasta 10 on 6 October. A few days later, on 11 October 1916, he was transferred to Jagdstaffel 11. However, his stay with Jagdstaffel 11 was to be fairly short. He was commanded to Seeflugstation (Naval Air Station) Tondern just four weeks later.[21]

On 14 October, another experienced airman joined Jasta 11. Lt. Franz Anslinger, who was born on 1 May 1887 in Freiburg im Breisgau. After entering the Prussian Fliegertruppe, he trained as a pilot from June through December 1915 with Flieger-Ersatz-Abteilung 9 and Fliegerschule (Aviation School) Freiburg. On 21 December 1915, Franz Anslinger was assigned to Kampfgeschwader der Obersten Heeresleitung (Combat Squadron of the Army High Command) III's Kampfstaffel 15, where he was awarded his Pilot's Badge on 28 February 1916.[22] Shortly thereafter, on 10 March, he was reassigned to Kampfstaffel 14 of Kagohl III and promoted to Leutnant on 25 August 1916. He then made his way to Jagdstaffel 11 via Armeeflugpark 6 on 14 October.

Above: Oblt. Rudolf Emil Lang, the first Staffelführer of Jagdstaffel 11. (R. Kastner)

Six days later, on 20 October 1916, Oblt. Rudolf Lang was appointed as Jasta 11's Staffelführer (Staffel leader or Commanding Officer) in accordance with Feldflugchef Nr. 53826. Born on 16 May 1889 in Munich, Rudolf Lang pursued a military career and entered Königlich Bayerisches 12th Infanterie-Regiment 'Prinz Arnulf' as a Fahnenjunker (junior officer candidate) on 14 July 1908. It took him more than five years to become a Leutnant (on 23 October 1913), and by early 1914, recognizing the airplane's military potential, he applied for entry in the Fliegertruppe. His application was approved, and he was sent to Bavarian Flieger-Ersatz-Abteilung 1 in Schleißheim as a student pilot from 1 April through 23 September 1914. He was placed in charge of Militär-Fliegerschule I at Schleißheim on 25 October 1914 but was then sent into the field with Bavarian Feldflieger-Abteilung 6 on 12 December 1914. Promoted to Oberleutnant on 1 June 1915, he served with Feldflieger-Abteilung 54 for a little more than a month from 1 August 1915 to 10 September 1915 before going to Bavarian Feldflieger-Abteilung 9 on 1 November 1915. Just one month later, he joined Bavarian Feldflieger-Abteilung

Lt. Konstantin Krefft, personal record OP 11364, Bayerisches Hauptstaatsarchiv, Abt. 4, München

№

Personalbogen

des Leutnant d. Res. Konstantin Krefft.

Geboren	wann (Jahr, Tag, Monat.)	1	1892. 18. Januar.
	wo (Ort, Bezirksamt, Regierungsbezirk, Bundesstaat.)	2	Olden i/W. Preußen.
Größe		3	...... m ...... cm ...... mm
Religion		4	Kath.
Stand der eigenen Familie	Frau	5	—
	Söhne	6	—
	Töchter	7	—
	Stiefkinder	8	—
Eltern	Vorname und Stand des Vaters	9	Julius, Lokomotivführer.
	Geburtsname der Mutter	10	Anna, geb. Müller, Nürnberg
Stiefeltern		11	—
Studien vor dem Eintritt		12	cand. ing. Nürnberg.
Zivilverhältnis und Wohnsitz		13	
Hervorstechendes körperliches Geschick		14	—
Hervorstechende Talente und Kenntnisse		15	—
Bestrafungen		16	—

17				18
Zugang, Ernennungen, Beförderungen, Versetzungen 2c. 2c.				Orden und Ehrenzeichen: a. vaterländische, b. fremde
Jahr	Tag	Monat		
1914	8.	Oktober	Kriegsfreiw. für Schleißheim.	M. V. Kr. 2 b.
1915	6	Februar	Unteroffizier. befördert.	17. 6. 1915.
1915	26.	Februar	z. F. Flg. Abt. 5. versetzt.	Flugzeugf. Abz.
1915	8.	März	Vizefeldwebel ernannt.	Jan. 1915.
1915	28.	September	Offz. Stellv.^tr gen. Tapferkeit v. d. Feind.	E. K. II. Kl.
1915	5.	November	z. Flg. Abt. 5 Kdt.	9. 4. 1915.
1915	28.	November	z. Ltn. d. Res. d. Flg. Truppen.	E. K. I. Kl.
1916	5.	Februar	z. Flg. Abt. 18 Kdt.	5. 1916.
1916	8.	Februar	z. K. E. Kdo. Krag Kdt.	M. V. Kr. 4 b.
1916	21.	Februar	z. Ers. Schleißheim versetzt.	31. 5. 17.
1916	12.	April	z. bay. F. Flg. Abt. 5 vers. bis auf weiteres z. K. E. Kdo. II Kdt. verbleibt zunächst b. Fokkerstaffel Buzancy	M. V. O. 4. Kl. 28. 5. 17.
				b.
1916	25.	März	z. b. Flg. Abt. 2 vers.	
1916	25.	August	z. A. Fl. P. 6 vers. u. z. K. e. K. II. d. F. Flg. Abt. 18. Kdt.	
1916	19.	September	z. F. Flg. Abt. 5. b. vers. u. z. Kek III. Kdt.	
1916	11.	Oktober	z. Jagdst. 11 versetzt.	
1917	25.	Juni	als techn. Offz. z. [illegible] I. Kdt. Gen. Kriegsmin. Nr. 73073 v. 7. 11. 18. d. Idflieg. bis z. 1. 5. 18. z. Verfügung gestellt.	
1918	6.	August	z. Ers. 2. Fürth vers. (Kriegsm. Verf. 45549 F III. v. 9. 8. u. z. [illegible] Speyer Kdt. (gem. Kriegsmin. v. 6. 8. 18. Nr. 179535 Abt. II.)	
1919	1.	Januar	z. Ers. 1 Schleißheim vers.	
1918 (8 written over 9)	22.	Dezember	aus dem akt. Militärdienst ohne Versorgung entlassen z. Bez. Kommando Nürnberg.	

Above: Lt. Konstantin Krefft, personal record OP 11364, Bayerisches Hauptstaatsarchiv, Abt. 4, München

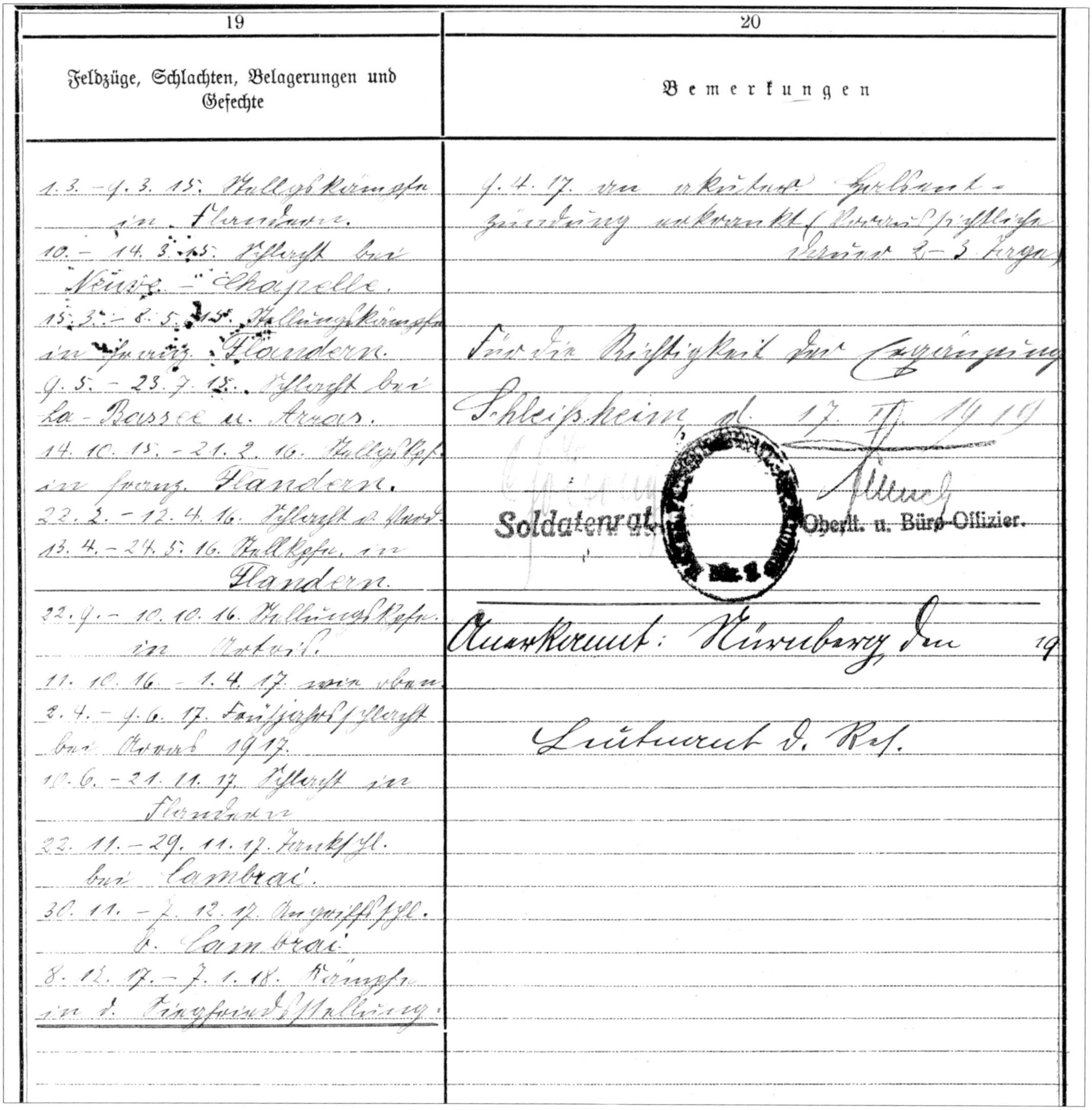

Above: Lt. Konstantin Krefft, personal record OP 11364, Bayerisches Hauptstaatsarchiv, Abt. 4, München

6 as an Kampfeinsitzer-pilot, and by 1 May 1916 was flying a Kampfeindecker with Bavarian Artillerieflieger-Abteilung 103. Shortly after Bavarian Kampfgeschwader VI's Kampfstaffel 31 was formed, he was appointed its Staffelführer on 15 July 1916 – a post he held until he became Staffelführer of Jasta 11 on 20 October 1916. He therefore had considerable experience in both the capacity of fighter piloting and serving as a Staffelführer.[24]

The Staffel's airfield at La Brayelle was located on even ground just south of Cuincy, about two kilometers west of the gates of the city of Douai.[25] It was a spacious, well-developed airfield that had been established by the French some seven years earlier. Included among its facilities were large wooden hangars, well-equipped workshops and storage depots, fuel tanks, a shooting range, well-constructed shelters, and even a lighting system for nighttime takeoffs and landings. Ground personnel were housed in clean, spacious barracks, while the pilots were quartered in houses in Douai. Special barracks were also provided close by the airfield for those pilots who were tasked with early morning patrols.

Above: Lt. Karl Kleinhenz in front of his Halberstadt D V, November 1916.

Above: Konstantin Krefft.

Several other Abteilungen shared this site with Jagdstaffel 11, and it was therefore busy all day long. The Staffel was stationed in the south-west corner of the airfield, close to "La Blancherie Ferme" (Blancherie = laundry and bleachery).

Jasta 11's operational area was in front of the German 6th Army's lines that in the autumn of 1916 ran about 60 kilometers from Armentières in the north to Quéant about 7 km north of Bapaume. Their flight time to the front was a mere 10 minutes, and in addition to the city of Douai, Jasta 11's pilots could see several major landmarks from the air to aid in their orientation of the general area: the Douai—Arras and Ostricourt—Hénin-Liétard—Lens railway lines, the Scarpe River, the Canal de la Haute Deûle, and the straight Roman Road from Cambrai to Bapaume. Located to the north of the German 6th Army was the German 4th Army, and to the south it was joined by the German 1st Army.

In the autumn of 1916, the 6th Army's sector of the front was relatively quiet. Soldiers on both sides had settled into well-developed positions and there were few major clashes. Accordingly, flying activity in the air above them was rather low.

To understand the military situation in the air it should be noted that at that time the opposing sides were following different strategies with regard to the use of their fighter units. The British had adopted an offensive strategy and ordered their fighter pilots to invade German airspace and force German aircraft to fight within it whenever possible. This strategy, which was only possible due to the numerical superiority of the Allied air units, placed the German aviation troops facing British sectors under constant pressure, but also led to significant losses of men and materiel among the British. Whereas German crews could at any time make an emergency landing in their own territory in the event of injury, technical failure or damage to their aircraft, British airmen facing a similar dilemma would inevitably be captured. German crews could also escape an unfavorable situation in aerial combat by spinning down. Experience had shown that British aircraft would not generally pursue them below an altitude of 1500 meters altitude in order to avoid coming under fire from German anti-aircraft guns. The majority of "out of control" aircraft claimed by the British as kills were in fact not shot down, rather, they had only spun down out of sight of their opponent.

In contrast, the Germans were pursuing a more conservative strategy which put less strain on their limited resources. According to the Kommandierender General der Luftstreitkräfte (Commanding General of the Air Forces), Gen. Lt. Ernst von Hoeppner, the German Fliegertruppe on average was outnumbered three to one, and the numerical disadvantage of such proportions made the adoption of an offensive strategy impossible.

In the summer of 1916, German flying units, as ordered by Army Headquarters, were conducting so-called 'Luftsperre,' or "air barrier (flights)", which consisted of airplanes flying along the front at fixed intervals to try to block any enemy intrusions into their airspace.

General von Hoeppner wrote about this strategy:

"One could understand how our infantry and artillery wanted to see themselves freed of enemy aircraft and believed this could be achieved by having their own aircraft flying up and down the front. Enemy air activity could not be eliminated by this.

Above: Vzfw. Alfred Prehn together with his two mechanics in front of his Fokker E III as single-seat-fighter pilot of Kampfeinsitzer-Kommando III. (R. Kastner)

The speed with which an enemy airplane could evade, the ability to change altitude without disturbing its observation work, the difficulty in recognizing an enemy in the air from afar – all of these resulted in this type of Luftsperre achieving nothing other than a tremendous squandering of resources to the detriment of our own reconnaissance opportunities."(26)

General von Hoeppner concluded that the"'Luftsperre", which spread quickly across the entire Western Front, *"had the most disastrous influence on the way our airmen were employed."* But in the autumn of 1916, before this was known, pressure from higher authorities forced Hptm. Maximilian Sorg, Stabsoffizier der Flieger (Stofl, or Aviation Staff Officer), the consultant on aviation matters for the 6th Army, to deploy Jasta 11 mainly in conducting 'Luftsperre' along the front.

Oswald Boelcke and other successful combat pilots took the different view that the use of one's own fighter aircraft should focus on responding to enemy air activity. Incoming enemy units should be immediately reported from the front lines, whereupon the fighter units would ascend to attack them. Luftschutzoffiziere (air raid officers) were posted right behind the front lines who, like the Flak Group commands, submitted continuous reports regarding the current position of the intruders, their flight direction, the number of airplanes, and their altitude to the Jagdstaffeln. In order to enable passing on this crucial information to the right recipient as fast as possible, a telephone system was installed that connected the air raid officers and Flak Group commands to the Stofl and to the Jagdstaffeln. According to contemporaneous German fighter pilots, British aircraft crews generally oriented themselves over German territory by keeping to prominent landmarks such as railway lines, river courses, or "Roman Roads". So based on the incoming reports, the Staffelführer was able to estimate the likely flight path of the British aircraft and react accordingly.

Initially, the primary task of Jasta 11's Staffelführer, Oblt. Rudolf Lang, was to familiarize his pilots with their role in the unit. Kampfeinsitzer-pilots, as well as those from Feldflieger-Abteilungen, were used to flying most of their missions alone and had to become accustomed to Staffel or Kette (a "chain" of 3 to 4 aircraft) coordination. The principle of flying at the front as a coherent Staffel with Ketten assignments was just as new to them in air combat as the formation of numerous single-seat fighters into a single permanent unit. In the autumn of 1916, there were no official instructions yet on how to run and organize a Jagdstaffel, and each Staffelführer was more or less left to his own devices in determining the way to do so. In modern day parlance, the entire process would be termed as "learning by doing". The individual leadership qualities of the Staffelführers,

Left: Vzfw. Leopold Anslinger, the brother of Franz Anslinger, visits Jasta 11. He stands in front of 'K'-marked Halberstadt D V, supposedly flown by Konstantin Krefft, December 1916 or early January 1917.

and the decisions they made in the early stages of their new positions, explain why the initial successes of many of the first Jagdstaffeln varied greatly in the last months of 1916.

Lt. Konstantin Krefft took one of the Halberstadt D Vs on a solo flight to the front on 20 October 1916. Upon his return, he reported having shot down an enemy airplane across the lines, but it went unconfirmed due to a lack of witnesses. His opponent may have been from No. 3 Squadron, RFC, which reported that one of its Moranes was forced to land within French lines after combat with an enemy airplane that day. Its pilot, 2Lt. W.B. Young, had been wounded and its observer, 2Lt. R. Davis, killed.(27) It is also possible that Krefft had attacked F.E.2d A6 of No. 20 Squadron, which was damaged in combat south of Lille and returned with a wounded pilot.(28)

The next day, three aircraft made their first flight at the front under the leadership of Staffelführer Oblt. Rudolf Lang. No appreciable results were achieved other than the beneficial determination that several pilots knew the area quite well due to having flown previous missions there with other units.

On 25 October 1916, the Staffel received another machine, Halberstadt D V 416/16, while on the same day Halberstadt D V D.412/16 had to be returned to the Armeeflugpark for repair due to insufficient elevator functions. In order to bolster the strength of the new

Above: Lt. Franz Anslinger, third from right, in front of Albatros D I D.425/16, the first Albatros single-seater to be delivered to Jagdstaffel 11. (R. Zankl)

formation, Halberstadt D V 417/16, D.418/16 and Halberstadt D II 107/16 were supplied by the Flugpark between 26 October 1916 and 31 October. Likewise, Halberstadt D V 412/16 came back repaired from the park during this period. Thus, the single-seater complement of the Staffel now consisted of seven operational aircraft.(29)

With these aircraft Jagdstaffel 11 conducted eight "Kriegsflüge" (war-flights) on three days during the week of 22 October to 29 October 1916, accumulating a somewhat modest total of seven flight hours. No significant events were reported, however. Over the same period of time, Jagdstaffel 10 conducted 20 flights amounting to a total of 15 hours.(30)

Right: Hptm. Maximilian Sorg, Stabsoffizier der Flieger (Aviation Staff Officer) of the German 6th Army. (P. Kilduff)

Nr.

K. B. Flieger - Abteilung (A) 297 O. U. 5.11.18.

Personal - Auszug

des Herrn Oberleutnant Lang Rudolf, Emil

Geburtsort : München Datum: 16.5.89.

Truppenteil : 12.b. Jnf. Regt.

Dienstgrad : Oberleutnant aktiv

Patent: 1.6.15.

ausgebildet als Führer

Abzeichen : 23.9.14.

Besondere Ausbildung : als techn. Offizier

Erstes Kommando zur Fliegertruppe : 1.2.14.

War in folgenden Stellen verwendet:

1.2.14. - 23.9.14. als Flugschüler
25.10.14.- 12.12.14. " Führer der Fliegerschule 1. Schleißheim
12.12.14.- 1.8.15. bei Feldflieger - Abt. 6b
1.8.15.- 10.9.15. kommandiert zu Feldflieger - Abt. 54
1.11.15.- 30.11.15. " " Abt. 9b
1.12.15.- 30.4.16. bei Feldflieger - Abt.6b als Eindeckerflieger
1.5.16.- 5.7.16. " Artl. Fl. Abt. 103b " "
15.7.16.- 20.10.16. " Kampfstaffel 31, Kagohl 6 als Führer
20.10.16.- 20.1.17. " Pr. Jagdstaffel 11 " "
20.1.17.- 27.4.17. " Württ. " 28 " "
1.5.17.- 15.12.17. " B. Fliegerschule 1 " "
5.12.17.- 28.12.17. " Flieger-Abt. 288. kommandiert
29.12.17.- 10.2.18. kommandiert zur Pr. Jdflieg zur Information
10.2.18.- 26.4.18. " " Fl. Abt. 287 " "
und " " 295
27.4.18.- 10.6.18. stellv. Führer der B. Fl. Abt. (A) 295
ab 22.7.18.(3.8.18.) Führer der B. Fl. Abt. (A) 297

Verwundungen:

17.6.18. Rippenbruch durch Überschlagen mit Flugzeug bei Notlandung.

Lang

Oberleutnant u. Abt. Führer.

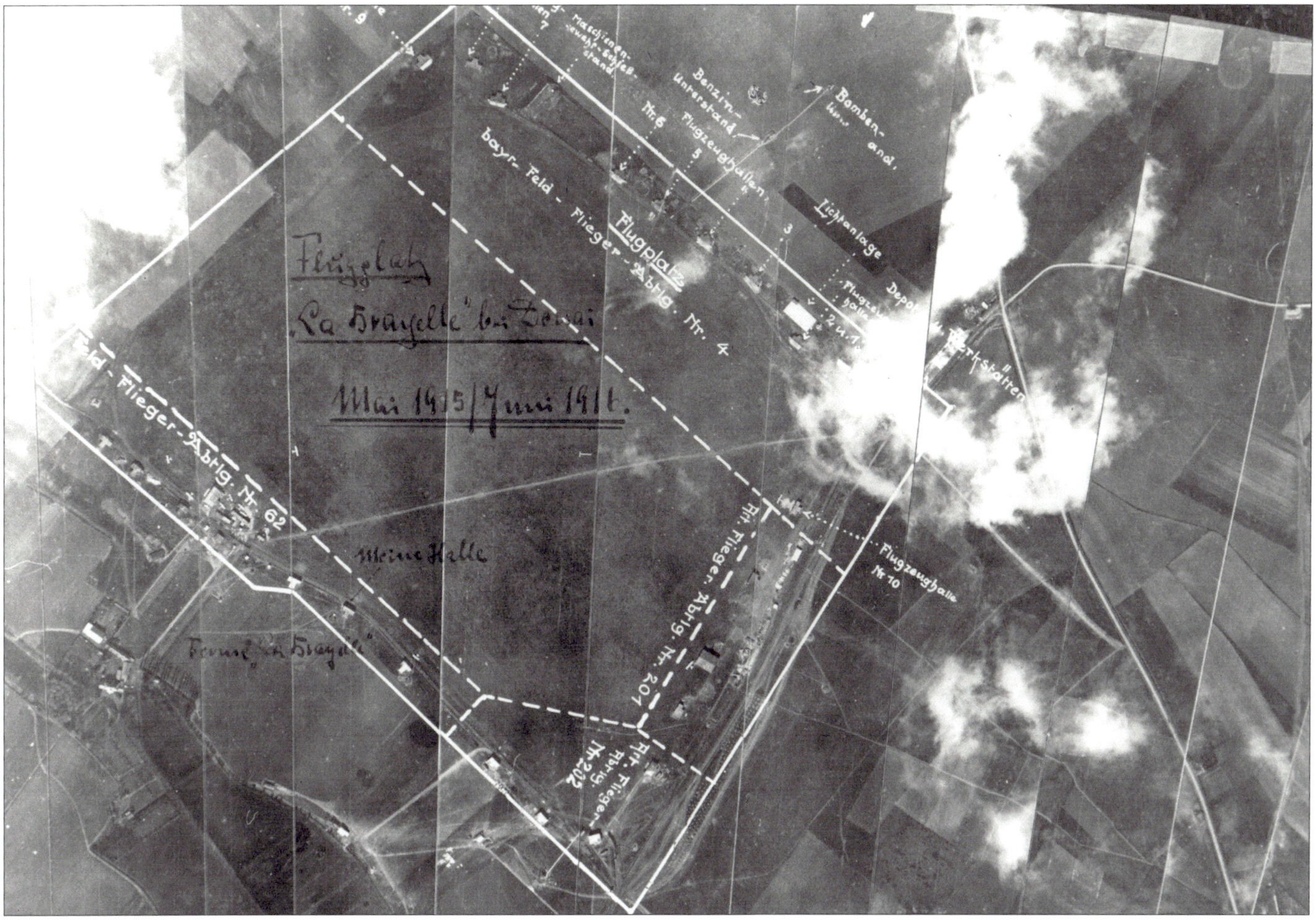

Above: Aerial photograph of La Brayelle airfield near Douai taken by Oblt. Ernst von Gusnar, observer in Feldflieger-Abteilung 62 in the summer of 1916. (www.anciens-aerodromes.com)

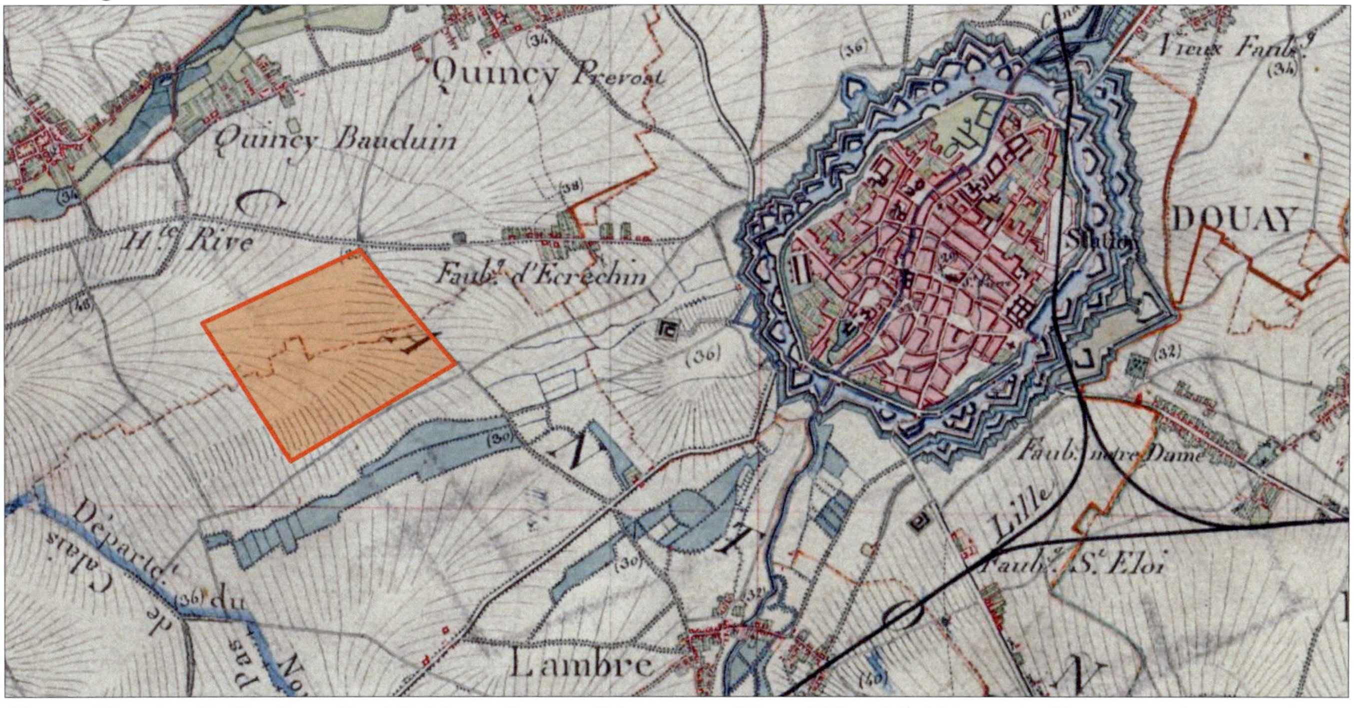

Above: Map detail of La Brayelle airfield, southwest of the gates of Douai. The airfield is marked in orange. (www.anciens-aerodromes.com)

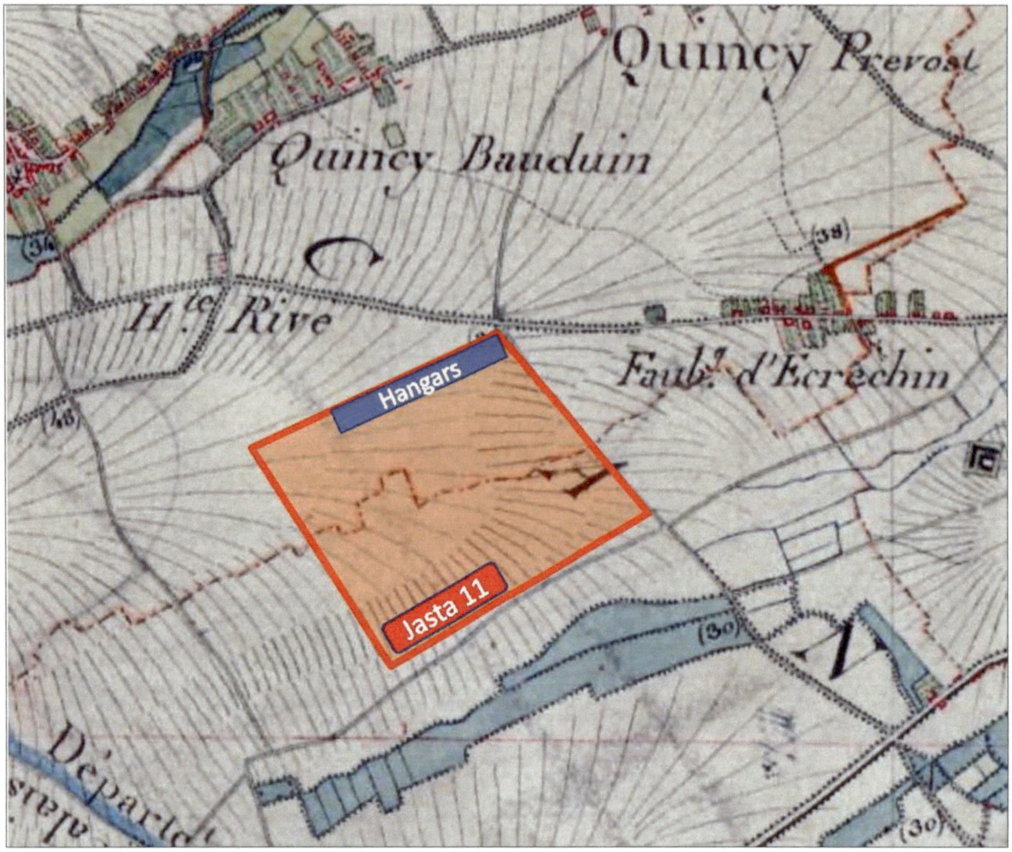

Right: La Brayelle airfield. Jasta 11 was located to the south-west of the airfield opposite La Blancherie Ferme. (www.anciens-aerodromes.com)

Below: Aerial view of part of La Brayelle airfield; clearly visible are aircraft hangars and aircraft tents. (www.anciens-aerodromes.com)

Above: Main area of operation of Jagdstaffel 11 in the area of the German 6th Army.

Following Page: The 6th Army Stofl's weekly report of activities for 22 to 29 October 1916. Point 2 lists the number of war flights, flying hours, and photo reports for each aviation unit. It recorded that Jasta 11 undertook 8 Kriegsflüge (war flights) on 3 days that totaled 7 hours. (Bayerisches Hauptstaatsarchiv, Abt. 4, München)

Nr. 19926.   A.H.Qu.,den 30.10.1916.

Stabsoffizier der Flieger

Truppen der 6. Armee.

Woechentlicher Taetigkeitsbericht

vom 22. mit 29.10.1916

Die beiderseitige Fliegertaetigkeit war durch Regen und Sturm stark beeintraechtigt:

1.) Feindliche Fliegertaetigkeit:

18 Geschwader mit 140 Flugzeugen (Vorwoche 40 Geschwader mit 264 Flugzeugen, nur am 22.10. sehr lebhaft (107 Flugzeuge)
Bombenangriffe auf Bhfe. Séclin, Hénin - Liétard, St. Leger;von 5 Angriffsgeschwader mit 79 Flugzeugen 93 Bomben. (Vorw.573)
Fliegerschiessen hat Gegner durchgefuehrt gegen 1 Batterie und 1 Beobachtungsstelle.
Das herausfordernde Fliegen feindl. Einzelflieger hat stark nachgelassen.
Feindl. Flieger gaben oefters Trompeten- und Hupensignale.
Angriffe auf unsere Ballone und Artillerieeinschiessflugzeuge dauern an.

2.) Eigene Fliegertaetigkeit:

	Flugtage	Kriegsfluege	Flugstunden	Bildmeldgn
F.Fl.A.5	1	2	6	6
" " " 18	1	3	7	24
A.Fl.A.227	4	11	12	9
F." " 5b	3	7.	7	14
A." " 240	4	9.	7	11
A." " 235	2	6	7	12
A ". " 202	4	13	21	19
J.St. 10	3	20	15	--
" 11	3	8	7	--
		79	89	95

3.) Fernaufklaerung:

am 22.10.16. Schiffsverkehr auf Kanal Aire - Béthune,
Bahnverkehr auf Strecke Béthune - Lillers u. Béthune - St. Pol ,sowie auf Strecke Doullens - St. Pol in Richtung St. Pol sehr lebhaft.
Bhf St Pol u. Béthune vieles rollendes Material. (5, 18 u. 5b,)
Feindliche Lager und Flughaefen unveraendert,
Feindl. Flakfeuer stark und gut bei Arras,Avenes le Comte u St. Pol

4.) Taktische Nahaufklaerung:

Belegt durch 65 Bildmeldungen,keine wesentlichen Veraenderungen, ausserdem Aufnahmen von eigenem rueckwaertigen Gelaende durch Reihenbildner fuer Vermessungszwecke. (5)
Feuernde fdl.Batterien wurden vom Flugzeug aus festgestellt:

gegenueber	II. B. A. K.	4 Batterien	(A.Fl.A.227)
"	XII. R. K.	2 Flakbatterien	(" " " 235)
"	Korps Arras	2 Batterien u. 1 Flakbatterie	(" " " 202)

Fdl. Batterien und M.W. stellten bei Annaeherung unserer Flugzeuge das Feuer ein. Die Zahl der sicher erkannten fdl. Batteriestellungen ist von 199 auf 151 zurueckgegangen (nach A.N.)

Frontline Operations of Jagdstaffel 11 November 1916 – Beginning of January 1917

Above & Below: Halberstadt D V D.420/16 was delivered to Jasta 11 on the 1 November 1916. After having served with the Jasta for almost three months, the aircraft was returned to Armeeflugpark 6 on 24 January 1917. It was overhauled and sent to Jasta 30 on 8 February. Just a week later, on 15 February 1917, Gefr. Schneider became disoriented and landed on Allied territory, suffering damage to his aircraft. The aircraft was repaired by its new owners, decorated with British cockades, and exhibited as a piece of booty. (J. Leckscheid)

Flight operations for the week of 30 October to 5 November 1916 were severely hampered by rain and storms, such that only 40 enemy aircraft were reported as having entered 6th Army airspace. Jasta 11 undertook five "Kriegsflüge", totaling 19 flight hours, on four days that week with no appreciable results.(1)

On 1 November 1916, the Staffel received three more Halberstadt D V fighters with military serial numbers D.413/16, D.420/16 and D.421/16.(2)

On 5 November, despite poor weather, elements of the Royal Flying Corps appeared over La Brayelle airfield and dropped bombs on it. Only one aircraft, Halberstadt, D V

Right: Lt. Kurt Wolff in typical flyer´s attire, with a cigarette on his lips. (L. Bronnenkant)

D.413/16, was damaged. It was sent to Armeeflugpark 6 for repairs and was returned to Jasta 11 on 12 November 1916. Halberstadt D II D.801/16, license-built by the Hannoversche Waggonfabrik AG, was also delivered to the Jasta in early November along with Halberstadt D II (Han.) D.802/16, D.803/16, and D.805/16 at or around the same time.(3) D.803/16 was damaged during one of its first flights on the night of 5 November, and Halberstadt D V D.412/16 made a crash landing the same day. Both machines were handed over to Armeeflugpark 6 for repairs.(4)

Additional pilots also began to arrive at Jasta 11 in November 1916. Lt. Kurt Wolff, formerly with Kampfstaffel 40 of Kampfgeschwader VII, came to the unit on 4 November 1916 after a short course of single-seat fighter training at Armeeflugpark 6, where he got accustomed to the Halberstadt D-types. He was followed by Lt. Georg Simon from Kagohl II's Kampfstaffel 12.

Kurt Wolff was born on 6 February 1895 in Greifswald in the Prussian province of Pomerania. His father died in 1901 and he lost his mother in 1911, so he and his sisters

Below: Lt. Kurt Wolff as a student pilot in the pilot's seat of an AEG (possibly a B.II) at the Fliegerschule Nieder-Neuendorf of Flieger-Ersatz-Abteilung 1. (L. Bronnenkant)

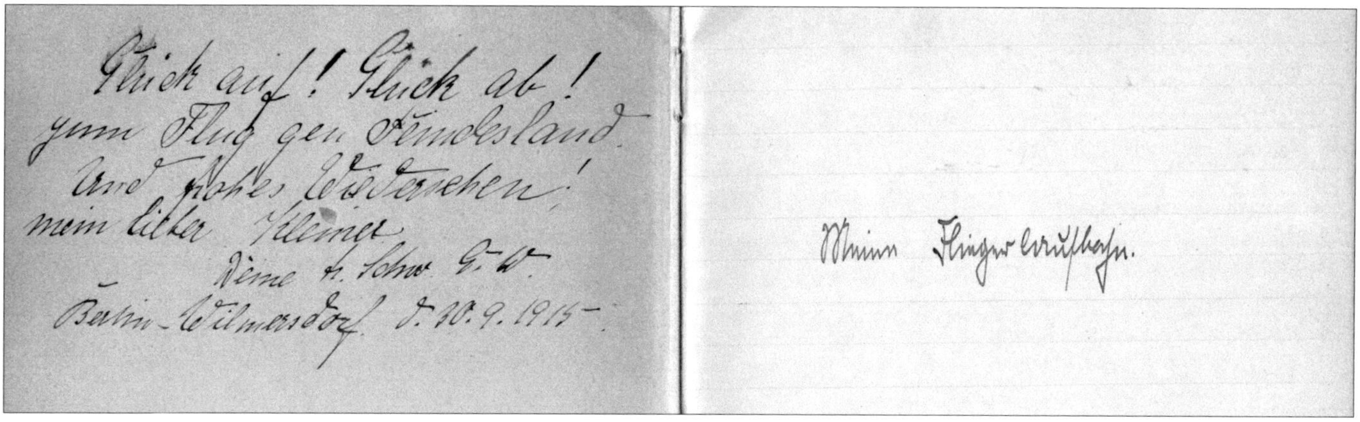

Above & Below: Excerpt of the handwritten diary of Kurt Wolff. (Private property)

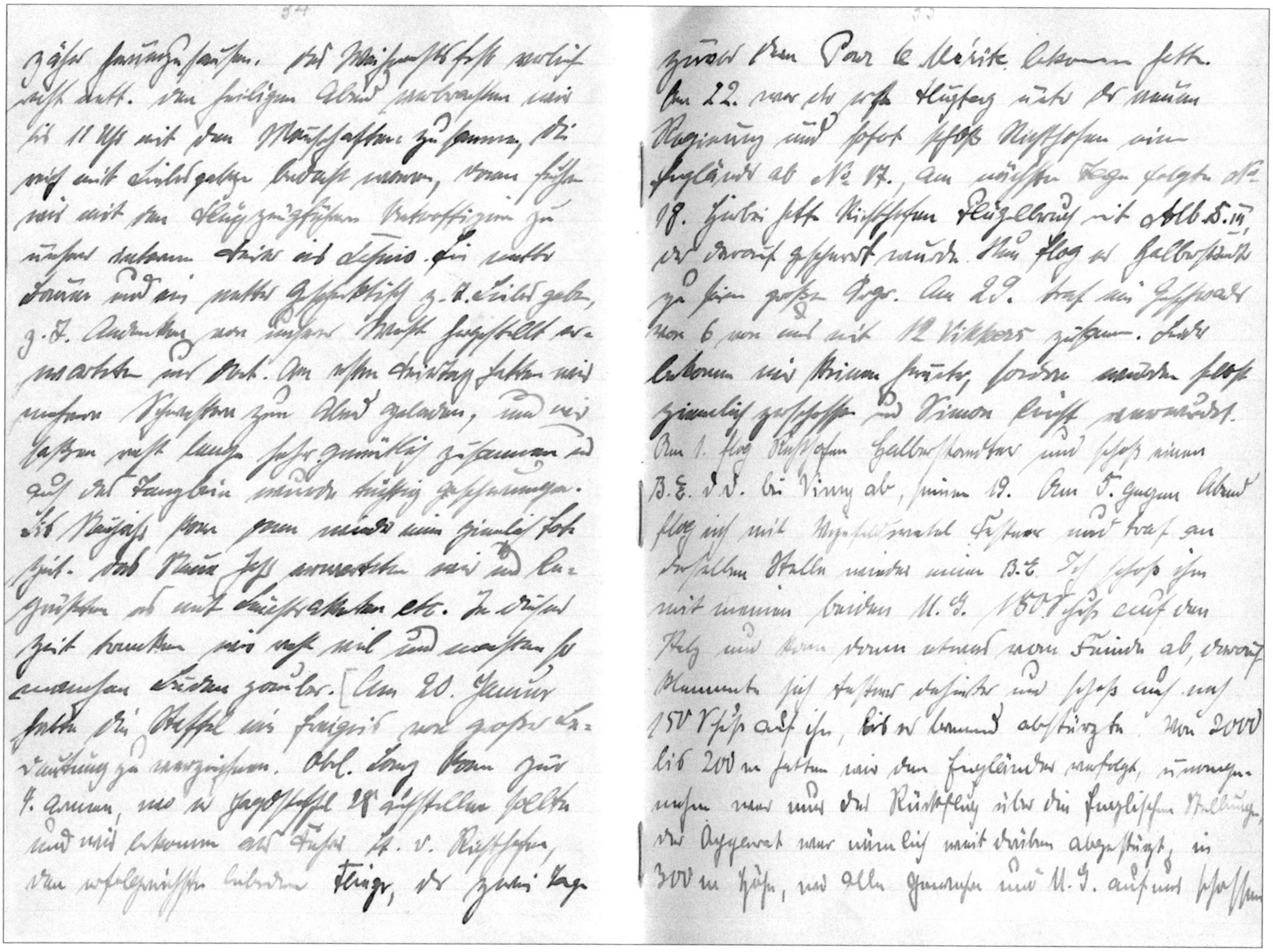

Elisabeth and Gertrud grew up with relatives in Memel, East Prussia (today Klaipėda, Lithunia). Following his graduation in March 1912, his technical interests led him to enlist in Schöneberg's Eisenbahn-Bataillon Nr. 4 (Railway Battalion 4) as a Fahnenjunker. He went to war with that unit and was promoted to Leutnant on 18 April 1915. Soon afterwards he applied for the Fliegertruppe, was accepted, and posted on 15 June 1915 to Flieger-Ersatz-Abteilung 1 at Döberitz before being ordered to attend the Militär-Fliegerschule at Nieder-Neuendorf.[5] It was said that his very first flight in an A.E.G. trainer there ended in a crash, injuring him and killing his instructor.[6] Following his recovery and completion of training, Wolff was first sent to Armeeflugpark Gaede on 1 October 1915 and then to Kampfstaffel 26 of Kampfgeschwader V on 18 December 1915.[7]

It was there, on 8 March 1916, that he and his observer, a

Above: Lt. Kurt Wolff in front of the "K"-marked Halberstadt D V, the personal plane of Lt. Krefft, December 1916 or early January 1917.

Above: Ground crew, military personnel, and a large number of curious local kids surround Kurt Wolff's Halberstadt D II (Han.) D.802/16 after a hard landing in November 1916, which resulted in a damaged horizontal tailplane. The removed component rests on the ground in front of the mechanics on the right side of the picture. (R. Kastner)

Lt. von Hartmann, experienced their first air combat:

"*...Hartmann* [and I] *went on patrol from here* [Metz airfield] *to Verdun in beautiful weather on the 8th. We were circling at 3,000 meters, unspared by the B.A.K.* [Ballon-Abwehr-Kanonen, or Flak guns] *fire over the hotly-contested positions and fortress, when we suddenly saw two points coming toward us that turned out to be Farman biplanes. When we saw the blood-red stripes on the rudders and the cockades on the wings, Franz* [the German nickname for an observer] *grabbed the machine guns, loaded both as calmly as if he were at the shooting range, and when the first Farman was quite close, he put a nice load into him and right after that, the second. Now the battle went back and forth, we always somewhat higher than the Frenchmen, right turn, left turn, alternating fire between the right and left machine guns, sometimes to the front, sometimes to the rear. A third Frenchman appeared, a Voisin biplane, who was also heavily attacked. I quickly pulled the machine up, quickly pushed down like crazy, quickly stood on my wings in a turn. Then one machine gun belt was used up;*

Right: Kurt Wolff together with one of his sisters, Elisabeth, called Liz.. The photo was taken during his leave after being awarded the Order Pour le Mérite at the end of May, beginning of June 1917. His mechanic Hermann Lohmeyer received the photo as a gift from Kurt Wolff.

Above: Kurt Wolff together with Konstantin Krefft, left, and Carl Allmenröder, right. He liked to be the joker if he felt comfortable around his comrades.

quick! the next ammunition case, which slipped from the hand. Let it lie – next one! One, two, the gun is freshly loaded again. Suddenly, another Aviatik machine appeared, set his sights on a Farman and a Fokker dove at him from above. That Farman did not want anything more to do with us. The Voisin quickly fled at an opportune moment; now only the other Farman remained with us. No more going away from him, but at him! We are behind him, he can turn as much as he wants but we are always close above him. Then we are no more than 50 meters apart. 'Shoot, shoot, like there's no tomorrow just like old people know,' Franz hears behind him. Now he was cleanly targeted and a series of 30 rounds pounded into the Farman. Suddenly – what's that? – the Frenchman somersaults and hurtles down like greased lightning. Catching him was impossible so we stayed where we were. We followed our prey with our eyes and saw him land on a forest meadow at Sommedieue, unfortunately in the French lines, where he still was 45 minutes later when we flew over the same place again. 'Bravo!' says the observer proudly, still somewhat worked up over the fight. The pilot puts his hand on his crash helmet. Congratulations. A strong handshake at 2,800 meters, the defeated opponent lies below, the winners above laugh and shake their right hands in the swaying aircraft. The first, fortunate air combat that both had struggled through, the first downed Frenchman. The whole fight with the three apparatuses had taken 20 minutes."[8]

Their claim went unconfirmed due to a lack of witnesses on the ground. Their victim may have been a Farman from Escadrille MF 35 manned by Sgt. Maurice Collet (pilot) and Soldat Pierre René Beauvais. Soldat Beauvais was killed but the wounded Sgt. Collet managed to land in the vicinity of Fleury-sous-Douaumont, just north of Sommedieue, which was in French territory. This would account for the lack of German eyewitnesses.

Three months later, on 17 July 1916, Wolff was moved to Kagohl VII's Kampfstaffel 40. Enthusiastic about the successes of single-seat combat pilots, he requested a transfer to fighter piloting and was ordered to Armeeflugpark 6 on 28 September 1916.[9] There, on 2 November, he was able to fly a 100-hp Fokker biplane (probably a D II) with a rotary engine for the first time:

"Although I flew with a control stick and a rotary engine for the first time, everything went brilliantly; the machine was marvelous and climbed like a wild monkey."[10]

Just three days later, on 5 November 1916, he was transferred to *Jasta* 11:

"On 4 November, I went to Jagdstaffel 11 at Douai, Oblt. Lang commanding. There I made some flights around the field in a Halberstadt, shot the machine guns, and then flew with the Staffel."[11]

In his book, *Jagd in Flanderns Himmel* (*Hunting in Flanders' Skies*), Karl Bodenschatz, Jagdgeschwader I's Adjutant, characterized Wolff in this manner:

"At first glance, you could only say 'delicate little flower.' A slender, thin, little figure, a very young face, whose entire manner is one of shyness and more shyness. He looks as if you could tip him over backwards with one harsh word."[12]

A journalist for *Die Woche* magazine who visited Jasta 11 in the spring of 1917 wrote his impression of Wolff in somewhat dramatic terms that were customary at the time:

"Smooth shaven like almost all the young gentlemen there with him, Wolff is tall, narrow, and slim, with dark hair and dark eyes – an almost delicate appearance. His extraordinarily refined and reserved nature further masks the fact that he is a man of action nonetheless."[13]

Hermann Lohmeyer, who began serving with Jasta 11 in January 1917 and was one of Wolff's three mechanics, said that his demeanor was that of a modest and reserved man. He required conscientious work, discipline, and commitment from his mechanics but was friendly with them and always listened to what they had to say. Lohmeyer could not recall any instance where Lt. Wolff bellowed at him or the other mechanics in a commanding tone – which incidentally almost never happened with the other pilots too.

He recalled that Kurt Wolff had a very close relationship with his two sisters, since they had grown up without parents, and that they wrote each other regularly. Humor was another of Wolff's attributes. Although he appeared rather shy, he could be quite the joker and be very funny once he grew comfortable with the company he kept. But behind the rather inconspicuous, shy being was a man with tremendous willpower, ambition, and much self-discipline. He was an aggressive fighter in action but was known to disengage from a promising pursuit to come to the aid of comrades in need. His talisman was a pointed stocking cap that he always wore under his leather flight cap. Hermann Lohmeyer further recalled that Wolff was engaged and planned to marry and start a family after the war.[14]

The other new pilot, Lt. Georg Simon, was born on 4 January 1885 in Weidebrunn near Schmalkalden in the Prussian province of Hesse-Nassau. At age 31, he was one of the older pilots of the Staffel. He volunteered for the Fliegertruppe at the beginning of 1915 and was sent to Hannover to train as a pilot with Flieger-Ersatz-Abteilung 5 on 22 February 1915. On 15 May, he began a three-month stint with the Militär-Fliegerschule at Halberstadt and returned to Flieger-Ersatz-Abteilung 5 on 21 August. Now an Unteroffizier, he was transferred to Kagohl II's Kampfstaffel 12 on 2 January 1916. Further promotions occurred to Vizefeldwebel on 14 April, and to Leutnant on 29 June. He was transferred to Armeeflugpark 6 on 12 October 1916 and to Jasta 11 one month later on 7 November.[15]

Above: Lt. Georg Simon on the steps of Roucourt Castle in April 1917. Aged 31, he was at this time the oldest pilot in Jagdstaffel 11.

Weather conditions improved during the week of 6 - 12 November 1916 and enemy air activities increased accordingly. During this period, 212 incoming enemy aircraft were counted, and British aircraft were said to have dropped 245 bombs on German positions. Jasta 11, the only Jagdstaffel in the area of the German 6th Army, flew sorties on four days that yielded 38 Kriegsflüge and 42 total flight hours.[16] Aviation units under the 6th Army reported 20 air combats, but the weekly report did not break them down by unit and we do not know how many involved Jasta 11. In any event, Jagdstaffel 11 had no successes.[17]

The weather deteriorated again during 12 - 18 November and halved the number of enemy aircraft intrusions into 6th Army airspace. Jasta 11 had 26 Kriegsflüge totaling 27 flight hours over the course of four days, but again without any noteworthy results.[18]

Alfred Prehn was commanded to Seeflugstation (Naval Air Station) Tondern on 13 November 1916. He would not return because his assignment became an official transfer on 1 March 1917.[19]

On 14 November 1916, three Halberstadt fighters including one flown by Kurt Wolff set out on a sortie. Wolff, however, experienced technical problems:

"On 14 November, three of us made a flight during which my engine croaked, so that I had to make an emergency landing at Billy-Montigny near the tethered balloon 'Hohenzollern'. I was pleasantly received there, put the machine within the balloon hangar, and then drove back again with the car the mechanics had brought. The crate was back in shape again the next day and I took off. Just as I was 300 meters high, the engine failed again, I had to make an emergency landing, and overturned terribly. Fortunately, I had on my crash helmet so that nothing else happened to me. The propeller, rudder, right upper wing, and a strut had had it. Afterwards, I flew a new machine over from the Park. And just when I was about to take off from Valenciennes, my old Franz, Hartmann, suddenly appeared. While I flew early in the morning, he drove in our car to Douai, where we spent a very nice day."[20]

On 15 November, Albatros D I D.425/16 was sent back to Armeeflugpark 6 in exchange for Albatros D II D.473/16. Halberstadt D Vs D.1099/16 and D.1100/16 were also added to the unit roster on 16 and 27 November, respectively.[21]

The week of 12–18 November 1916 was characterized by heavy haze and low clouds. On four days of this period, Jasta 11 undertook 26 Kriegsflüge amounting to 37 flight hours.[22]

Additions to the pilot roster continued the next week with Lt. Carl Allmenröder joining on 20 November and Vzfw. Josef Baierlein on the 26th.

Carl Allmenröder, the sone of a pastor, was born on 3 May 1896 in Wald, near Solingen in the Rhineland (today a part of Solingen). His brother, Wilhelm, was two years older and would also become a fighter pilot. Their neighbor's daughter, Helene Kortenbach, who was engaged to Carl but married Wilhelm after Carl's death, recalled that the boys had been raised in a home that was "liberal-protestant" and strongly influenced by humanism. One of their father's standard sayings was the Kantian categorical imperative: *"Act according to the maxim that you would wish all other rational people to follow, as if it were a universal law."* Their parents placed great value on honesty, compassion, faithfulness, faith in God, and patriotism.[23]

Carl, in keeping with his humanistic education, decided to study medicine after graduating from secondary school. His parents were quite proud when he attended the University of Marburg an der Lahn, but the outbreak of war interrupted his plans. He, like so many other of the world's young men, volunteered at age 18 for military service and went into the field with Ostfriesisches Feldartillerie-Regiment Nr. 62 when it was deployed in Galicia. The spring of 1915 brought him several recognitions: his Leutnant commission on 22 March, the Grand Duchy of Oldenburg's Friedrich August Cross, 2nd Class on 18 May, and Prussia's Iron Cross, 2nd Class on 27 May.[24]

Carl's company commander at Feldartillerie-Regiment 62 wrote in a performance review:

"In no time at all, he earned the high regard of his superiors and comrades not only by his superb military achievements but also by his youthful, open nature."[25]

He felt a deep bond with his former service arm even as a member of the Fliegertruppe. He would be seen time and again flying low and wagging his wings over artillery positions to greet the gunners.[26]

Carl, like his brother Wilhelm before him, applied for service in the Fliegertruppe in early 1916. On 29 March, he was accepted into Flieger-Ersatz-Abteilung 5 to undergo pilot training at the Militär-Fliegerschule at Halberstadt. He returned to Flieger-Ersatz-Abteilung 5 on 30 May to complete his training and was sent to Armeeflugpark 6 on 25 August to await further disposition in the field, which occurred on 25 September when he was commanded to Feldflieger-Abteilung 18. His next stop was Jasta 11, where he was sent on 20 November.[27]

Frau Helene Allmenröder remembered Carl in these words:

"Of the two brothers, Carl was the wilder and more daring, always on the lookout for adventure, but at the same time considerate to children and adolescents who were weaker. When a larger child struck a smaller or weaker one, Carl was there to help the weaker one. He even let us girls participate in games of Indians. When we were in what would today be called our 'teenage' years, most of the young girls in our area secretly fell in love with Carl because of that and his carefree and humorous nature as well. Of course, I was in love with him too. Wilhelm was the calmer, the prudent, and more reflective one of the two."[28]

Allmenröder invited fellow artilleryman Otto Fuchs to Jasta 11 for a few days in early June 1917 before Fuchs had to move on to his actual assignment, Jasta 30. Fuchs remembered this about Carl:

"I was at the Jagdfliegerschule Valenciennes, when I had the misfortune to come so close to the ground during an evasive maneuver with another airplane that the flight instructors had to throw themselves to the ground. As a result, I was immediately

Above: The Halberstadt D II (Han.) D.801/16 "S" of Lt. Georg Simon, marked with his personal identification, a white "S" on a black background, around December 1916 or early January 1917. The aircraft was exchanged for an Albatros D III in late January 1917 at Army Air Park 6. From there, it later went to Jasta 28.

Above: Albatros D II D.473/16 was delivered to Jasta 11 on 15 November 1916. On 11 December 1916, the aircraft was returned to Army Air Park 6, from where it was then sent to Jagdstaffel 10. On 2 May 1917, Jasta 10 pilot Gefr. Lemke landed the aircraft in Allied territory after losing his orientation. The photo shows the aircraft with British cockades being applied after capture. It is noticeable that only the number "4", an identification marking then typical for Jasta 10, had been applied to the aircraft. No traces of a removed previous marking that could be ascribed to its time with Jasta 11 can be made out. (G. Van Wyngarden)

Above: Carl Allmenröder as a war volunteer shortly before going to the front. (L. Bronnenkant)

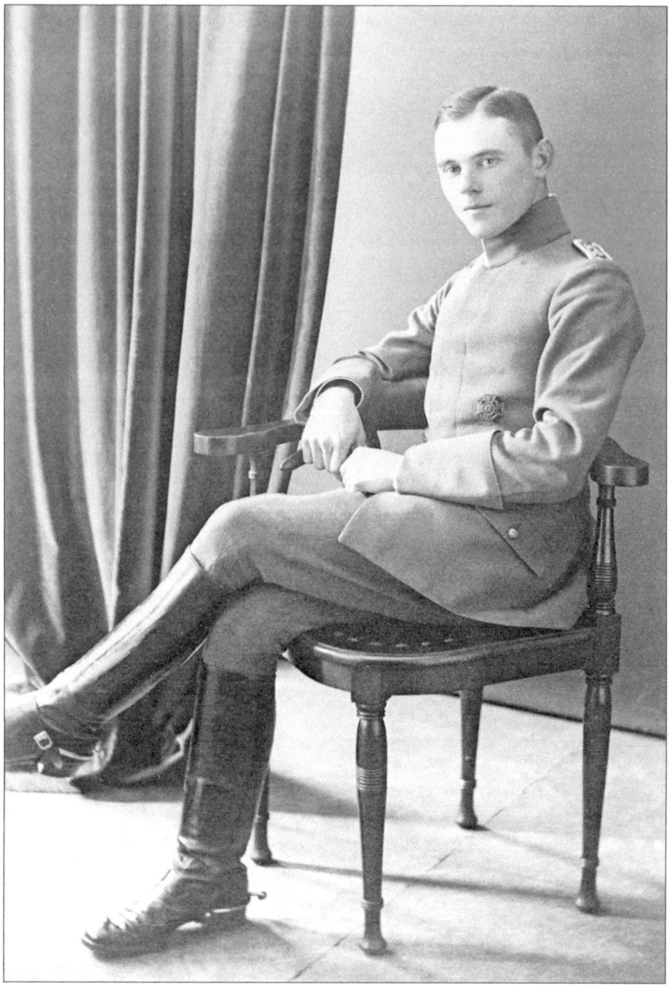

Above: Lt. Carl Allmenröder after his promotion to Leutnant d. Res. in 1915.

thrown out of the school. Just at that moment Allmenröder landed. I gathered all my courage and asked him to put in a good word with my instructors. He sat down with me and asked me what's happened. When I told him that I had flown a whale [a Roland C II - that was difficult to fly], *he seems to be impressed. As a result, he took me to his Jagdstaffel 11, without asking the school to give me a chance to become a fighter pilot.*

He was very pleasant, very friendly, very casual, and devoid of any arrogance – a real comrade. When I arrived with him at Jasta 11, he introduced me to all the gentlemen at dinner. The next day, he showed me the airfield and aircraft. On the first Jagdflug [hunting flight], *he impressed upon me to stay close to him and instructed another pilot to look after me. He seemed very concerned about the safety of his airmen. When I learned of his death some three weeks later, I was quite saddened.*"(29)

Josef Baierlein, a native of Munich, was born there on 16 November 1886. At 30 years of age, he was the second-oldest member of *Jasta* 11 when he joined it on 26 November. As early as 1912, he had acquired pilot's license No. 187 on an Otto-Doppledecker. On 4 August 1914, he was drafted into the Bavarian Flieger-Ersatz-Abteilung at Oberschleissheim but was deferred from military service on 1 November 1914

Above: The Allmenröder family in the summer of 1916. From left: Ilse Hendrichs (née Allmenröder), Wilhelm, Frau Allmenröder, Carl.

and worked there as a civilian engineer. During 10 - 27 July 1915, he was assigned to Armee-Abteilung von Strantz's

Right: Carl Allmenröder at the Halberstadt Military Aviation School, May 1916.

Armeeflugpark to check aircraft bracing. On 21 January 1916, he was transferred to Kampfeinsitzer-Abteilung Mannheim for training as a single-seater pilot. On 13 May 1916, he was transferred to Armeeflugpark 6. On 5 July 1916, he was assigned to Bavarian Feldflieger-Abteilung 2 as a single-seat fighter pilot but returned to Armeeflugpark 6 on 30 July 1916. On 26 November 1916, he was ordered to Jagdstaffel 11 which became an official transfer on 22 December 1916.(30)

The week of 19–25 November 1916 consisted mostly of poor weather with haze and fog. Enemy activity was primarily confined to their side of the lines and only 38 hostile aircraft entered the 6th Army's airspace. Jasta 11 compiled 17 Jagdflüge totaling 20 hours over three days that included a lengthy dogfight involving 27 aircraft between Jasta 11 and an enemy formation on 23 November.(31)

Kurt Wolff wrote this about it in his diary:

"I had my first fight with a single-seater on 23 November, and it was vigorous. Seven of us took off with Österreicher in command and encountered an English Geschwader of about 20 apparatuses. We took it on and luckily, some other single-seaters came there. I myself attacked a lattice-tail that soon went down

Left: Carl Allmenröder on 10 September 1916 in Valenciennes; at this time he was a member of Feldflieger-Abteilung 18. The aircraft is an LVG C.II.

in spirals, wrapped in thick smoke. Unfortunately, I could not pursue him any further because I was seized from behind by several Englishmen. Then I spun through the entire Geschwader, attacking first this one and then that one, whereby my engine was shot through by an air cannon. I landed with that hit and four machine gun hits in the wing."(32)

Jasta 11 had tangled with two flights of No. 25 Squadron FE.2s, as reported in the British communiqué for that day:

"Returning from a bombing raid against Brebieres village and station, two offensive patrols of 25 Sqn were heavily engaged with about twenty hostile Scouts. The rear patrol, under Capt Tedder, was attacked by eight Scouts, who dived at the three highest machines. Lt Morris was hit in the arm at once and having his aileron control cut at the same time, he planed back to the lines and landed safely near Arras. Sgt Matheson's radiator was hit, and he was forced to return to the lines and landed safely. The second patrol, under Capt Chadwick, turned to assist Capt Tedder, and a general mêlée ensued, the eight German Scouts being reinforced by about twelve more. At this period of the fight, 2Lt Moller was hit in the back, but continued fighting until his radiator was hit and his engine seized up. He landed safely near St. Eloi. Towards the end of the fight, two De Havillands of 29 Sqn came up and one of them damaged and drove down one of the enemies. Lt Knight, 29 Sqn, reports seeing four hostile machines driven down in damaged condition. One of these was shot down by Sgt Horrocks and 2Lt Woollven and was seen to crash early in the fight. Another was driven down by Sgt Brown and Cpl Bower, 25 Sqn, and yet another, engaged by Sgt Green and Cpl Bower, 25 Sqn, at close range, went down in a spinning and uncontrolled nose dive. 2Lt Johnson and Lt Heald drove down another. After fifteen minutes' close fighting the hostile Scouts drew off and the F.E.s returned towards the lines in formation." (33)

Evidently, Wolff had been one of those seen to have been "driven down in damaged condition," but it turned out that no German pilot was actually lost or had to make an emergency landing.

Above: Carl Allmenröder (left), with his observer Lt. Hein Adalbert Lubarsch in October 1916 in front of Roland C II 1003/16 – one of the aircraft they flew at Fliegerabteilung 18.

During the following week, 26 November - 2 December 1916, Jasta 11 only flew on two days due to heavy fog conditions. The unit conducted "Luftsperre" patrols and recorded 14 Kriegsflüge totaling 14 hours.(34)

But it was not just the bad weather that affected the Jasta's operations. The engines of the Halberstadt fighters turned out to be less than reliable, as Kurt Wolff noted in his diary:

"On the 27th, Uffz. Festner and I had the job of protecting an artillery airplane directing fire onto Bethune. My propeller suddenly stopped at 2,000 meters and I had to go down and landed safely near Henin-Liétard. Simon made an emergency landing for the same thing that day."(35)

The weather deteriorated further during 2–10 December 1916, such that Jasta 11 was only able to mount 8 frontline flights totaling 8 hours on one day without any success.(36)

Kurt Wolff wrote about this time:

"On 4 December, three of us flew in a Geschwader. The other two went down over Douai while I could not find the field and kept wandering around. I got to Cambrai, landed there to ask where I was, then went to Valenciennes, again landed there and then happily got home. I received my Iron Cross, 1st Class on 6 December 1916. At the time, there was little flying because of the bad weather, so I usually amused myself by roaring around without an rpm indicator."(37)

During the following week, 9-16 December, the weather did not improve. Only limited flight activity occurred in the area on 11 and 16 December, and Jasta 11 was able to conduct only 8 frontline flights totaling 10 hours without any measurable results.(38)

In December 1916, two aircraft were repaired by Armeeflugpark 6 and returned to Jasta 11: Halberstadt D V D.116/16 on the 4th and D V D.413/16 on the 11th. Albatros D II D.473/16 was in turn sent to Armeeflugpark 6 which left Jasta 11 exclusively supplied with Halberstadt fighters. As of 31 December, it had a total of 15 Halberstadt D II, D V, and D II (Han.) airplanes at its disposal.(39)

The poor weather continued into 17–20 December, during which Jasta 11 made 15 uneventful frontline flights on one day that totaled 17 hours.(40)

The weather then improved dramatically during the final week of December and the Jasta was able to carry out 33 frontline flights totaling 30 hours over the course of three days. No dogfights were reported, however.(41)

On 20 December, the Staffel came very close to suffering its first loss, and Kurt Wolf described the event in a letter to his sister Elisabeth:

"Yesterday we celebrated the miraculous salvation of a comrade. He came back in the dark, and while coming down in a controlled glide hit against a mast on the edge of our airfield and crashed. The plane was completely smashed, and we all expected a dead man, but nothing had happened to him."(42)

Unfortunately, it is not clear from the available documents which pilot survived the emergency landing unharmed.

He also states in his letter:

"There has been little going on here lately because the Lords have been flying very little and very timidly in recent days. They seem to be very afraid of us, as they always avoid an air fight. In this way, of course, you can't achieve anything, and you can't shoot down an Englishman."(43)

A lack of activity allowed Jasta 11 to spend a quiet Christmas and New Year:

"The Christmas party was quite nice. We spent Christmas Eve

Above: Carl Allmenröder (right) together with his fiancée Helene Kortenbach during a visit to the Altenburg an der Lahn monastery site in August 1916.

Above: The brothers Wilhelm and Carl Allmenröder. The photo was taken on 10 September 1916 on the occasion of Carl's visit to his brother at Fliegerabteilung (A) 227.

together with the men in the ranks, who were generously gifted with 'Liebesgaben', until 11:00 p.m. Then we went with the Unteroffizier pilots into the Kasino for our internal celebration. A nice tree and a nice gift table, part 'Liebesgaben', part mementos made by our repair and maintenance men awaited us there. The next day, we invited our nurses for the evening's festivities, and we sat together quite comfortably for quite a long time and our dance legs were quite heartily brandished. It was a rather happy time again up to the New Year. We waited for the New Year and welcomed it with signal rockets, etc. We drank quite a lot and held many jamborees at this time."[44]

Jasta 12, based in the neighboring 1st German Army sector at Riencourt from 1 November 1916 to 23 December 1916, but which had occasionally been deployed to the 6th Army sector, was shifted to Armee-Abteilung A on 23 December 1916 due to the lack of air activity, which meant that Jasta 11 was the 6th Army's sole fighter unit at the end of the year. [45]

Jasta 11 had had an inauspicious beginning under the leadership of Oblt. Rudolf Lang. In its first three month of existence, it had carried out over 200 combat flights resulting in only one unconfirmed victory claim. On the positive side, however, it had suffered no losses. Still, of the 12 Jagdstaffeln established before 6 October 1916, it ranked in last place in terms of success. Although poor weather hampered its operations, this cannot be considered to be the sole cause of its lack of success. Other fighter units, such as Jastas 10 and 12, that had been active in the 6th Army region then, had secured several air victories in that sector during the same period.

The New Year began with mostly unsettled and rainy weather. For the week of 1 – 7 January 1917, the region's Stofl reported only three flying days that involved continuous enemy patrols and artillery bombardments along the front, with 69 enemy aircraft penetrating German airspace. Jasta 11 recorded 25 Kriegsflüge totaling 23 hours. The regional Stofl also reported 12 air battles during this period, but the extent to which Jasta 11 may have been involved is unclear from the documents.[46]

On 14 January 1917, the Staffel received Kogenluft Order I. Flieg. Nr 57439, in which Oblt. Lang was appointed the Staffelführer of the newly-formed Jagdstaffel 28 and was instructed to go to Flieger-Ersatz-Abteilung 10 at Böblingen to organize it. Lt. Manfred von Richthofen, formerly with Jagdstaffel 2 'Boelcke', was named to replace him as the new leader of Jasta 11.[47]

Above: Pilots of Jagdstaffel 11 in November 1916 in front of Albatros D I D.425/16, from left: Gefr. Sebastian Festner, Lt. Konstantin Krefft, Lt. Karl Kleinhenz, Lt. Kurt Wolff, Lt. Georg Simon, Lt. Carl Allmenröder. The identification of the individuals in the photo has been inscribed on the backside of the original print.

Above: The Halberstadt D II (Han.) D.805/16 flown by Uffz. Erwin Howe. The personal marking is a white "H" on a black background.

Above: The same group of Halberstadt D V and D II (Han) aircraft. Two Halberstadt D.Vs carry the same personal marking "C", the personal marking of Carl Allmenröder. It is very unlikely that two aircraft had the same personal marking except for the aircraft of the Staffelführer. The reason for this may be that one of the aircraft is a new Halberstadt D V delivered in November or December, while the other "C" aircraft is his old aircraft delivered in October 1916.

Above: Halberstadt D V and D II (Han) aircraft of Jagdstaffel 11 around the turn of 1916/17. Some personal identifications are clearly visible: C = Lt. Carl Allmenröder (D V), K = Lt. Konstantin Krefft (D V), H = Uffz. Erwin Howe (D II (Han.) 805/16) and W = Lt. Kurt Wolff (D II (Han.) 802/16).

Above: "Blow up" of the Halberstadt D V flown by Lt. Carl Allmenröder.

Above: "Blow up" of the Halberstadt D V flown by Lt. Konstantin Krefft.

Above: „Blow up" of the second Halberstadt D V with the "C"-marking of Carl Allmenröder.

Vzfw. Georg Braunmüller joined the Staffel two days later on 16 January. Braunmüller was born on 4 December 1887 in Zangelsdorf (today a part of the city of Dahme/Mark) in the Prussian province of Brandenburg. His parents were farmers there. Braumüller's pilot training began on 19 June 1915 at Flieger-Ersatz-Abteilung 2 at Adlersdorf. He received further training at Flieger-Ersatz-Abteilung 4 and Militärfliegerschule Schneidemühl until he was sent to Armeeflugpark Bug on 21 August 1915 and then transferred to Feldflieger-Abteilung 63 on 27 August. On 27 February 1916, he earned his Pilot's Badge. About three months later on 23 August 1916, he was reassigned to Armeeflugpark 9 at Bialystock as a pilot at its Versuchs- und Übungsflugpark (aviation depot for testing and training); but just four days later he was transferred to Feldflieger-Abteilung 64. On 8 December 1916, he was assigned to Armeeflugpark 13 to undergo fighter pilot training at Jagdstaffelschule I.(48)

It turned out that Oblt. Rudolf Lang did not depart for his new command until 20 January 1917, a delay that most likely was due to the inclement weather. As a final note, an order was issued that same day to establish the newly-formed Jagdstaffel 30 at Phalempin airfield, about 16 km north of Douai.(49)

Above: Uffz. Howe's Halberstadt D II (Han.) 805/16 marked "H" on the snow-covered airfield of Jagdstaffel 11 around December 1916. (R. Absmeier)

Above: The well-prepared Christmas party for the non-commissioned officers of Jagdstaffel 11 on 24 December 1916.

Dienststelle : Jagdstaffel 11

(Standort) : Douai, den 31. Dezember 1916.

eingetr.

Personal-Bestandsnachweisung.

Veränderungen zur Kriegsrangliste. (Nur für Jdflieg. auszufüllen.)
Abgang : (wohin ?), Beförderung, Patentverleihung, Umschulung, Prüfung.
Bei Zugang : (woher ?) Kriegsrangliste auf besonderem Bogen beifügen.

(Unterschrift und Dienstgrad :

Oberlt. und Staffelführer.

eingetr.

Above: Personnel inventory of Jagdstaffel 11 dated 31 December 1916. (Bayerisches Hauptstaatsarchiv, Abt.4, München)

Flugzeugführer:

Lfde. Nr.	Dienstgrad	Name	Geburtsdatum	Patent	Truppenteil	Zugehœrig oder kommandiert	Abzeichen seit	Bei der Dienststelle seit	Woher zur Dienststelle gekommen	Bemerkungen
1	Oblt.	Lang	16.5. 1889	1.6. 1915	12.b.JR	zug.	23.9. 1914	22.10. 1916	bay. Kgohl 6	
2	Lt.	Kleinhenz	28.8. 1892	1.8. 1914	b.5.FAR	»	25.7. 1916	11.10. 1916	KEK III	
3	»	Allmenroeder	3.5. 1896	30.3. 1915	FAR.62	»	.	21.11. 1916	F.A.18	
4	»	Wolff	6.2. 1895	18.4. 1915	Fea. 1	»	Juni 1916	5.11. 1916	A.Fl.P. 6	
5	Lt.d.	Oesterreicher	1.2. 1890	28.9. 1915	Fl.Btl. Schl.	»	Aug. 1914	11.10. 1916	KEK III	
6	»	Krefft	18.1. 1892	28.11. 1915	Fea. Schl.	»	Jan. 1915	11.10. 1916	» »	
7	»	Simon	4.1. 1885	29.6. 1916	Fea 6	»	29.2. 1916	7.11. 1916	A.Fl.P. 6	
8	»	Anslinger	1.5. 1887	25.8. 1916	Fea.9	»	28.2. 1916	16.10. 1916	A.Fl.P. 6	
9	Vzf. d.R.	Baierlein	16.11. 1886	.	Fea. Schl.	»	10.8. 1914	26.11. 1916	»	
10	Vzf.	Prehn	20.4. 1894	.	Fea 2	»	25.12. 1915	16.10. 1916	Jgst. 10	
11	»	Braumüller							A.Fl.P.6	bis 15.1. z.J.St. Schule kdt.

~~Beobachter~~===

Lfde. Nr.	Dienstgrad	Name	Geburtsdatum	Patent	Truppenteil	Zugehœrig oder kommandiert	Abzeichen seit	Bei der Dienststelle seit	Woher zur Dienststelle gekommen	Bemerkungen
12	Uffz.	Festner	30.6. 1894	.	Fea. Schl.	»	.	11.10. 1916	KEK III	
13	»	Howe	28.3. 1894	.	Fea.1	»	6.11. 1915	27.10. 1916	Jgdst.	

Jsta 11 205

ıerku
ef.
mand

Lfde. Nr.	Diens grad
1	U'zhl

Above: Personnel inventory of Jagdstaffel 11 dated 31 December 1916. Vzfw. Braumüller is commanded to the Jagdstaffelschule until 15 January 1917. The dates of transfer listed here differ in part from the dates on the Kriegsstammlists (KSt.L.) and Kriegsstammrolle (KSt.R.) (Bayerisches Hauptstaatsarchiv, Abt.4, München)

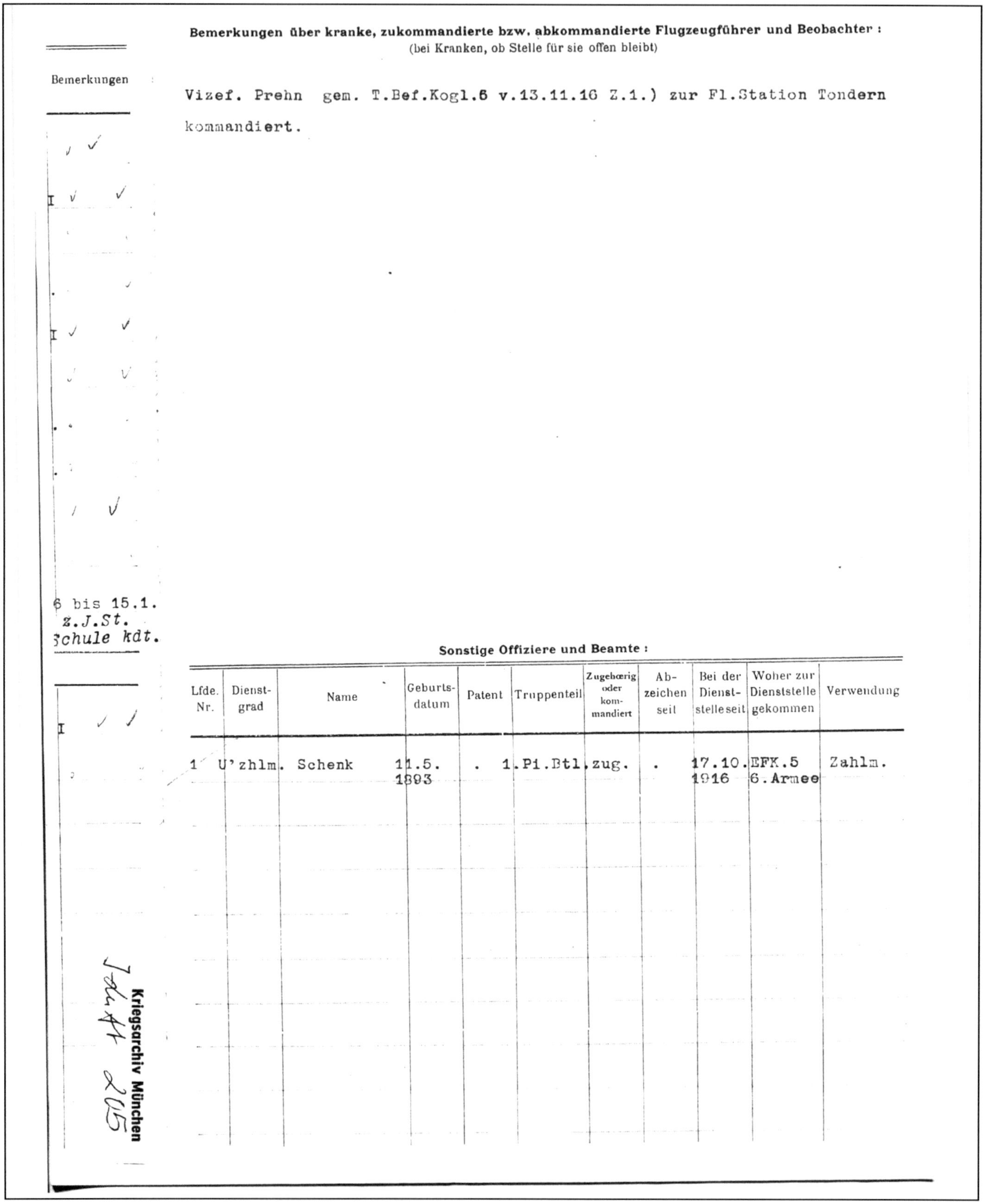

Bemerkungen über kranke, zukommandierte bzw. abkommandierte Flugzeugführer und Beobachter :
(bei Kranken, ob Stelle für sie offen bleibt)

Bemerkungen :

Vizef. Prehn gem. T.Bef.Kogl.6 v.13.11.16 Z.1.) zur Fl.Station Tondern kommandiert.

6 bis 15.1.
z.J.St.
Schule kdt.

Sonstige Offiziere und Beamte :

Lfde. Nr.	Dienstgrad	Name	Geburtsdatum	Patent	Truppenteil	Zugehœrig oder kommandiert	Abzeichen seit	Bei der Dienststelle seit	Woher zur Dienststelle gekommen	Verwendung
1	U'zhlm.	Schenk	11.5. 1893	.	1.Pi.Btl.	zug.	.	17.10. 1916	EFK.5 6.Armee	Zahlm.

Kriegsarchiv München

Above: Personnel inventory of Jagdstaffel 11 dated 31 December 1916. Above is noted: According to Kofl 6's daily order of 13 November 1916, Vzfw. Prehn is commanded to Fliegerstation Tondern. (Bayerisches Hauptstaatsarchiv, Abt.4, München)

Dienststelle: Jagdstaffel 11

(Standort): Douai, den 31. Dezember 1916.

Flugzeuge.

(Unterschrift, Dienstgrad): Lang

Oberlt. u. Staffelführer

Aircraft inventory report filed by Jagdstaffel 11, dated 31 December 1916. (Bayerisches Hauptstaatsarchiv, Abt. 4, München)

	Stand am 31. 12. 1916		Abgang vom 1. November bis 31. Dezember 1916.			
	Fabrik	Nummer	Fabrik	Nummer	Grund	Verbleib
R						
G						
D	Ha D II	107	Ha D III	116	Bruch bei Notlandung	A.Fl.P.6
	Ha D III am 4.12.16 von A.Fl.P.6. zurück	116	" D V	413	Besch.dch.Brandb.Wurf	"
			Alb.D I	425	telegr.Weisung des A.Fl.P. am 11.12.16 abgeliefert	"
	Ha D V	410	" D II	473		
	Ha D V	412	Ha D	803	bei Abendlandung stark beschädigt	"
	Ha D V am 11.12.16 zck.	413				
	Ha D V	414				
~~E~~	Ha DV	415				
	Ha D V	416				
	Ha D V	417				
	Ha D V	418				
	Ha D V	420				
	Ha D V	421				
	Ha D	801				
	Ha D	802				
	Ha D	805				
	Ha D V	1099				
	Ha D V	1100				

Aircraft inventory report filed by Jagdstaffel 11, dated 31 December 1916. (Bayerisches Hauptstaatsarchiv, Abt. 4, München)

96

Kommandeur der Flieger 6. A.H.Qu., den 23.1.1917.

No. 22400

Wöchentlicher Tätigkeitsbericht
vom 14.1. mit 21.1.17.

1.) Organisatorische Veraenderungen:

Jagdstaffel 30 wird ab 20.1.17. bei 6. Armee aufgestellt (O.H.L. Ic 44332 op. v. 14.1.17.)
Flughafen Phalempin. Fuehrer Oblt. Bethge
Die Fuehrung von Jagdstaffel 11 uebernahm am 20.1.17. Lt. Frh. von Richthofen (Jagdstaffel Boelcke).
Bisheriger Fuehrer der Jagdstaffel 11, Oberlt. Lang, uebernahm am gleichen Tage die Fuehrung von Jagdst. 28 bei 4. Armee.
(Kogenluft. I Flieg. No. 57439 v. 14.1.17.)
Hauptm. Moosmair, Fuehrer bay. Flieg. Abt. (A) 292, zu Fea Schleissheim versetzt.
Hauptm. Hailer, Fea. Schleissheim, zum Fuehrer bay. Fliegerabt. (A) 292 ernannt, (Kogenluft. I Flieg. 57343 v. 13.1.17.)-Bay. J.J. IVa No. 566 v. 10.1.17.)
Feldfliegerabteilung 40 tritt zur 2. Armee zurueck (H.Gr.Krprz.v. B. Ic No. 7889 v.17.1.17.)

2.) Allgemeine Wetterlage:

Meist tiefhaengende geschlossene Wolken u. Schneegestoeber.

3.) Feindl. Flugtaetigkeit:

Sehr gering. 2 Bomben von 1 Flieger auf La Passée (kein Schaden)

4.) Eigene Flugtaetigkeit:

Jm ganzen 34(65)Kriegsfl. 25(67)Flugstd. 36(49)Bildmeldungen:

Right: Weekly activity report of the Kommandeur der Flieger:
Under "Organizational Changes," one can read, among other things:
"Jagdstaffel 30 is established at 6th Army on 20.1.17 (O.H.L. IC 44332 po. v. 14.1.17. Phalempin airfield, Staffelführer Oblt. Bethge.
The command of Jagdstaffel 11 was taken over by Lt. Frh. von Richthofen (Jagdstaffel Boelcke) on 20.1.17.
Previous leader of Jagdstaffel 11 Oblt. Lang took over leadership of Jagdstaffel 28 on the same day. At 4th Army. (Kogenluft I. Flieg. No 57439 v. 14.1.17.
General weather conditions:
Furthermore, the general weather situation reads: Mostly low-hanging, closed clouds and snow flurries.
Enemy aviation activity:
Little enemy flight activity is reported.
Own flying activity:
In total, 34 (65) war flights, 25 (67) flight hours, 36 (49) photo reports. [In parentheses the data of the previous week].
(Bayerisches Hauptstaatsarchiv, Abt. 4, München)

The New Staffelführer Lt. Manfred Freiherr von Richthofen

The Book *Der Rote Kampfflieger*

Above: Formal Portrait of Lt. Manfred Freiherr von Richthofen, the new Staffelführer of Jasta 11, with him wearing no other decorations besides the "Pour le Mérite". (L. Bronnenkant)

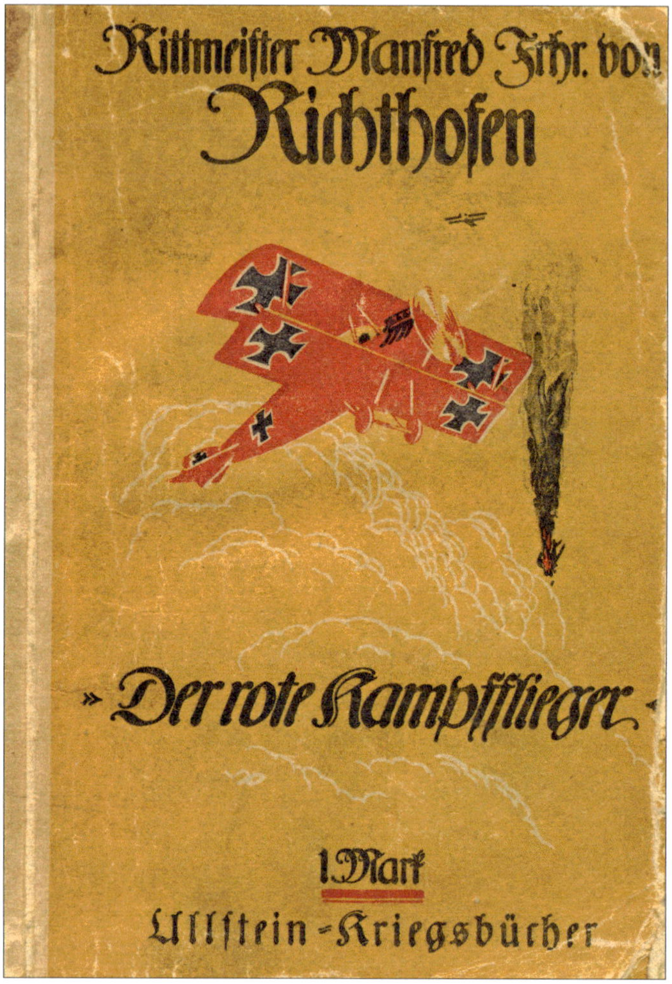

Above: "Der rote Kampfflieger" was published in October 1917, and seen here is the original cover art.

Much has been written about Manfred von Richthofen over the course of the past century, and his alleged autobiography, published under the title *Der rote Kampfflieger*, has been regularly treated as a firsthand source of information. But it must be pointed out that quotes ascribed to Manfred von Richthofen in this book should not necessarily be treated as such.

Certain individuals were idolized as heroes by the nations involved in the First World War, just as had occurred in previous conflicts. High-ranking commanders were usually offered for this purpose, as they were already known to a certain degree by the general population. At the beginning of World War I, this role was primarily filled for the Germans by Paul von Hindenburg, the commander-in-chief of the troops in East Prussia. But as the war progressed and became prolonged, a certain amount of 'war fatigue' grew among the populace and new heroes had to be offered to strengthen the people's will to persevere. The use of the airplane – a new weapon employed for the first time in a conflict on a broad scale – captured the public imagination and brought attention to its individual pilots, which naturally progressed into their being presented as national heroes. In particular, the newly-evolved role of fighter pilot came to the forefront. It was comprised of daring young men who used their

Left: The ghostwriter behind the book "*Der rote Kampfflieger*", Erich von Salzmann, pictured here during his service as officer of the Schutztruppe in German Southwest Africa (now Namibia) during the uprising of the Herero and Nama in 1904.

new weapon to bravely and directly confront their enemy in the air; and their success was easily demonstrated and comprehended by the number of victories they could attain. Books and newspaper articles that featured photographs of these pilots and the wreckage or captured examples of their conquered foes could be constantly produced that had immeasurable propagandistic value. They also gave the readers an opportunity to visualize and vicariously experience the successes of these new heroes.

The German Army's leadership, like the French High Command, recognized the growing popularity of successful fighter pilots and quickly began to take advantage of their propagandistic value to increase war morale. In France, it was pilots such as Roland Garros and Jean Navarre. In the German Empire, the first aviators to be awarded the Order Pour le Mèrite, Oswald Boelcke and Max Immelmann, initially fulfilled this role. Accordingly, the book *Hauptmann Boelckes Feldberichte* (Captain Boelckes´s Field Reports) was published by Friedrich Andreas Perthes in Gotha in 1916, and publisher August Scherl produced the book *Immelmann, meine Kampfflüge* (Immelmann, My Combat Flights) in Berlin the same year. Both works primarily were compilations of the airman's letters and combat reports. Max Immelmann had died in a crash on 18 June 1916 and Oswald Boelcke had lost his life in a midair collision with a comrade on 28 October 1916. So, both men were no longer alive when these works were published, and their stories had ended. As a result, the German Army's leadership looked for a suitable replacement who could carry on the dead heroes' propagandistic value. In early 1917, Manfred von Richthofen was the most successful living German fighter pilot and they obviously turned to him. He began to be featured in newspaper articles and plans to publish a book about his exploits were undoubtedly set in motion. As far as we know, however, von Richthofen was quite unenthusiastic about such a project. He was extraordinarily busy leading a Jagdstaffel at a hotly-contested sector of the front, personally flying up to five missions per day, and training his pilots for air combat – not to mention his numerous administrative duties. It would not have been possible for him to both fulfill his military duties and concentrate on such a work.

Nevertheless, the Supreme Army Command wanted it done and it appears that by March/April 1917, Manfred von Richthofen acceded to their wishes. A compromise had been reached and a ghostwriter was hired for him since he was too busy to take on the project himself. The man chosen to be von Richthofen's ghostwriter was the military officer and journalist Erich von Salzmann, an employee of Ullstein Verlag.(1) Erich von Salzmann held his first meeting with von Richthofen in the spring of 1917 in Douai, and the men met again during von Richthofen's leave the following May.(2)

When analyzing the book's contents, one may assume that von Salzmann received copies of von Richthofen´s air combat reports during the course of one of the early meetings in order to use them as a basis for the book, since the combats and events described in the book only cover the period up to the end of April 1917.(3)

While Manfred von Richthofen was at home in Schweidnitz during his leave from the front in May 1917 a shorthand typist from Ullstein Verlag recorded memories provided by von Richthofen. In this context it should be mentioned that, according to former Staffel members, Manfred von Richthofen did not keep a diary.(4) Another meeting between von Richthofen and von Salzmann is recorded to having taken place in August 1917.(5) Any further communication between them with regard to the book's contents is not documented. The book's manuscript must have been very close to being finished by this time, however, since the finished work made its first appearance on the shelves of book dealers in October 1917.

It is important to emphasize here that the contract with Ullstein, still available at the Ullstein archive, stipulated that Erich von Salzmann could revise the manuscript not only linguistically and stylistically, but also in terms of content. It is also significant that von Richthofen gave von Salzmann his power of attorney in the contract:

gs-Bücher

3 —
3 —
29 Juni 1917

Vertrag

Zwischen Herrn Oberleutnant Freiherr Manfred von R i c h t h o f e n

und der Firma Ullstein & Co, Berlin, ist folgendes Abkommen getroffen:

§ 1. Herr von Richthofen

überträgt der Firma Ullstein & Co ~~das Recht der alleinigen Buchausgabe seines Werkes~~

das alleinige und ausschliessliche Verlagsrecht seines in Arbeit befindlichen Werkes über seine Fliegererlebnisse.(Die Festsetzung des Titels bleibt gegenseitiger Vereinbarung vorbehalten).

Die Firma Ullstein & Co nimmt dieses Werk in ihre Bibliothek „Kriegs-Bücher" auf.

Den Zeitpunkt der Veröffentlichung bestimmt der Verlag.

Die Vergebung des Werkes zum Abdruck in Zeitungen und Zeitschriften ist nur ~~mit Genehmigung~~ ~~des~~ dem Verlage Ullstein & Co gestattet.,und zwar nur für Bruchstücke des Werkes.Sollten für derartige Abdrucke Honorare eingehen,so ist die Hälfte dieser Honorare an den Verfasser abzuführen.

Das Manuskript soll möglichst Ende April dieses Jahres im Besitz des Verlages sein.

§ 2. Der Herr Verfasser erhält von jedem abgesetzten Exemplar, für das ein Ladenpreis von M. 1,— festgesetzt ist, eine Tantieme von 8 Pf. (8 % des Ladenpreises). Die Tantieme wird dem Herrn Verfasser mit Mark ~~8~~10 000.-(~~acht~~zehntausend Mark) garantiert. Diese Garantie ist zahlbar, nachdem das Plazet der Zensur erteilt ist.

Weitere Abrechnung über tantiemepflichtige Exemplare erfolgt für jedes Kalenderjahr bis spätestens

geändert mit Einverständnis
Firma E. v. Salzmann

zum 15. April des folgenden Jahres. Die Firma Ullstein & Co läßt auf Wunsch die Richtigkeit dieser von ihr für das abgelaufene Kalenderjahr gegebenen Tantiemeabrechnung durch einen gerichtlichen Bücherrevisor bestätigen.

§ 3. Sollte der Umfang des Werkes den Raum von 256 Seiten im Format der Kriegs-Bücher überschreiten, so verpflichtet sich der Verfasser zu entsprechenden Kürzungen, oder erteilt für den Fall, daß er sie nicht rechtzeitig oder nicht im ganzen Umfange ausführen sollte, der Verlagsbuchhandlung die Genehmigung, diese Kürzungen durch eine ihr geeignet scheinende Persönlichkeit bewirken zu lassen.

§ 3. Die endgiltige Festlegung des Textes mit der Verlagsbuchhandlung zu vereinbaren, ist Herr Hauptmann von Salzmann ermächtigt.

§ 4. Der Verlag ist berechtigt, die Uebersetzungsrechte des Werkes nach fremdsprachlichen Ländern zu verkaufen sowie Abdrucksrechte des deutschen Textes nach dem Ausland zu vergeben; jedoch ist er verpflichtet, die Hälfte der hierfür eingehenden Honorare an den Verfasser abzuführen.

§ 5. Der Verfasser hat das Recht, sein Werk 10 Jahre nach Abschluß dieses Vertrages in eine Gesamtausgabe seiner Werke aufzunehmen; jedoch darf der Ladenpreis des betreffenden Bandes nicht unter M. 2,— festgesetzt werden.

§ 6. Der Verfasser hat Anspruch auf dreißig Freiexemplare.

§ 7. Der Verfasser räumt für seine nächsten 3 sich auf den Krieg beziehenden Werke unter den in diesem Vertrag festgelegten Bedingungen dem Verlag Ullstein & Co das Vorkaufsrecht ein. Der Verlag hat sich spätestens 4 Wochen nach Erhalt des Manuskriptes zu entscheiden, ob er von diesem Vorkaufsrecht Gebrauch machen will oder nicht.

Frhr. v. Richthofen
Rittmeister.

Berlin den 29. 6. 1917
Ullstein & Co
[illegible]

Vollmacht für Herrn Hauptmann a. D. Erich von Salzmann, Berlin W., Meinekestrasse 16.

= =

Ich bevollmächtige hierdurch Herrn Hauptmann a. D. Erich von Salzmann, gegenüber der Firma Ullstein und Co. in Angelegenheiten des von der Firma herauszubringenden Buches " Der rote Kampfflieger" irgend welche ihm notwendig erscheinenden Aenderungen oder Verbesserungen selbständig und ohne mich vorher zu befragen, vorzunehmen, ebenso, meine eigenen Interessen gegenüber der Firma Ullstein & Co. oder solchen Personen, die mit Bezug auf die Fertigstellung des Buches in Frage kommen, wahrzunehmen.

Herr von Salzmann ist berechtigt, in meinem Namen alles, was das Buch anbetrifft, selbständig zu regeln.

Frhr. v. Richthofen

Rittmeister und Führer der Jagdstaffel 11.

Above and 2 Preceding Pages: The actual contract between Firma Ullstein and Richthofen, laying out all of the details, payment, and the notice that von Salzmann was empowered to write the book and make any editing decisions he wanted. As noted in the text, this raises serious doubts about the veracity of much of the material in "*Der Rote Kampfflieger.*"

"*...with respect to Ullstein & Co. in matters relating to Der Rote Kampfflieger to be published by said company, to make all changes or improvements deemed necessary by it without having to question me in advance and, in addition, to represent my interests towards Ullstein & Co. or such other relevant persons regarding the completion of my book. In this regard, Mr. Salzmann is entitled to act on my behalf.*"[6]

Manfred von Richthofen also stipulated that:

"*Herr Hauptmann von Salzmann is authorized to settle upon the final version of the text with the publisher...*"[7]

Therefore, upon the contract's execution, von Richthofen relinquished any rights he otherwise might have had to influence the book's final version.

Der rote Kampfflieger therefore differed significantly from Boelcke's and Immelmann's 1916 publications in that they primarily consisted of direct quotations of letters and reports. In the case of *Der rote Kampfflieger*, von Salzmann largely had a free hand in what he wrote. We can be certain that he, as a Hauptmann, ensured that the expectations of the Supreme Army Command were met as well as those of Ullstein and Co. We must also suspect that he presented what he believed the reader wanted or needed to hear.

Considering the many factual and chronological inconsistencies and errors in the book, Erich von Salzmann must have at times altered the original combat reports and the interviews that were available to him. We must also consider that von Salzmann was under a lot of pressure to deliver a finished product within a short period of time and that this might have contributed to some of its factual inconsistencies and errors. In any event, the book, published in October 1917, was greeted by a great deal of enthusiasm by the public. It became an overwhelming commercial success, selling 560,000 copies of its 1917 edition. In 1920 extended versions were published.

The German Army Command's intended goal of achieving a work of high propagandistic value was certainly met. But from today's point of view, it is impossible to distinguish which parts faithfully presented the "real" facts and events described by von Richthofen from what might have been altered, rewritten, embellished, or even invented by von Salzmann; and we can only speculate how much of von Richthofen's original content found its way into the final product rendered by von Salzmann.

Therefore, any quotations from this book that are directly attributed to von Richthofen must be treated with a high degree of caution. We also have contemporary evidence that von Richthofen himself was not pleased with the final result. Former Jasta 11 pilot Carl-August von Schoenebeck informed the author that a pilot who had arrived at Jagdgeschwader I in the late autumn/early winter of 1917/18 had intended to ask von Richthofen to autograph a copy of *Der rote Kampfflieger*. When Wilhelm Reinhard learned of this, he took the pilot aside and advised him to get rid of the book if he *"möchte sich keinen Schiefer bei Richthofen einziehen"* ("didn't want to get on the wrong side of Richthofen"). Carl-August von Schoenebeck also related that he could not recall the book ever being a topic of any conversation at the Staffel. It seemed as if there was a taboo against mentioning it and that it was treated as if it simply did not exist by the members of Jagdstaffel 11 and Jagdgeschwader I. This assessment was confirmed for the author by Hans-Georg von der Osten.[8]

Consequently, passages from the book should not necessarily be taken at face value. In particular, it should be understood that any assessment of Manfred von Richthofen's personality or psyche based only on passages from *Der rote Kampfflieger* (as done by some authors who were likely unfamiliar with its genesis) lacks any secure foundation.[9] For all of these reasons, passages quoted from *Der rote Kampfflieger* are clearly identified herein and should be viewed accordingly.

Manfred von Richthofen's Military Career Before his Appointment as Staffelführer

Lt. Manfred Freiherr von Richthofen arrived at Jasta 11's airfield La Brayelle on 20 January 1917. After having served for four and a half months with Jagdstaffel 2 'Boelcke' and with 16 aerial victories to his credit, an appointment to become the Staffelführer of another Jagdstaffel could not have been unexpected. Nevertheless, he did not seem to be enthusiastic about his new position, i.e., according to his alleged response as related in his ghostwritten autobiography *Der rote Kampfflieger*:

"*I have to say I was annoyed. One was so used to working wonderfully with one's comrades at Jasta Boelcke. To have to start all over again with settling in, etc., was for the birds.*"[1]

But duty was duty and 'the new man' arrived at Jasta 11's La Brayelle airfield a few days later. So, who was this 'new man'?

Manfred Freiherr von Richthofen was born on 2 May 1892 in Kleinburg near Schweidnitz (today's Borek, now a suburb of Wroclaw) in Lower Silesia as the second child and first son of career officer Major Albrecht Freiherr von Richthofen and his wife, Kunigunde.[2] The Richthofen children were raised in a strict household with their father deciding early on that Manfred and his brothers, Lothar and Bolko, should pursue military careers just like him. As Eddy Lübbert (the son of Jasta 11's Friedrich-Wilhelm Lübbert) once told the author, the children behaved with respect and politeness toward their parents, whose relationship with them was more formal than cordial. Still, their relationship with their mother was much closer than with their strict father.[3]

Obedience, conscientiousness, self-sacrifice, love for the

Above: Major Albrecht von Richthofen, father of Manfred and Lothar von Richthofen. (L. Bronnenkant)

Above: The house of the von Richthofen family in Kleinburg near in Schweidnitz, where Manfred and his siblings grew up. (E. Wischnowski)

homeland, and loyalty to the royal family were the iron principles of the brothers' upbringing and molded them from their early youth on. The summer holidays that Manfred and Lothar routinely spent at their aunt's estate compensated for their strict upbringing at home and at cadet school. She always greeted them with the words: "*You can do whatever you want here.*" Those holidays were times of freedom and adventure with fishing, hunting, living in tents, playing games of Indians, and campfires. The brothers loved their aunt for this carefree time which encouraged their enjoyment of adventures and challenges. Accordingly, it should be little

Right: Manfred von Richthofen photographed at the age of seven, four years before he had to join the cadet corps. (E. Wischnowski)

Right: The Richthofen brothers, from left: Bolko, Manfred, and Lothar von Richthofen. Manfred is now wearing the uniform of the Cadet-Corps, and the wintry setting suggests the photo may have been taken during his Christmas holidays in 1903. (E. Wischnowski)

Above: Three-quarters of the Richthofen family in one picture. From left: Kunigunde von Richthofen (mother), Manfred,, his sister Ilse, and the youngest brother Bolko, now wearing the Cadet-Corps uniform. (Taktisches Luftwaffengeschwader 71 "Richthofen")

Above: Manfred von Richthofen and his sister Ilse in front of the family home in Schweidnitz. (Taktisches Luftwaffengeschwader 71 "Richthofen")

surprise that the brothers shared traits of iron discipline, conscientiousness, self-sacrifice, and patriotism, mixed with a love of adventure and the pursuit of challenges, which became the cornerstone of their military success.

Manfred had private education lessons until he was nine years old and then went to school in Schweidnitz. His military career began at an early age. He was enrolled in the Cadet Corps at Wahlstatt (today's Legnickie Pole, Poland) in 1903 when he was 11, and as was stated in *Der rote Kampfflieger*:

"I was not overly fond of becoming a cadet, but it was my father's wish, so I wasn't consulted."(4)

The book related also that, after having spent six unpleasant years in Wahlstatt, which was known for its strict discipline, he attended the Senior Cadet Academy at Gross-Lichterfelde at age 17:

"I liked it a lot better at Lichterfelde. You weren't so isolated from the world, and I began to live somewhat more like a human being."(5)

By entering his sons in a cadet school, Major Albrecht von Richthofen was following the tradition of aristocratic families living under European monarchs where the nobility saw themselves as the political and military pillars of the state and monarchy. Noblemen were an important, if not dominant, part of the Officer Corps, not only in Imperial Germany but also in other countries such as Great Britain, Austria-Hungary, Russia, Italy, and the Republic of France. The German military offered excellent training for its Officer Corps. It not only included military disciplines and skills, but also potentially provided strong qualifications for later civilian life. Moreover, a successfully completed military career often promised later success in business or politics.

The quest for national and economic greatness, which included colonialism, was common among European states in the 19th and early 20th centuries and routinely led to tension and conflict between those powers. The military therefore figured prominently in such efforts and therefore had a highly respected role in society, seeing itself as part of the social elite.

After his time as a cadet, Manfred von Richthofen – a passionate horseman – joined Ulanen-Regiment Kaiser Alexander III. von Russland (Westpreußisches) Nr.1, based at Militsch (now Milicz, Poland) as a Fahnenjunker (Ensign) in 1911, and received his officer's patent on 19 November 1912. During his time there he participated in a number of horse riding events. One of them was 1913's Kaiser Prize race, in

Above: Manfred von Richthofen, far left, pictured during a jump at a horse race in Breslau, before the war. (E. Wischnowski)

which he was the only participant to complete the entire cross-country course without any mistakes, despite the fact that he had broken his collarbone in a fall 70 km from the finish.[6]

After the outbreak of the war, his Uhlan regiment initially served in Belgium. As the initial war of movement in the west quickly turned into trench warfare, the cavalry's traditional role was greatly diminished and he was posted to the Eastern Front to serve with Infantry Regiment No. 154.

In January 1915 he was posted to the 18th Infantry Brigade, back on the Western Front, as an ordnance officer. This role, however, did little to satisfy his quest for action, and he began looking elsewhere for a greater challenge. The increased use and notoriety of military aircraft attracted his attention and he applied for entry in the Fliegertruppe.

He was sent to Flieger-Ersatz-Abteilung 7 in Cöln (Cologne) for observer training on 27 May 1915 and then to Flieger-Ersatz-Abteilung 6 at Grossenhain on 10 June 1915. This was because at that time, an observer was usually the senior member of a crew who, as an officer, had been trained to identify and interpret military positions, materiel, and movement. After completing his course, he was transferred to Feldflieger-Abteilung 69 on 21 June 1915. From there, he joined Brieftauben-Abteilung-Ostende (B.A.O., or Carrier Pigeon Unit Ostende), the cover name for Germany's first bombing unit (later known as "Kampfgeschwader I") in late August 1915 that was under the direct control of Supreme Army Command.[7]

Above: Clothes make the man! This very rare view of Manfred von Richthofen in civilian clothing leaves him almost unrecognizable. (E. Wischnowski)

Left: Manfred von Richthofen's great passion was hunting, as it was for many aristocrats. (Taktisches Luftwaffengeschwader 71 "Richthofen")

On 1 September 1915, he experienced his first air combat with Lt. Georg Zeumer as his pilot.

Der rote Kampfflieger related the events of that day as follows:

"We flew five to six hours a day without ever seeing an Englishman… suddenly, I spotted a Farman who was unabashedly trying to fly a reconnaissance mission. My heart pounded as Zeumer approached him. I was eager to see what would happen next. I'd never seen an air combat and only had a vague concept of it… Before I knew it, both of us, the Englishman and I, were charging past one another. I had fired four shots at most when the Englishman suddenly gave us a full load from behind. I have to say I had no feeling of danger because I had no conception of how such a fight would end up. We turned around each other a few more times until finally, much to our surprise, the Englishman did an about turn and flew away from us. I was greatly disappointed, and so was my pilot."[8]

B.A.O. was redeployed from Ghistelles (Gistel) in West Flanders, Belgium, to Rethel in the northeast of France on 1 October 1915, and during the train ride to Rethel, von Richthofen had an encounter that would have the most profound impact on his future, as described this way in *Der rote Kampfflieger*:

"A young, unpretentious Oberleutnant sat at the next table in the dining car. There was no particular reason for him to attract any attention except for one outstanding fact: he was the only one of us who had already shot down an enemy flier – and not just one, but four. He'd even been mentioned by name in the Army Reports. He impressed me because of his quite dashing experiences. Despite my great efforts, I had nothing to show for them, that is, nothing that had been officially acknowledged. So, I wanted to learn how this Oblt. Boelcke really did it. So, I approached him and asked: 'Tell me honestly, how do you really do it?' He laughed, quite amused, even though I had posed the question in all seriousness. Then he answered me: 'Well, good heavens, it's really quite simple. I fly right up to him, aim well, and then he just falls down.' I just shook my head and thought that I'd done that too, only they hadn't fallen down for me. The difference was, however, that he flew a Fokker and I my big Kampfflugzeug."

I made an effort to get to know this nice, unassuming man, who had made such an impression on me. We often played cards together, went for walks, and I asked him many questions. Thus, I gradually came to the conclusion: 'You must learn to fly a Fokker yourself, then things will go much better.'"[9]

Manfred von Richthofen had run into Oswald Boelcke, who, with four victories to his credit, had been temporarily loaned out to B.A.O. by its sister unit, Brieftauben-Abteilung Metz (B.A.M.). As it turned out, von Richthofen had already talked Georg Zeumer into giving him flying lessons by this time. But after his first solo attempt on 10 October had ended in a crash, it was decided that he should undergo formal training elsewhere. On 8 November 1915 he was transferred to the Inspektion der Fliegertruppen (Inspectorate of Aviation) and then on to Flieger-Ersatz-Abteilung 2 and the Fliegerschule (aviation school) at Döberitz on 15 November where he completed the last of three flight examinations on 25 December 1915. This was followed by a transfer to Kagohl II`s Kampfstaffel 8 on 23 March 1916.

There he was given the opportunity to fly a Fokker Eindecker in action. But since only limited numbers of the monoplane were available at the time, he had to share the machine with Lt. Reimann; and Reimann destroyed the rare bird as the result of an emergency landing between the lines. A Fokker E III arrived in due course as a replacement, but then it was von Richthofen´s turn to crash it, after it suffered

Above: A.E.G. G II G.6/15 saw service with Kagohl II and was marked with a prominent cat symbol. It was presumably flown by Lt. von Richthofen and Lt. Zeumer. (L. Bronnenkant)

Above: Lt. Georg Zeumer flew a number of frontline missions as pilot along with his observer Manfred von Richthofen. (L. Bronnenkant)

Above & Below: Two snapshots of Manfred as a beach guest at Ostend beach. Being stationed in nearby Ghistelles had its pleasant side. (G. VanWyngarden)

Above: Manfred von Richthofen with his former observer Lt. Alfred Gerstenberg. (L. Bronnenkant)

Above: Hptm. Oswald Boelcke, the master teacher of the German Jagdflieger, in a casual pose that was typical for him. (L. Bronnenkant)

Above: Kampfstaffel 8 of Kagohl II at Mont Airfield. The Roland C II on the far left may have been the aircraft flown by Manfred von Richthofen and Alfred Gerstenberg.

Above: Manfred von Richthofen as a member of Kampfgeschwader II, Kampfstaffel 8. (G. VanWyngarden)

Above & Below: Manfred von Richthofen's crash-landed Fokker E III in a cornfield. It's hard to believe that he remained unharmed. (Taktisches Luftwaffengeschwader 71 "Richthofen")

The ground crew marvel at the crashed Fokker E III. (L. Bronnenkant)

The ground crew pose with the crashed Fokker E III. Left is von Richthofen´s dog "Moritz". (J. Miller)

The Fokker E III is recovered by the mechanics for transportation. (Taktisches Luftwaffengeschwader 71 "Richthofen")

Above: Parade formation of Kampfgeschwader II at Kowel on the Eastern Front, on the occasion of the visit of the Commander-in-Chief of the Eastern Front Generalfeldmarschall Prinz Leopold von Bayern.

Above: The pilots of Kampfstaffel 10 are introduced to Prinz Leopold von Bayern; Manfred von Richthofen is 6th from left. (L. Bronnenkant)

Above: Oblt. Wilhelm Boelcke, the man who recommended Manfred von Richthofen to Oswald Boelcke. (L. Bronnenkant)

a mid-flight engine defect during its third sortie to the front.

Von Richthofen claimed his first victory as a pilot (he was flying a two-seater on this occasion) on 25 April 1916, and although it was actually mentioned in the Heeresbericht (Army Report), it went unconfirmed in the end. Another important change happened a little more than two months later, as Kagohl II was moved to the Eastern Front on 28 June 1916, where it settled into its new base outside of Kowel (now Kovel, Ukraine). It was there in mid-August 1916 that Oswald Boelcke, visiting his brother Wilhelm who was the Staffelführer of Kagohl II`s Kampfstaffel 10, ran into Lt. von Richthofen once again. Boelcke, who was on his way back from an enforced leave, was now slated to return to the Western Front to take command of and staff the soon-to-be established Jagdstaffel 2. Before he did so, he visited his brother and asked Wilhelm if he could recommend any pilots to him. Wilhelm named von Richthofen, among others, and Boelcke promptly invited him to join his unit. Opportunities to engage in aerial combat against the Russian forces were rare, and so it comes as no surprise that Manfred eagerly accepted the unique offer and arrived at Jasta 2 on the

Above: Oswald Boelcke, center, together with his brother Wilhelm, right, during his visit to Kovel.

Western Front on 1 September 1916.

Jagdstaffel 2 was stationed at Vélu airfield near Bertincourt within the German 1st Army's zone when von Richthofen arrived there. Just one day later, its Staffelführer, Hptm. Oswald Boelcke, scored the first air victory for the newly-formed Jagdstaffel; and he was the only one to gain air victories (his 20th through 26th) during the two weeks that followed. In this context it must be taken into account that, for a little over two weeks, only three or four fighter aircraft were available to the Staffel. But the early sorties with the Jagdstaffel gave von Richthofen the opportunity to learn from Germany's best fighter pilot – and he learned very quickly and achieved his first confirmed victory on 17 September 1916. He forced F.E.2b 7018 from 11 Sqn. RFC to land on German territory near Villers-Plouich. The pilot 2/Lt. L B F Morris was wounded and died of his wounds. The observer, Lt. T. Rees was killed during the air fight.

He was successful for the second time just six days later. In an aerial fight involving Ketten from Jagdstaffeln 2 and 4 versus six Martinsyde G. 100/102s of 27 Sqn. RFC, he shot down one of the British aircraft. A week later, on 30 September 1916, he recorded his third air victory.(10)

By the time of Oswald Boelcke's death as the result of a mid-air collision on 28 October 1916, Manfred von Richthofen had six aerial victories to his credit. Following his eighth victory, von Richthofen was awarded the Royal Hohenzollern House Order, Knight's Cross with Swords, on 11 November 1916.

On 23 November 1916, he and Lts. Reimann and Collin attacked an Airco D.H. 2 south of Bapaume, after which he recorded in his combat report:

"I attacked, together with 2 planes, a Vickers one-seater at 3,600 meters alt. After a long dogfight of 3 – 5 minutes I had

Above: Manfred von Richthofen´s first victory at Jagdstaffel 2, F.E.2b 7018, is surrounded by German soldiers after making a forced landing in German territory. (G. VanWangarden)

pressed down my adversary to 500 meters. He now tried to escape, flying to the front. I pursued and brought him down after 900 shots."[11]

His opponent was the famous English fighter pilot Major Lanoe George Hawker, commander of 24 Squadron and recipient of Britain's highest military decoration, the Victoria Cross. He had fatally crashed in Airco D.H. 2 5964.

By year's end, Manfred von Richthofen was tied with Wilhelm Frankl as Germany's highest-scoring living fighter pilot with 15 confirmed successes. He downed his 16th victim on 4 January 1917 and was subsequently informed that he was being given command of Jasta 11.[12]

Above: Lt. Manfred von Richthofen in front of his Albatros D II which had a red painted fuselage and a black and white "ring" marked on the nose of the airplane. The color combination references the black-white-red German flag. (L. Bronnenkant)

Above: Pilots of Jagdstaffel 2 in November 1916, from left: Staffel commander Oblt. Stefan Kirmaier, Lt. Hans Imelmann, Lt. Manfred von Richthofen and Lt. Hans Wortmann.

Just days later, he learned that he had been awarded the Order Pour le Mérite on 21 January 1917. And so it was that he arrived at Jasta 11's La Brayelle airfield with Prussia's highest award for bravery dangling from his Ulanka's (Uhlan tunic's) collar.

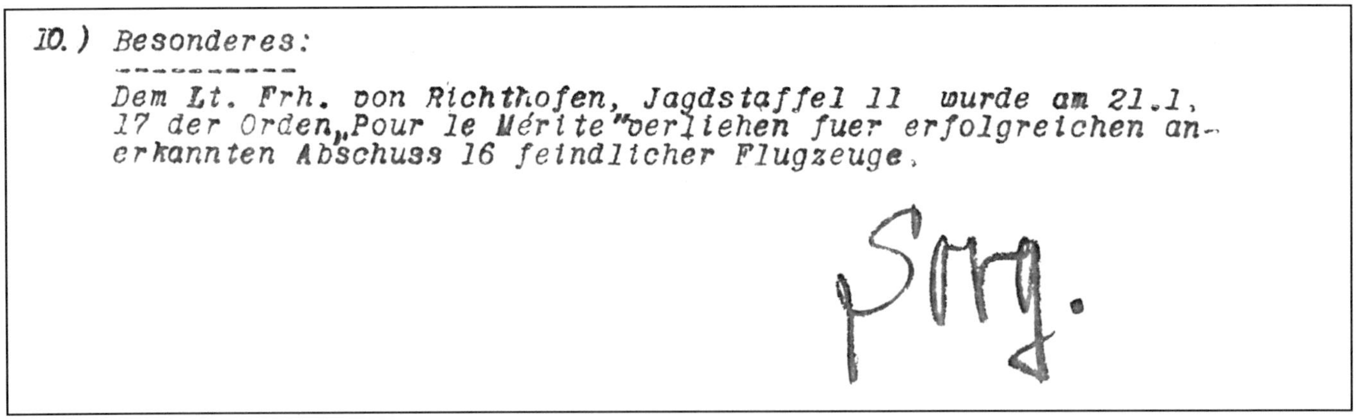

10.) Besonderes:

Dem Lt. Frh. von Richthofen, Jagdstaffel 11 wurde am 21.1.17 der Orden „Pour le Mérite" verliehen fuer erfolgreichen anerkannten Abschuss 16 feindlicher Flugzeuge.

Sorg.

Above: Weekly activity report of the Kommandeur der Flieger from 14 January 1917 to 21 January 1917. No.10. Special: Lt. Frhr. von Richthofen, Jagdstaffel 11 was awarded the Order "Pour le Mérite" on 21 January 17 for successful confirmed shooting down of 16 enemy aircraft. Signed Sorg (Bayerisches Hauptstaatsarchiv, Abt. 4, München)

Manfred von Richthofen in Contemporary Reports

Above: Lt. Otto Brauneck, whose letters serve as an important contemporary source of information for this book. (M. Schmeelke)

Above: Busso von Alvensleben met Richthofen as student pilot, and from June 1917 to November 1917 served as a member of Jagdstaffel 4.

Much has been written about the personality and character of Manfred von Richthofen, ranging from glorifying him as a god-like war hero to demonizing him as a psychopathic killer. Many authors have been influenced by their ideological orientation or the political context of their times. During and after the First World War, von Richthofen was stylized as a glorious, almost superhuman hero in Germany, while over the course of the last ten to fifteen years some authors in Germany have presented Manfred von Richthofen as a psychopathic, murderous Prussian militarist, who greatly enjoyed seeing enemy aircraft crews crashing down in flames.

Accordingly, depending on the ideological and political preferences of the respective authors, interpretations of Manfred von Richthofen's character have often been based on his alleged words from *Der rote Kampfflieger* which, as we have seen, may not actually have been his. In light of this, it is preferable to allow those who personally knew or interviewed von Richthofen to be the true authorities on the man. To do so, we have written statements of contemporary witnesses as well as interviews that Alex Imrie and this author conducted with surviving members of Jagdstaffel 11 and Jagdgeschwader I, all of whom knew von Richthofen personally.

A reporter from *Die Woche* magazine, who visited Jasta 11 in April 1917, provided this portrait of Richthofen:

"Richthofen is medium-sized, muscularly built, almost stocky. Bright eyes peer out under blond hair from a good, intelligent face with a strong profile; eyes that have something special as Boelcke's did. It probably is that his eyes are full of modest simplicity because the young cavalry Rittmeister is reserved, yes, almost embarrassed, such that one thinks he could easily blush, but these eyes are clear and large. What is unforgettable about them is that the blue pupil, even looking straight ahead, does not touch the lower eyelid and the white of the eye remains visible underneath.

He is no friend of many words; indeed, a certain awkwardness that comes from his young age holds him back somewhat until he gets to know a person better and gets on closer terms. Then everything changes quickly and all kinds of Silesian friendliness bubbles out from him. His handshake is short and firm. His behavior is that of a well-bred young man of a good family. His manner is that of a Prussian officer who has matured under the discipline of an officer corps."(1)

Carl Allmenröder described von Richthofen in a letter to his fiancée, Helene Kortenbach, with the following words:

Above: Lt. Carl-August von Schoenebeck, Jagdstaffel 11, with whom the author had a number of conversations about Manfred von Richthofen.

"Richthofen is a remarkable person who cast a spell over us. He is able to inspire us and mold us into a very close-knit group. Each looked out for the other and Richthofen looked out for all of us. He demanded courage, dedication, self-sacrifice, and loyalty and conducted himself as a personal example. You could always count on him to break off a dogfight and come to the aid of a comrade who had difficulty in another fight. This instilled confidence and a sense of security among us. He pays personal attention to each of us, even if he wasn't very successful at first like me. Everyone got a chance to prove himself. In private, he is quite a comrade, humorous, and anything but a stiff officer of the nobility".(2)

Otto Brauneck, who was transferred to Jasta 11 on 9 April 1917, wrote to his parents on 20 April 1917:

"Everyone [at Jasta 11] *is very nice, especially Richthofen, who you don't address as Rittmeister but only by his name, who is of a small, strong build and brimming with health, with very light blond hair and large blue eyes..."*(3)

Busso von Alvensleben, a flight student at the Army Air Park La Briquette near Valenciennes at the time, wrote in a letter after his encounter with Manfred von Richthofen:

"I met the 'older' Richthofen at Jasta 4 in Douai on the 17th. But that of course is the wrong expression and he made a great impression on me. He is actually quite young, 24 years old, and charming to be around. Once you got to know him, you could hardly believe that he practices such a rough trade. But he inspires you so much that your only wish is to have him as your teacher.

Above: Of all members of Jagdstaffel 11, Oblt. Wilhelm Reinhard was probably the closest to von Richthofen, after his brother Lothar, and may well have known him better than most people serving alongside him.

He is very energetic and has achieved his fabulous successes by training his Staffel to work together brilliantly."(4)

Carl-August von Schoenebeck, who served with von Richthofen in Jagdstaffel 11 from July 1917 through January 1918, characterized his Geschwaderführer in this way during a conversation with the author:

"The most striking thing about him was his charisma. I've seen very few people who've had such a personality as Richthofen. He radiated supreme authority. He had an ability to inspire and win over people. He was quite clear in his statements and orders. We all had tremendous respect for him and made every effort to fulfill our military responsibilities well and win his favor.

At first, he was restrained around new airmen, sometimes even harsh. You got the impression that he first wanted to get a good picture of the man and his character. Character meant for him: comradeship first, then loyalty, honesty, and the willingness to be self-sacrificial. If you passed the 'character test,' he became more approachable. You could go to him with any problem, and he took time to listen and to support you as much as he could.

As a combat pilot, especially a young one, you inevitably found yourself in situations where you were afraid of being shot down. Such experiences were nerve wracking. One of Richthofen's skills was recognizing when a pilot had problems or fear after such an experience. He would discreetly take him aside and do everything

Above: Nurse Käte Oltersdorf, who attended Manfred von Richthofen´s head wound in July 1917. Following his recovery, she was later invited to his Jagdgeschwader for a visit and received some personal gifts from him in gratitude for her care. (Taktisches Luftwaffengeschwader 71 Richthofen)

he could to help him regain his courage and self-confidence. As a rule, he then ensured that the pilot flew at his side during the next few missions to help him feel more secure. One could be sure that no word of such a conversation ever leaked out.

In an air fight you could be absolutely sure that he would come to your aid if he recognized you were in trouble. This gave as all a safe feeling.

He took his time and tried hard to pass on all his experience and knowledge to us young airmen. In the reviews that followed missions, he could be quite direct and sharp in his criticism but always fair, and some would leave the meeting with 'red ears' [i.e., very embarrassed]. *On the other hand, he was also ready to give out credit and praise. What he hated most was if a pilot showed 'laurig'* [cowardice] *in air combat or thereby left a comrade in serious danger. Such a pilot didn't stay long with the Staffel.*

He despised 'Duckmäusertum' [moral cowardice] *and demanded that everyone speak their mind freely even if he didn't agree with their opinion.*

Above: Friedrich-Wilhelm Lübbert, fighter pilot of Jagdstaffel 11.

After his brother, [Oblt. Wilhelm] *Reinhard was the next closest to him."*(5)

Wilhelm Reinhard, the man who Carl-August von Schoenebeck named as von Richthofen's second-closest confidante, wrote this about him in the spring of 1918:

"Our Rittmeister was quite a man. A well-educated personality, but difficult to understand at first for someone who's only known him for a short time. In the typical Silesian way, he came out of his shell slowly around people he didn't know. But once he'd grown fond of you, he would have gone through fire for you. There were many, particularly in the Fliegertruppe, who held the opinion that our Rittmeister was proud and unapproachable. It's probably true that when he first looked at people, he actually judged them not by their appearance but only for what they'd done. He was a completely military soldier. To him, an infantryman who stuck it out through daily artillery barrages or had participated in many difficult attacks was certainly as worthy as a pilot with air victories. He was reserved, and he opened up only to those whose character he had learned to respect over time.

During leisure time, he sought to continue his education; but he could never have been described as a bookworm. Conversations with him were always stimulating… I often saw him creating in-depth military reports. Shortly before his death, he made notes for us on what he thought a capable fighter pilot did, how he fought in the air in small and large groups – in short, what a fighter pilot should and should not do.

Above: Lt. Wilhelm-Gisbert Groos. (G. VanWyngarden)

He was a very purposeful character. Whatever he asked of others, he demanded first of himself. He tried to carry us along and make us into capable pilots through his example. He was often the first at the airfield and we would take off after him, ashamed. He also knew what he wanted with his superiors. If he knew something was right, he'd push it through with iron determination. He wasn't a one-sided character. He was interested in everything and of course particularly in military matters. He liked to talk with infantrymen who'd just come from a battle in the trenches and let them tell him how they had fared, and what they had seen of German and enemy aircraft. If he drove past artillery observation posts, he liked to get out and look at the battlefield from there with binoculars himself... He particularly liked going to the Luftschutzoffiziere [air raid officers], *getting detailed reports from them, and observing with his binoculars himself...He was not solely interested in fighter aviation; he was also familiar with the concerns and trials of the Fliegerabteilungen – the infantry airmen, the assault aviators, the artillery airmen – and he liked to talk with those gentlemen about the efforts of these individual air arms. In short, he wasn't a specialist in fighter aviation only, but had an interest in the entire military structure and was a born general staffer. Once something was firm in his mind, he carried it through to the end. When he recognized that something was ineffective, he made a quick decision and didn't withhold his opinion. His favorite people were those superiors to whom he could speak freely and did not begrudge frankness. He also loved when his subordinates spoke as frankly as he did. He never had a problem with those gentlemen who expressed contrary opinions and quite sincerely said: 'I like it that way.' When he realized that someone wasn't able to deal with the truth, he shut them out, which was sometimes misinterpreted. He did not like idle chatter. If you confided a secret to him, you could count on him being as silent as a grave. He was a beloved superior who looked after you with his entire being. The result was that we would go through fire for him.*

Above: Lt. Johann Janzen, Pilot and Staffelführer of Jagdstaffel 6 in Jagdgeschwader I. (A. Imrie)

He displayed another side while among comrades in the Officers' Mess. Then he was amused by the youths and in the mood for joking around. He never took offense to a friendly joke. In the Officers' Mess, he was our equal, drank, and was funny, but never to the extreme. He enjoyed conviviality and welcomed guests; but he didn't like those – especially reporters – who wanted to question him. Then he became introverted, closed up, and didn't say a word. He put his whole heart and soul into Jasta 11. To him, the best was just good enough. He told us how to perform air combat, showed us in the air, and demanded we do the same. If we didn't,

Above: Vzfw. Franz Hemer pilot of Jagdstaffel 6.

if we lacked the nerve or bravado that he demanded of every fighter pilot, then he told us frankly that he thought it would make more sense to take up another military occupation."(6)

The noteworthy thing about this description of von Richthofen's personality is that Reinhard wrote it in the spring of 1918, two years before it was published in the book *Ein Heldenleben*.(7)

The comradely behavior that Wilhelm Reinhard said Richthofen displayed was not limited to just members of his unit. Nurse Käte Oltersdorf, who cared for Richthofen at Courtrai hospital after he had suffered his head wound in July 1917, wrote in her diary how he looked after his fellow wounded comrades there:

"Suddenly, I was called [over to Richthofen]*:*

"Sister ["Schwester", or "sister", was how a nurse was addressed in Germany], *would you like autographed photos of me for your seriously ill patients? I would much rather have them go there than to a bunch of teenagers back home.'*

"Yes, Herr Rittmeister, gladly, quite gladly!"

I was given 50 and then another 100. My patients fell over themselves with joy."(8)

Käte Oltersdorf also described how Richthofen, as soon as he was able to leave his bed, visited the hospital's wards to encourage the wounded. The first time he did so, she and her staff had been overwhelmed by their duties and dirty laundry and litter had piled up in her ward:

"Richthofen climbed over all the dirty piles. I was horrified."

"Good God, Herr Rittmeister, you're visiting my ward now when I'm up to my neck in filth?"

"Well, Sister, I have to check if you're working!

"Well, if that's it, go ahead."

I was extremely busy. Richthofen went to each of the men, all of whom had suffered a head wound, and chatted with them. My

Above: Lt. Walter Böning, Staffelführer of Jagdstaffel 76.

seriously ill patients were happier today than I had ever seen them – especially the one shot in the neck and spinal cord, who always told me so much about the 'Red Devils' [Jasta 11]. *He ate quite well again and said:*

"Sister, I'm growing fat now that I've seen the best of our Red Devils in person!"(9)

Friedrich-Wilhelm Lübbert, a pilot with Jasta 11 from December 1917 to February 1918, wrote of his personal experience with von Richthofen:

"Richthofen was a pilot through and through. Over time, he became one of the most popular men in Germany. One would think that a man with such demands as were placed upon him by one of the most strenuous activities there is, and with the great popularity that he enjoyed, would have no more room for friendship and comradery. The opposite was true, however. Richthofen was just as good a comrade as he was a superior to the officers of his Jasta and Geschwader. Off duty, he was like any other comrade with us. He'd play hockey with us, if there was no flying, and often participated in a card game in the evening. You could go to him with any question and any concern and be sure to find sympathy and help when you needed it.

Richthofen was unsurpassable as a teacher. I've been with various Flieger-Ersatz-Abteilungen and at the Jagdstaffelschule. I never met a teacher who could make the theoretical technique of air combat as clear as Richthofen did. He was always ready to answer any questions that were asked. He actually loved it when his pilots were inquisitive. He never became impatient even when our questions must have been amateurish and foolish. He accepted each one with

Above: Otto Fuchs, as a member of Fliegerabteilung (A) 292, witnessed Jagdstaffel 11 being engaged in aerial combat on several occasions, and was with Jasta 11 for a few days in June 1917.

Above: Vzfw. Alois Heldmann, a pilot of Jagdstaffel 10.

the greatest of patience. Each young pilot who came to his Staffel first had to fly with Richthofen alone to the front a few times. After these flights, details of what was seen and experienced were promptly discussed with the beginner.

Richthofen was quite decided upon one point: he tolerated only those pilots who actually achieved something in the Staffel. He would watch the beginner for some time; if he then came to the conclusion that the person concerned did not meet the requirements he needed in a combat pilot – be it in terms of his moral qualities or in terms of his technical abilities – then the person concerned would certainly be sent packing again. But that was really great for us, that Richthofen would evaluate you simply for your achievements and not for external reasons.

As a superior, Richthofen was loved by all. The men in the ranks, particularly the mechanics, who generally had an especially close relationship with their pilots, loved and admired him above all else. It was only natural that such a man would have an ideal relationship with his officers as their superior. The calmness with which he treated the officers under his command, even when he must have been internally upset, was admirable…

Despite his extremely strenuous job as a fighter pilot, superior, and teacher, Richthofen didn't neglect his intellectual and sport interests. He often read at night, mostly beautiful literature of a serious nature but often scientific literature as well… Richthofen was never idle. If there was no flying, he practiced shooting at the machine gun station – with fabulous skill, by the way – or he went riding, which from time to time was an absolute necessity for him as an enthusiastic cavalryman, or he went hunting…He had all the qualities that a successful fighter pilot must have: fly well, shoot well, see everything, always stay calm and get up to the enemy quickly… He never had an unhealthy ambition that cost so many other good combat pilots their lives… If his Staffel or Geschwader was involved in a dogfight, he saw everything and everybody. Not only did he deal with his own adversary, but he monitored his pilots at the same time, either to provide them with timely assistance or to be able to critique them later on how they shouldn't have done something."(10)

It should be noted that Friedrich-Wilhelm Lübbert, as well as his brother, came from a Jewish family who had converted to the Protestant faith. He held Manfred von Richthofen in high esteem throughout his life.

Wilhelm-Gisbert Groos, who served with Jagdstaffel 11 as a pilot from 24 May 1917 to 16 September 1918, wrote to Alex Imrie:

"He gave you the feeling that he was never afraid, and that gave us courage and also helped us overcome some of our own fear. He protected every member of the Staffel, if ever possible, but there was no pardon if a pilot snuck away from the fight. That pilot would be transferred immediately."(11)

Johann Janzen, pilot and later Staffelführer of Jagdstaffel 6 in Jagdgeschwader I from 16 October 1917 to 9 June 1918, reported to Alex Imrie:

"Richthofen had an incredible sense for the flight maneuvers of our opponents. It seemed as if he knew them in advance. Sometimes he performed maneuvers that at first seemed quite strange to us, for example, to lure enemy aircraft over our lines and then cut off the way back. In the vast majority of cases, his

Above: Oblt. Karl Bodenschatz, Adjutant of Jagdgeschwader I.

Above: Hans-Georg von der Osten in flying dress. (A. Imrie)

maneuvers were successful. Sometimes it was almost scary [how he anticipated enemy flight maneuvers]. *When I later became a Staffel leader, I considered myself lucky to have learned so much from him."*(12)

Franz Hemer, a pilot in Jagdgeschwader I's Jasta 6 from September 1917 until the end of the war, explained during a long telephone conversation with the author in 1976:

"Richthofen was a Rittmeister and Geschwader commander as well as a nobleman. I was just a simple Vizefeldwebel. However, he treated each pilot equally, whether he was an officer or in the ranks. The decisive factor was how he performed as a fighter pilot. There were no differences between ranks, and we NCO pilots ate and socialized in the Officers' Mess. It was very different from what I had experienced before in the 'Kommiss' [slang term for the military]. *He never flaunted his nobility either, unlike other noble officers who treated us like third-class beings during training. Our Staffel leaders Reinhard, Kirschstein, and Neckel, were the same as Richthofen."*(13)

Walter Böning, the Staffelführer of Jagdstaffel 76, first met Richthofen on 14 April 1918. Jasta 76's airfield at Liéramont was located only about 30 km away from the Cappy airfield that Jasta 11 had occupied two days earlier. Böning had gotten into his Albatros D Va and made the short hop over to Cappy where he was greeted by Richthofen and his pilots. Richthofen personally showed him the Staffel's Fokker Dr.I aircraft. At this point, Böning had 15 air victories to his credit, which seemed to impress Richthofen, who invited Böning to lunch and said that he would be happy to request Böning for his Jagdgeschwader.(14)

When asked about Richthofen as a person, Böning said:

"I only met him twice, but both times I was deeply impressed by his personality. People were strongly attracted to him, and he radiated confidence and reliability. He was a man that you trusted implicitly and that gave you the impression of being 'a man of his word.' He was charismatic and was a natural-born leader without being overbearing or aloof. On the contrary, he seemed completely down-to-earth. I also saw that he had a sense of humor when we were at lunch together. When I learned a few days later that he had not returned from a flight against the enemy, which was then followed by the news that he had been killed, I was deeply affected even though I had only known him

Abschrift.

April 1918.

Geschwaderflüge.

Bölke teilte seine 12 Piloten im Herbst 1916 in zwei Ketten. Er machte jede 5 – 6 Flugzeuge stark. 6 – 7 Flugzeuge lassen sich von einem Führer am Besten führen, übersehen und sind am Beweglichsten. Jm Allgemeinen genügt diese Kampfstärke noch heute. Der Engländer hat in Bezug auf Geschwaderfliegen die grössten Erfahrungen und ist meist ebenso gegliedert.

Bei sehr starkem englischen Flugbetrieb ist man jedoch gezwungen mit stärkeren Geschwadern zu arbeiten; ich starte mit 30 – 40 Maschinen, also ein Geschwaderflug (Grund : Das unterlegene deutsche Jagdflugzeug, oder starke Geschwadertätigkeit). Die Gliederung bei einem so grossen Geschwader ist folgende :

Der Geschwaderkommandeur am weitesten Voraus und tiefsten,
Staffel 1 – links
" 2 – rechts
" 3 – hundert m über dem Kommandeur
" 4 – in Höhe der Staffel 3 als Letzte hinder dem Kommandeur. Abstand 150 m. Staffeln richten sich nach ihrem Staffelführer, die Staffelführer nach ihrem Kommandeur.

Vor jeden Start mus unbedingt das was man machen will, (z.B. die Richtung, in der ich zuerst fliegen werde) besprochen werden. Die Besprechung vor dem Start ist mindestens ebenso wichtig, wie die nach dem Flug.

Zu jedem Geschwaderflug gehört eine grössere Vorbereitung als wie zum Flug innerhalb einer Staff el. Daher ist es erforderlich, dass man vorher den Geschwaderflug ansagt. Also sage ich z. B. am Abend, dass am nächsten Morgen von 7° ab das Geschwader sich startbereit halten soll. Unter startbereit verstehe ich in diesem Falle : Zum Fluge komplett angezogen, jeder Flugzeugführer neben oder in seiner Maschine und nicht etwa in einem Starthaus ohne angezogene Fliegersachen.Die Monteure sind bereit an ihren Maschinen.Die Maschinen aufgebaut fix und fertig zum Start.Da ich nicht wissen kann, ob um 7 Uhr der feindl.Flugbetrieb einsetzt,so ist es möglich,dass das ganze Geschwader angezogen,eine oder mehrere Stunden auf dem Platze wartet.

Durch telefonischen Anruf (wenn auf verschiedenen Plätzen), durch Glockenschlag (auf einem Platz) wird der Start befohlen.Es startet jede Staffel für sich,als letzter ihr Staffelführer,sammelt die Staffel in niedrigster Höhe (100 m)über einem Punkte,rechts oder links von der vorher angegebenen Flugrichtung des Kommandeurs.Dann startet der Kommandeur und schlägt gleich die von ihm vorher befohlene Richtung ein.Der Kommandeur fliegt,bis sämtliche Staffelführer die vorgeschriebenen Plätze eingenommen haben,stark gedrosselt.Damit die Staffeln nicht durcheinander kommen,ist es zweckmässig jeder Staffel ein Staffelabzeichen zu geben. Das Flugzeug des Kommandeurs muss stark auffallend angestrichen sein.Während des Sammelns darf der Kommandeur keine Kurven fliegen.Er fliegt also möglichst langsam,meist der Front entgegen.Hat sich der Kommandeur überzeugt, dass das Geschwader geschlossen ist und keine Maschine mehr abhängt, so kann er allmählig beginnen,die Leistungsfähigkeit seiner Maschine auszunutzen.

Die Höhe,in der der Kommandeur fliegt,ist die Höhe,in der das Geschwader zu fliegen hat.Grundfalsch ist,ein Flugzeugführer fliegt 200 m höher oder 50 m tiefer. In so grossem Verband (30 – 40 Maschinen) muss der Platz der Staffelführer während des ganzen Fluges innegehalten werden.Es empfiehlt sich,besonders bei Anfängern auch innerhalb der Staffeln eine Platzordnung zu bestimmen.Die Platzordnung innerhalb der Staffel kann so mannigfaltig sein,dass man dafür schlecht eine bestimmte Vorschrift gibt.Bei einer gut eingeflogenen Kette erübrigt sich eine genaue Platzeinteilung.Ich führe Jagdstaffel 11 am liebsten wie das Feld einer Reitjagd,dann ist es gleichgültig,ob ich kurve,stürze,drücke oder ziehe.Ist die Staffel jedoch nicht so gut eingeflogen,so ist eine Platzordnung angebracht.

Glückt der Geschwaderflug nicht,so ist in 99 Fällen fas Führerflugzeug daran schuld. Es richtet sich mit seinem Tempo nach dem langsamsten seines Geschwaders. Die dem Kommandeur am nächstenflieg den Staffelführer dürfen nicht > eng auffliegen,dass dem Kommand

ei

Right: Copy of the original "Geschwaderflüge" document, written by Manfred von Richthofen as late as April 1918. These instructions were given to the Staffelführers of the Jagdstaffeln of Jagdgeschwader I to serve as as the basis for training their pilots. (A. Imrie)

Above: Lt. Carl Deilmann, member of Jagdstaffel 6 until September 1917, was one of very few people to provide a more critical view on Manfred von Richthofen´s personality. (A. Imrie)

Above: Vzfw. Kurt Delang, pilot of Jagdstaffel 10, in early 1918.

for a short time."(15)

Hans-Georg von der Osten, supplemented:

"Richthofen wanted to fulfill his military role and serve his country in the best way possible. He was a charismatic leader. He combined courage with flying skill, military intelligence, and the ability to think tactically."(16)

Otto Fuchs, a sensitive observer of his surroundings, had seen Jasta 11 pilots in action as a member of Fliegerabteilung (A) 292 and was at Jasta 11 for just a few days in June 1917. He described the atmosphere in the Jasta, which he was quite unaccustomed to, as "Heldenluft" ('an air of heroes'):

"The pilots regarded themselves as the elite of fighter pilots and the personal effort of the pilots to always give everything they had was quite high. Despite all the Staffel leader's prudence, this could mean a quick end for an aviator with modest flying skills. I also have the impression that the 'air of heroes' led some pilots to overestimate themselves and take risks that they just could not handle."(17)

Alois Heldmann, who flew with Jagdstaffel 10 from November 1916 until the end of the war, confirmed von Richthofen's relationship with NCOs in a conversation with this author in 1975:

"I of course was not as close to Richthofen as others were because I was with Jasta 10. But what always impressed me was his charisma and his ability to lead people. His very presence radiated security and confidence. He treated every pilot the same way, whether he was an officer or an NCO. His instructions and orders were clear and intelligible. He always brought the leaders of the four Staffels together [after Jagdgeschwader Nr. I was formed] *to discuss the missions with them and listened to their opinions. He had a particularly friendly relationship with* [Lt. Werner] *Voss* [Staffelführer of Jasta 10]. *His death must have hit him quite hard, but he didn't show it."*(18)

Karl Bodenschatz, Jagdgeschwader Nr. I's adjutant, confirmed Heldmann's statement about the loss of friends when the author, together with US aviation historian Neal O'Connor, visited him in 1976:

"Richthofen had been deeply affected by the loss of his friends [Kurt] *Wolff and* [Werner] *Voss but outwardly displayed no signs of his grief, at least in public."*(19)

As interviewed members of JG I reported in unison, von Richthofen showed his respect for the loss of a Staffel member but did not let any grief overtake him or be expressed to the outside world. With regard to this, Hans-Georg von der Osten, member of Jagdstaffel 11 from 11 August 1917 to 16 March 1918 and briefly Staffelführer of Jagdstaffel 4, told the author:

"Once you got to know him better, you noticed that his sometimes somewhat brash manner was really for self-protection. When he was internally affected by the loss of one of our pilots, he did not show it externally. 'Keep your composure and fulfill your duty' was an iron principle of his personality.(20)

Maintaining an iron demeanor and not publicly displaying emotions was a principle of his aristocratic origins. As a result, he could be perceived by those around him as unemotional or aloof.

Of all the statements and recollections known to the author by former members of Jagdgeschwader I that concerned von Richthofen, only one was critical of him. Lt. Carl Deilmann, a pilot who flew with Jasta 6 in the spring

Above: Richthofen Museum in Schweidnitz, around 1934. The silver cups, which Manfred von Richthofen had made for each air victory he scored, are lined up under a portrait of the man. The cup "Dem Sieger im Luftkampf" (The victor in aerial combat), which was awarded for his first confirmed victory, takes the center. (E. Wischnowski)

Above: An amused-looking Manfred von Richthofen.

and summer of 1917, wrote the following in 1978:

"I was a member of Jagdstaffel 6, which was assigned to Jagdgeschwader I, for about a year. I went on leave in July/August 1917. Before my leave, I handed over three indisputable victory claims along with the usual documentation (which I still have today) to my Staffel's office to pass on to the Geschwader. When I returned, the victory claims had been assigned to others. I therefore took the relevant documents under my arm and reported to the commander. I wrote the conversation down word for word and still remember it. I spoke with Rittmeister von Richthofen and asked him to review the decisions on the kill allocations. He snapped at me without looking more closely at my intended report: 'How dare you question my decision!' I replied: 'Herr Rittmeister, I am not questioning your decision and am here in order to bring you documents for review.' The reaction was: 'Perhaps you never enjoyed a proper upbringing?' I quickly saluted and did an about turn."(21)

It should be noted that following a submission by the Geschwader, it was the office of the Commanding General of the Air Forces that actually confirmed or denied such claims. Once an air victory was confirmed by that office, it was only rarely withdrawn or awarded to another pilot. It should also be mentioned that Carl Deilmann himself pointed out that the disagreement occurred not long after von Richthofen had suffered his head wound in July and that he wondered if the wound had affected the Rittmeister, resulting in his harsh reaction.

Carl-August von Schoenebeck highlighted another important aspect of Manfred von Richthofen as Staffelführer – the fact that he invested a great deal of effort in the training of new pilots. He would take such "Häschen" ["bunnies"] along with just him on one or two front flights in order to observe how the new man behaved. On such occasions, he was a teacher rather than a fighter pilot. He also observed mock dogfights over the airfield. Richthofen placed particular importance on one training exercise where the pilot was told to approach his dogfight opponent up to within 50 meters. Richthofen wanted his new pilots to be unafraid of getting that close. Carl-August von Schoenebeck said that this was an exercise for which all the newcomers had a healthy degree of respect:

"Airplanes have no brakes! If you didn't build up enough momentum, you'd never reach the opposing airplane – then you

Right: Manfred von Richthofen with his Great Dane Moritz. (E. Wischnowski)

Left: Excerpt from motion picture footage filmed at Marckebeeke Castle. Manfred von Richthofen is seen enjoying the joint affection of his Great Dane Moritz as well as other dogs of the Staffel. (L. Bronnenkant)

could count on getting rebuked by Richthofen after landing. If you built up too much momentum, then you shot over the opposing airplane into the direct line of fire of its machine guns. In the worst case, you rammed the opposing airplane. I was bathed in sweat after these first exercises and understood what it meant to be in air combat. We learned in this way to get the correct speed and to gain better control of our aircraft. By the way: Richthofen attached no importance to aerobatics."[22]

This is also reflected in Manfred von Richthofen's discourse on "Geschwaderflüge" ("Geschwader Flights"). In the section titled "Wie bilde ich mir Anfänger heran" ("How I Train Beginners") it said:

"...Daredevils, and not aerial acrobats, are needed. I require in-flight target practice and tight turns at high altitude. If a pilot meets my requirements, he flies 50 meters to the left and behind me during his first few flights and pays attention to his leader... I attach less importance to the skill of flying itself. I shot down my first 20 while flying was still quite difficult for me; but it doesn't hurt if one is a skilled pilot. Incidentally, I prefer the pilot who can only turn left but still goes straight for the enemy versus the diving and turning flier from Johannisthal who attacks too cautiously."[23]

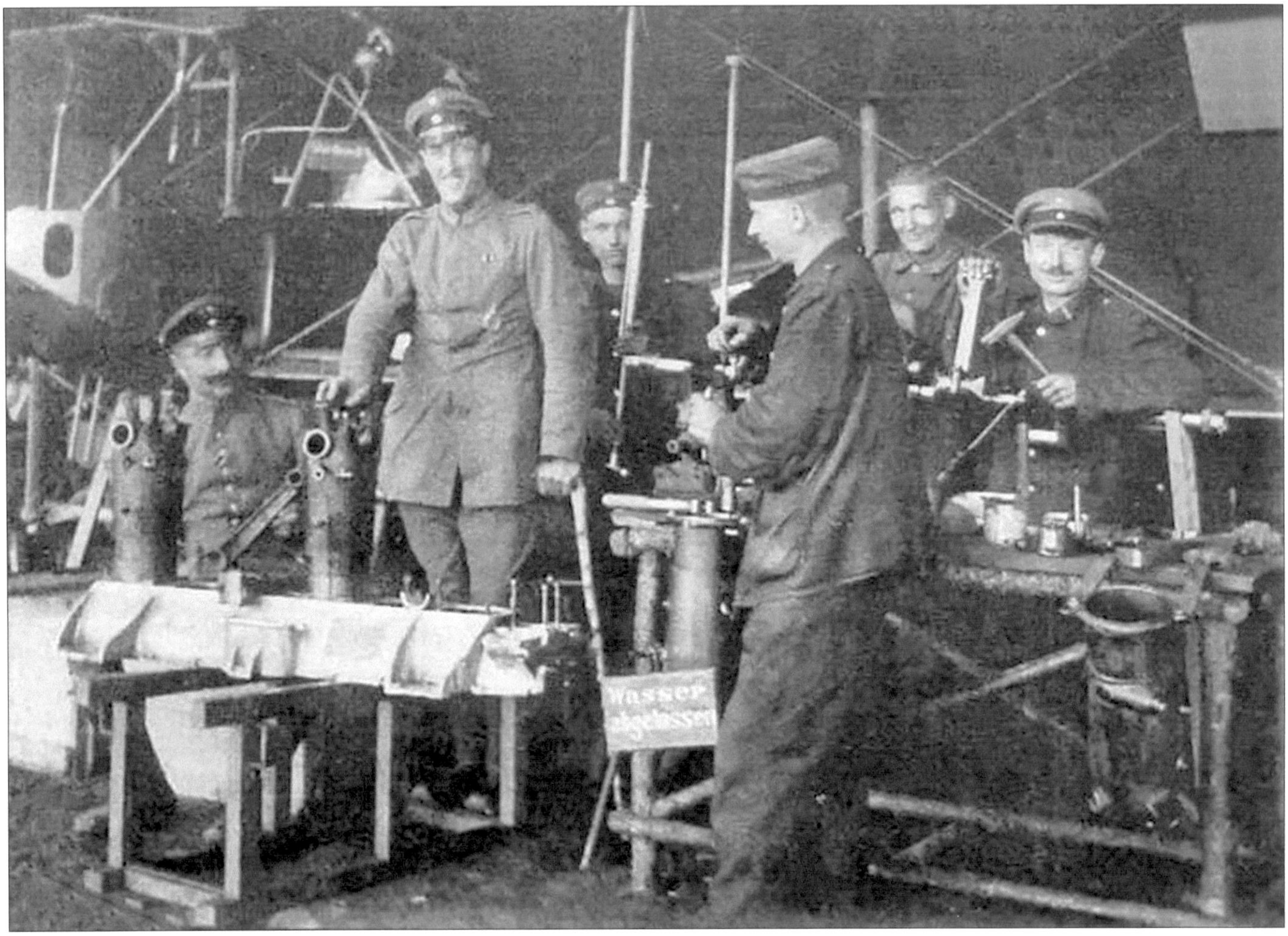

The pullman wagons used as quarters for the pilots and observers of Kampfgeschwader II. Between the wagons there is a small garden with paths and a wooden hut where the flight crews can make themselves comfortable. However, it got very hot in the pullman wagons in the summer.

Facing Page, Above: Josef Holzapfel, Manfred von Richthofens first mechanic, is the tall man in the middle, seen here while servicing a Mercedes engine. The sign reads "Wasser abgelassen", indicating that the water was drained from the engine.

"Geschwaderflüge" also provided very specific instructions on how pilots were expected to behave in certain situations:

"For a beginner, it is at least just as important to know what he has to do to keep from being shot down... The enemy chooses the best moment to attack the hindmost airplane in a Kette. Coming out of the sun, he dives at him and can get a kill with only a few shots. It is imperative for everyone to devote their main attention to what's behind them... If a beginner is surprised from behind, he should under no circumstances try to escape by diving. The best and, in my opinion, only correct response is to suddenly make a very tight U-turn and then go on the attack as quickly as possible."[24]

Kurt Delang, a pilot with Jagdgeschwader I's Jasta 10 from January to April 1918, saw Richthofen in combat and described his mode of attack in a letter to the author:

"Richthofen's successes were primarily because he got quite close up to his opponent! He gave each of his pilots a 'Zigarre' [dressing down] *if he saw that he had started to open fire from 150 meters! Richthofen would go up to within a few meters of his opponent, look through the gunsight, and shoot! Therein lay his success."*[25]

Manfred von Richthofen was also personally proud of his air victories. For each of his aerial victories he had a small silver cup made, inscribed with the date of the victory – a habit that might be interpreted as a sign of vanity due to his ambition. He also collected components of the aircraft he shot down, such as the serial numbers marked on the rudders or the manufacturers' company badges, as personal memorabilia. We might speculate that this practice sprang from his hobby as an animal hunter, where trophies from one's victims were routinely collected and displayed. In this way he differed from his teacher Oswald Boelcke, who rarely kept any mementos from his air victories as von Richthofen and others did. However, Manfred von Richthofen was far from alone in collecting souvenirs of Allied aircraft shot down or forced to land. A large number, if not the majority, of fighter pilots from all air forces involved in this war collected such "victory trophies".[26]

The recognition of his military achievements through honors and decorations was well received, but should not be overvalued as Wilhelm-Gisbert Groos reported, it was not the impetus behind von Richthofen's actions:

"Richthofen flew and fought for his Kaiser and our Fatherland, not for decorations, but from a sense of duty."[27]

But what was Manfred von Richthofen like between combat missions? Carl-August von Schoenebeck provided the following answer:

"On the ground, and especially when we had 'Fliegerwetter' [pilot's weather], *we had a level of personal freedom that I had never before experienced in other military units. Richthofen would then be in the mood for practical jokes and festivities. On the other hand, there was iron discipline during times of action.*

When Richthofen was off duty, he had a dry and sometimes raucous sense of humor. He loved jokes and boisterous celebrations and participated in pranks himself. I was the victim of pranks he was involved in at least twice. I think he was most amused by the pranks from us young airmen. Then he was relaxed, the military severity fell away from him, and he was a young man like us."[28]

Wilhelm-Gisbert Groos was part of one of these pranks by Manfred von Richthofen:

"On the ground, after duty, Richthofen was like a boy with a wonderful sense of humor. While Jagdgeschwader I was stationed at Marckebeeke, Richthofen nicknamed [Wilhelm-Gisbert] *Groos 'Count Groos of Marckebeeke'* [the reason was that Groos, a tall man, had a very elegant, almost aristocratic appearance]. *Major Albrecht von Richthofen, Manfred's and Lothar's father, visited the unit and Manfred introduced Groos by this joking title. The elder Richthofen, thinking he was talking to a fellow nobleman, began a long conversation with him. The talk got around to politics and the Major said: 'Politics are not for us* [meaning the nobility]. *We are only suited for military services, nothing else.' When Groos told Manfred of this, the younger Richthofen nearly died of laughter and told his father that Groos was not 'noble'. The Major was very disappointed at first, but later he sent Groos a photo of himself with a special dedication to 'Count Groos of Marckebeeke.'"*[29]

A special place in the life of Manfred von Richthofen was taken by his Great Dane "Moritz", affectionately called "Sugar Pig" by him because Moritz loved sweets. He regularly joked and played with him on the airfield or in the park.

In *Der rote Kampfflieger*, he is quoted as follows:

"The most beautiful creature that the world has ever created is the genuine Ulm Great Dane, my 'little lapdog', Moritz. I bought it in Ostend from a good Belgian... Zeumer took a second one and called him 'Max'. Max found a sudden end under a car. But Moritz has been excellent. He slept with me in bed and was excellently educated. He accompanied me from Ostend at every turn and grew very close to my heart."[30]

Jasta 11 mechanic Hermann Lohmeyer reported on his memories of Richthofen:

"Richthofen had a strong personality that made an immediate impression. There was a respectful distance between us mechanics and our Staffelführer, of course, but Richthofen always treated us with respect in return. When a Leutnant who was new in the Staffel complained to him one day that his mechanics weren't working fast enough, his curt reply was: 'They're doing their best, leave the men alone!' He had a very comradely relationship with his own mechanics, especially with his first mechanic [Josef Holzapfel]."[31]

Josef Holzapfel was assigned to Manfred von Richthofen as his First Mechanic at Kagohl II's Kampfstaffel 8. It resulted in a very trusting relationship that led to Holzapfel, at Richthofen's request, following him to Jasta 2 'Boelcke' and then eventually to Jasta 11.[32]

Left: Josef Holzapfel sits on the fuselage of a captured British Sopwith Triplane at Marckebeeke airfield in the second half of 1917. The aircraft was credited to Lt. Walter Blume of Jasta 26. FSLt. R. Kent of No. 10 (Naval) Squadron was flying this Triplane N5357 when he was captured on 11 July 1917 (G. VanWyngarden.

Below: Air Combat Report dated 18 March 1918. From the book *"Jagd in Flanderns Himmel - Hunting with Richthofen"* by Karl Bodenschatz.

(66. Sieg)

Jagdgeschwader 1 . 18.3.1918 11.15 Uhr vormittags

an der Strasse Molain - Vaux Andigny.

600 m südwestl. Molain.

Sopwith - Camel-1 Engländer;

Flugzeug Nummer: B 5243

Motor: Clerget 35751

Fok.Dr.I.152; rotes oberes Trag=
deck, rote Haube, rote Laufräder u.roter Schwanz.

1 Kanadier, gefangen.

Jch startete mit 30 Flugzeugen meines Geschwaders und flog ge=schlossen, alle 3 Staffeln führend, in 5300 m Höhe gegen den Feind. Wie wir uns der Front näherten, sah ich mehrere englische Geschwader die gerade unsere Linien überflogen, etwa Richtung Le Cateau. Das er=ste Geschwader das mir begegnete war etwa in 5500 m Höhe. Jch schoß mit Lt.Gussmann, Jasta.11 zusammen den letzten der Gegner, einen Bri=stol-Fighter ab. Er verlor die Tragflächen und Lt.Gussmann brachte ihn bei Joncourt zum Absturz.

Darauf sammelte ich meine 30 Flugzeuge, stieg auf 5300 m und folgte zwei bis nach Le Cateau durchgebrochenen Geschwadern. Als der Gegner den Versuch machte auszubiegen und nach der Front zurückzukom=men, griff ich an. Das mir am nächst fliegende Flugzeug, scheinbar ein Breguet oder Bristol-Fighter, wurde von mir und Lt.Loewenhardt,Jasta.10 beschossen, worauf dem Gegner der Benzintank kaputt geschossen wurde und ich sah, wie das Flugzeug senkrecht abstürzte.Lt. Loewenhardt brach=te es zum Absturz.

Von zwei englischen Einsitzer-Geschwadern griff ich ein Flugzeug mit Wimpeln an und zwang den Gegner bei Molain zur Landung. H.

<u>Zeugen des Luftkampfes:</u>

<u>siehe Anlage!</u>

Frhr. v. Richthofen

Rittmeister und Geschwaderkommandeur.

Holzapfel reported that Richthofen cared for his mechanics in touching ways. When they were in Russia, Richthofen himself drove to towns to swap foodstuffs for things that were scarce for his mechanics. Reservations about mingling with his subordinates were foreign to him. Holzapfel added:

"The officers [at Kampgeschwader II] *slept in railroad cars and the rest of us slept in tents near the airplanes. We had made ourselves quite comfortable, the main thing being that it wasn't as hot as it was in the railroad cars. One day Richthofen came to us and asked whether he could sleep in our tent. He wanted to know if he wouldn't be disturbing the others and if they might agree. Even though the others of course acted enthusiastically about it, Richthofen had the keen sense that the men wouldn't feel as free in his presence as when they were alone.*

This is what happened: one night the NCO on duty stumbled over the bracing rope in the dark and a puddle of water that had pooled in the bagged canvas spilled all over Richthofen, who had camped near the entrance. Wet as a cat, he called for his orderly, had his belongings brought back to the railcar, and said, laughing: 'These are OK guys, they got rid of me in a decent way.'

Holzapfel continued: 'Don't misunderstand this... he was 'the Baron' in our eyes, he was an officer, I the subordinate. But he didn't at all act like that. He had a splendid way of dealing with us. And that's what enabled us to perform at the highest level.'(33)

As Josef Holzapfel later found out from Richthofen's mother, he always wrote home that his mechanics were everything to him and that he had to take care of them. When Holzapfel arrived for the inauguration of the Richthofen Museum in Schweidnitz in 1933, Bolko von Richthofen, Manfred's youngest brother, greeted him with the words:

"Now go and see my mother. She wants to see you first!"(34)

In contrast to his relaxed behavior towards members of his Jagdstaffel, no matter which rank or position, Richthofen was somewhat reserved when he was visited by reporters or journalists. He behaved as though he was suddenly embarrassed by his fame. As former Jasta members confirmed, he kept such visitors at a distance and was relieved when, as he put it, they "left the field." He abhorred any expressions of flowery adulation.

Carl-August von Schoenebeck brought up another side of Richthofen's personal conduct in one of our conversations:

"In July 1917 he had a significant disagreement with our Kofl [Hptm. Bufe, Kommandeur der Flieger of the 4th Army] *regarding the use of the Geschwader. The Kofl pressured Richthofen to deploy it in accordance with his views. He* [Richthofen] *spoke about it with Reinhard* [Staffelführer of Jagdstaffel 11] *and Döring* [Staffelführer of Jagdstaffel 4] *and was extremely irritated by the man's incompetence. I was still a novice at the time, but the Kofl's instructions didn't make sense to me either. Calling on his authority and personality, Richthofen stood his ground and didn't give in one millimeter."*(35)

Hans-Georg von der Osten confirmed this when he said:

"Richthofen did not shy away from staunchly defending his opinion among higher ranks and contradicting his superiors. Although he always displayed the polite respect that was required for them, he was never under any circumstances willing to tell them what they wanted to hear. He was a man of moral courage.(36)

Hans-Georg von der Osten also recalled how Richthofen liked to encourage his pilots by awarding them aerial victories in which he had participated:

"If Richthofen had taken for himself all the victories he had been involved in – which he could have done as Geschwaderführer – then his tally likely would have stood at 100."(37)

Carl-August von Schoenebeck added:

"I myself witnessed how he attacked an enemy airplane with a comrade and eventually caused it to crash. In several such instances, he waived his claim and left the kill to the other pilot, especially if he hadn't gotten any kills yet but had proven himself through his courage and aggressive spirit in action."(38)

One example confirming these claims can be found in Richthofen's combat report of 18 March 1918, which stated:

"I, together with Lt. Gussmann, Jasta 11, shot down the last opponent, a Bristol Fighter. He lost his wings and Lt. Gussmann ***brought him down.****"*

The report went on to say:

"The airplane closest to me, apparently a Breguet or Bristol-Fighter, was shot at by me and Lt. Loewenhardt (Jasta 10), whereupon the enemy's fuel tank was shot apart and I saw the airplane fall vertically. ***Lt. Loewenhardt brought him down.****"*(39)

Richthofen could have easily claimed the two victories of this combat for himself, but instead he preferred to leave the credit for them to his two fellow pilots who were also involved in the fight.

Wilhelm-Gisbert Groos also confirmed this in an interview he gave to aviation historian Heinz Nowarra:

"In service matters [he was] *at all times fair, especially in the air over the front. He saw everything. He gave new men in the Jasta every chance to score a victory. He gave away many victories, if by doing so, a young pilot was able to score his first kill.*(40)

The theory promulgated in some publications that Richthofen was bent upon racking up as many air victories as possible in order to reach the magic number of 100 hardly conforms with these testimonies provided by the men who actually flew with him.

In other publications, Richthofen was accused of mainly picking on two-seaters because they supposedly were comparatively easy to shoot down. This amateurish view has no basis in reality and ignores the brutal realities of World War One air combat. Two-seaters were, in fact, highly dangerous opponents because they had a man in the rear seat armed with one or two machine guns who could fire in multiple directions at any approaching opponent as opposed to a single-seater that could only fire forward. When a formation of two-seaters were able to form a defensive circle, their attackers were met with a hail of projectiles.

In discussions with Alex Imrie, US aviation historian Bill

Nicht in das Flugzeug oder die vorderste Linie mitnehmen!
Geheim!

Vorschriften
für den Stellungskrieg für alle Waffen.

Teil 13.

Weisungen
über den
Einsatz von Jagdstaffeln

Vom 25. Oktober 1917.

Herausgegeben
vom Chef des Generalstabes des Feldheeres.

Above & Following Page: Extract from "Instructions for the Deployment of Jagdstaffeln" with the order to primarily attack enemy artillery cooperation planes.

— 4 —

II. Auftragserteilung für Jagdstaffeln.

5. Die Ausschaltung der feindlichen Flieger- und Ballonbeobachtung ist von entscheidender Bedeutung für den Artilleriekampf; daher besteht die Hauptaufgabe der Jagdstaffeln darin, durch immer wiederholte Vorstöße über die Front die hinter den feindlichen Linien einschießenden Flugzeuge zu vernichten und damit der feindlichen Artillerie die Augen zu nehmen.

Puglisi, and the author, the vast majority of former fighter pilots attested to the fact that fighting a two-seater with a capable crew was considerably more dangerous than fighting a single-seater. This can be seen in the fact that a good number of successful German fighter pilots lost their lives in combat with two-seaters. Naturally, combat with enemy single-seaters occurred frequently as well, as their task was to protect their two-seater comrades.

Air combat with enemy fighter planes, despite its knightly and romantic portrayal in many books, was not the primary job of fighter pilots unless it served to ward off attacks on their own reconnaissance, artillery, and bomber forces. If they succeeded in shooting down hostile, two-seat "working aircraft", the ground troops were effectively supported; that is, if enemy reconnaissance results, and particularly aerial photographs of German positions, failed to materialize, enemy artillery could not direct their fire on them and bombing missions were also thwarted.

By attacking two-seat "working aircraft," Richthofen, as Staffelführer, was following the *"Weisungen über den Einsatz von Jagdstaffeln"* ("Instructions on the Deployment of Jagdstaffeln") that read:

"The elimination of enemy aircraft and balloon observation is crucially important during an artillery battle; therefore, ***the principal task of the Jagdstaffel is to destroy the enemy aircraft spotting for their artillery from behind enemy lines by constant and repeated offensive action over the front****, thereby taking away the eyes of the enemy artillery."*[41]

Richthofen wrote about this in his discourse on "Geschwaderflüge" in April 1918:

"A two-seater should be attacked from behind at high speed, exactly in line with its flight path. The only way to avoid the machine-gun burst of a skilled observer is to remain calm and incapacitate him with one's first shots. If the enemy goes into a turn, you have to be careful not to overshoot him. ***A long dogfight with a fully combat-capable, small two-seater is the hardest...*** *I believe that attacking a two-seater from the front is very dangerous. You almost never completely incapacitate him. I am against first getting in the line of fire of the fixed, forward machine gun and then in the observer's...* [42]

Right: Manfred von Richthofen (left) poses with one of many two-seaters he brought down. F.E.2b A5446 of No. 18 Squadron was his 14th victory, downed on 20 December 1916. (Taktisches Luftwaffengeschwader 71 "Richthofen")

Approaching the Personality of Manfred von Richthofen

Above: Manfred von Richthofen with bicycle together with his pilots in 1917. From left: Lt. Plüschow, Manfred von Richthofen, Lt. Lothar von Richthofen, Lt. von Hartmann Offz. z.b.V., unknown, Lt. Karl Esser, Lt. Otto Brauneck, Major von Richthofen, unknown. (L. Bronnenkant)

So, what conclusion can we draw from the firsthand accounts and statements of Richthofen's contemporaries with regard to his personality? Based on their testimony, his personality might be described as follows:(1)

First, there are two personality traits that become clear: One was the love-of-country, discipline, comradeship, and devotion to duty that his father and cadet training instilled in him; and this trait was in no way exclusive to Manfred von Richthofen. Rather, it was typical for an entire generation of young Germans who were born after what was then Germany came into existence in 1870. The other was a strong love of adventure and challenges. This, in turn, can be directly attributed to his childhood experiences.

Due to his patriotic upbringing, which was common at the time, and his early entry into the Cadet Corps, Manfred von Richthofen was a patriot and military man in every respect. He had an unquestioning devotion to his military duties as an aviator, Staffel leader, and later Geschwader commander. This was only possible due to his strict, self-imposed discipline. He drank only moderately and usually went to bed at around 10:00 p.m. in order to be fit for the next day's activities. Celebrations or "Fliegerwetter", where weather conditions were unsuitable for flying, were exceptions to this rule. Then he enjoyed staying up later, followed by a really good night's rest.

His contemporaries stated that they were thoroughly impressed by Richthofen's personality, which they unanimously termed as "charismatic." People were strongly attracted to him in such a way that they quickly put their trust in him and were not afraid to turn to him with their problems. He was a born leader who made an impact just by his presence alone. This was combined with an ability to motivate any of his subordinates – from pilots to mechanics – to do their best. As Carl-August von Schoenebeck said: *"We were all ready to follow him through thick and thin."*

This included the knowledge by each of his pilots that

Above: Another photo of Richthofen on his bicycle in April 1917. (L. Bronnenkant)

Above: A relaxed and joking Manfred von Richthofen on the steps of Bethune Castle. (L. Bronnenkant)

he would never hesitate to immediately disengage from a potential victory in order to rush to the aid of a beleaguered comrade. He led by personal example, and it was only natural that he would do the same in the air as a Staffel and Geschwader leader. He never demanded from his pilots what he himself could not commendably do himself. In today's terms, he "walked the talk."

This put Manfred von Richthofen's style of leadership in complete contrast to that favored by the last wartime leader of the Geschwader, Hermann Göring, who accomplished only a single aerial victory as its commander over a course of roughly 4 months. Hermann Göring preferred leading the Geschwader from the ground and rarely exposed himself to the dangers of aerial combat during the closing stages of the war. However, this topic will be dealt with in more detail in a later volume of this series.

Richthofen devoted much time to providing the best training to his pilots, not only so that they would be successful in combat but also because it gave them the best chance of coming home safely. He had a talent for explaining complicated things in a clear, practical way. He very quickly recognized promising potential in new fighter pilots and was patient with and continued to support them even when it took them a long time to "get the hang of it".

Like his teacher and role model, Oswald Boelcke, he had the ability to read the flight movements of enemy units, to anticipate their subsequent maneuvers, and then advance to the most favorable position from which his own unit could attack. According to other Staffel leaders such as Josef Jacobs, Otto Fuchs, Johann Janzen, and Heinz Arntzen, this was one of the most important skills that a Staffel leader needed to have in order to be successful.(2)

Richthofen had great personal courage, coupled with a joy of accepting challenges and a thirst for adventure. He was not foolhardy, however, and tempered that thirst by taking calculated risks. He was never a gambler like some other fighter pilots. This was reflected in the fact that he refrained from "suicidal tethered balloon missions" because, according to his own statement, he had once been given a "bloody nose" during one such attempt. He had the greatest respect for the tethered balloon specialists; but observing a target for a long time and only then meticulously planning an attack – like Pour le Mérite pilots Heinrich Gontermann or Fritz Röth – was not in his nature. His style was to attack an enemy airplane, get up to point-blank range as quickly as possible, and fire. The fact that he was an excellent marksman certainly was of great help in that respect.

Another noteworthy aspect of his personality was his drive and ambition. As an officer and Staffel leader, he viewed military success as the ultimate fulfillment of his duty; thus, he and his men should bring down as many hostile airplanes as possible.

Nurse Oltersdorf took note of how the success of Richthofen's Geschwader took precedence even over his private life:

"Now [October 1917] *a rumor spread in the newspaper: 'Richthofen married!' I had to laugh when this was read to me. To me, the notion that Richthofen had married was simply ridiculous. I said: 'That just isn't like him.' He was far too sensible and knew that his life was always at risk, and once said that if he were to get married, he would become cowardly because his thoughts would then be back with his wife who would be worrying so much about him. It was quite enough 'that my dear mother and sister worry about me! Thank God I have the most understanding mother you could ever imagine. She is quite prepared that something could happen to us.'*(3)

Camaraderie and team spirit were of the highest priority to him. He was a comrade who was respected by the men under his command, but not a "pal". In between missions, Manfred von Richthofen was anything but a stiff aristocrat.

Above: The booklet *„Kreuz wider Kokarde - Jagdflüge des Leutnant Ernst Udet"*, published in 1918.

Above: Manfred von Richthofen, on the left, with Lt. Algernon Bird, who he forced to land on 3 September 1917, on the right. Bird was then a guest of Jagdstaffel 11.

He played hockey, billiards, or cards with his men and behaved like a comrade among comrades with a sense of humor – and he appreciated a good prank every now and then.

At the same time, he sensed when one of his pilots could not cope with a situation or was even afraid and needed his help. Then he took him aside inconspicuously. He talked to him and gave him every kind of support he could provide.

Personnel problems were rarely an issue because he had a keen instinct for knowing when to "turn a blind eye" and allow his men certain liberties or when iron discipline was needed instead.

Another prominent trait was his willingness to stand up for his beliefs. He was, like his men, unfailingly polite and militarily correct toward his superiors, but did not kowtow to them. If he had formed an opinion based on his experience regarding a military matter and judged it to be sound (such as the best use of an aviation unit at the front), he would steadfastly defend it – with a good dose of *"Silesian stubbornness"*, as Carl-August von Schoenebeck put it – to his superiors.

A fairly common criticism leveled at von Richthofen is based on one passage in the book by Ernst Udet, *Kreuz wider Kokarde - Jagdflüge des Leutnants Ernst Udet* (Cross against Cockade – Hunting Flights of Leutnant Ernst Udet), published in 1918, which mentioned:

"... he shot down all enemy aircraft in flames. He said that in his opinion, this was the only genuine aerial victory. Once I had shot down a Sopwith whose pilot had been killed by a head shot. The plane had not caught fire, but just hit the ground. When I reported the downing to Richthofen, I was almost ashamed to respond to his question: 'burning?' in the negative."[4]

Karl Bodenschatz, who served as the adjutant of Jagdgeschwader I from July 1917 until the end of the war, repeated this representation from Udet's book almost verbatim in his own work *Jagd in Flanderns Himmel*, published in 1935:

"There was seldom a time when his opponent was not shot down in flames. One of his first questions to his comrades filing a combat report was: 'In flames?' The fighter pilots of his Staffel were pretty embarrassed when they had to answer no to this question."[5]

This pair of identical descriptions is often considered as the main argument for a negative interpretation of Manfred von Richthofen's character. In fact, they are often referred to as an indication of his fanatical will to "wipe out" his opponents, or even interpreted as a murderous, sadistic trait.[6]

However, these often referred-to quotes from the Udet and Bodenschatz books raise questions that cast doubt on their authenticity. Ernst Udet received the news of his transfer to Jagdgeschwader I on 18 March 1918 and arrived there a week later on the 25th, after which he was promptly appointed deputy Staffelführer of Jagdstaffel 11.[7] He was not there for long, however, because he departed for home leave on 7 or 8 April and did not return until after von Richthofen's death on 21 April.[8] As such, Udet could only have personally witnessed von Richthofen's behavior over the brief period from 26 March through 6/7 April 1918. Additionally, it is by no means certain whether Udet was a witness to all the aerial victories scored by von Richthofen during this two-week period when von Richthofen was credited with 8 to 10 aerial victories, only six of which went down in flames.[9] But it is certain that Udet was not a

Fokker Halb D.I v
Iflieg

16

werden.

Document dated 25 Okt 1917

Auch beantrage ich, sämtliche von MAG zu liefernden Berg-Flugzeuge mit Fokker MG Steuerung liefern zu lassen.

Betreffend dem Baue der Kampfeinsitzer wären die Ansichten des Frh.v.Richthofen, die ich bei Einsicht der Fokker-Korrespondenz bei MAG gelesen habe, nicht uninteressant:

„Eine Hauptbedingung für ein Jagdflugzeug ist, dass in grösserer Höhe die Maschine in den schärfsten Kurven, mehrmals 360°, mit voll laufendem Motor nicht fällt. Ein Gewinnen an Höhe wäre ideal. Dass dies nicht der Fall,ist,eine Hauptschwäche beim Albatros D III.

Steigleistungen: Wie sind die Steigleistungen im einzelnen von 1000 zu 1000 m? Es ist wichtigt, dass die Leistung beim Umlaufmotor in grösserer Höhe nicht allzusehr nachlässt.

Empfindlichkeit der Steuer.

a) Verwindung: Bei Seitensteuer und Verwindung müssen kleine Ausschläge grosse Wirkung haben; also sehr empfindlich.Die Engländer haben doppelte Verwindungsklappen. Bei Albatros D III reicht Verwindung nicht ganz aus. Verwindung Hauptsache beim Jagdflugzeug.

b) Seitensteuer: Seitensteuer des Albatros D III gut

c) Höhensteuer: Dürfte nicht zu empfindlich sein. Bei Fokker Eindecker machte sich dieses im Luftkampf oft unliebsam bemerkbar. Bei Albatros D dagegen sehr angenehm.

Geschwindigkeit: Grosse Geschwindigkeit auch in grosser Höhe erforderlich.

Steigleistung: Die alten Fokker-D haben Steigleistung auf Kosten der Geschwindigkeit zu sehr bevorzugt. Ein Drosseln auf ganz geringe Geschwindigkeit ist auch bei einer sehr schnellen Maschine unbedingt erforderlich.

Aussicht: Aussicht oben, unten, seitlich muss einwandfrei sein. Albatros D III gut; Alb.D II besonders nach unten schlecht Alb.D I nach oben im Kurvenkampf nicht zu gebrauchen. Beim Halberstädter D ist oberes Tragdeck zu nahe an den Augen und genau in Augenhöhe. Etwas weiter ab vom Auge und etwas höher ist für den Luftkampf angenehmer.

This Page & Following Page: Richthofen's report to Anthony Fokker. The handwritten notes are by the late American aviation historian P. Grosz. (G. VanWyngarden)

Filed Halb. D.I ✓
Idflieg

17

Festigkeit:

Sturzflüge: Einen senkrechten Sturzflug von über 1000 m, dabei eine Verwindungskehrtwendung muss die Maschine unbedingt aushalten; bei Albatros D III nicht immer der Fall. Maschine muss Oberdruck auf Flächen aushalten /Verwindungs-Looping/; auch verunglückte Loopings und vielfache scharfe, überrissene Kurven mit voll laufendem Motor. Bei Sturzflügen ist die Möglichkeit, grosse Fahrt zu erreichen, dringend notwendig. Ein langsamer Stürzflug, wie beim Halberstädter, meistens zwecklos.

Falltank: Falltank für 20 Minuten Vollgas.
Drucktank für 1½ Stunden Vollgas.

Instrumente: So wenig, wie möglich; Tourenzähler, kleiner Höhenmesser, Druckmesser.

Drosselung am Steuergriff: MG Knöpfe in einem Griff für die rechte Hand; Drosselung für die linke Hand, beides am Steuergriff.

Hilfsvisier: Gewehre nahe bebeneinander in der Mitte.
500 Patronen pro Gewehr. Geschossart ist Geschmaksache.

Windschutzscheiben: Wenn man Ölen des Motors vermeiden könnte, keine. Sonst möglichst kleine.

Festschnallvorrichtung: Muss vorhanden sein.

~~[illegible]~~/ Einbau eines Spiegels: Unterstützt sehr die Faulheit und Unachtsamkeit.

Dieses sind nur private Ansichten von mir und den Herren meiner Jagdstaffel.

gez.v.R i c h t h o f e n m.p."

PKZ 1 → Seit einer Woche wird bei MAG noch an einem Schraubenfesselflugzeug mit 4 Hubschrauben gebaut.

In der Motorenabteilung sind die hiezu nötigen leichten Elektromotore auch schon vorhanden.

Die Fertigstellung dieses S.F.Flugzeuges ist für Ende Oktober vorgesehen.

III.

III.Lloyd.

Lieferungen: 48 Stück Bauart 46/18000 im Ausliefern.

LLOYD →
48 Stück Bauart 46.5/81000
45 " Bergzweisitzer 47/18000

im Bau.Lieferung beider Typen beginnt noch im Oktober paralell im Verhältniszahl 1:1

witness to von Richthofen's prior 68 victories.

Karl Bodenschatz's *Jagd in Flanderns Himmel* was published in 1935 and was clearly influenced by the Nazi ideology of Bodenschatz's then superior, General der Flieger Hermann Göring, who wrote the book's introduction.(10) Some of its contents, when compared to independent documents from Jagdstaffeln 4, 6, 10, and 11, present inaccuracies, omissions, misrepresentations and even factual errors. Despite Bodenschatz's obvious use of JG I's war diary, it is clear that his work was affected by other, later influences. Thus, the question arises whether Bodenschatz's statement was made from his own experience or simply borrowed from Udet's book.

Finally, the statements that "...[Richthofen] *shot down all enemy aircraft in flames*" (Udet) and "There was seldom a time when his opponent was not shot down in flames" (Bodenschatz) are simply untrue.

Karl Bodenschatz was the adjutant of Jagdgeschwader I from its formation on 1 July 1917 until the end of the war. If one examines von Richthofen's combat reports, he shot down 24 enemy aircraft during his time as its leader. Of these, only 11 went down in flames, namely victories No. 57, 59, 68, 69, 70, 72, 74, 75, 76, 79, and 80. This means that less than half of his victories while he was the leader of the Jagdgeschwader went down in flames.(11) Moreover, if one analyzes all 80 of his combat reports, only **23 state that the enemy aircraft burned after being hit by him, which amounts to only about 29% of his total victories.**(12)

In this context, it should be noted that the question of whether or not a machine that took hits in combat erupted in flames depended on where it was hit. Hits to the fuel tank, fuel lines, carburetor, or stored ammunition almost always resulted in a fire. Yet aircraft could also go down as a result of combat damage inflicted onto important structural components, such as the wing spars, struts, rigging wires, or control cables, or simply by disabling the engine. It was therefore more a matter of chance whether or not an aircraft went up in flames.

This therefore leads to the question of whether Manfred von Richthofen really placed such emphasis on a burning airplane as "the only genuine aerial victory" when almost three quarters of his own victories did not meet that criterion. As mentioned previously, he did not demand anything from his pilots that he did not fulfill himself in an exemplary way.

The former German fighter pilots of the First World War interviewed by the author, including Jasta 11 pilots such as Carl-August von Schoenebeck, Hans-Georg von der Osten, and Wilhelm-Gisbert Groos, explained the process related to the confirmation of an aerial victory. When an air combat report was filed by a pilot after having completed his mission, it was required to provide a detailed description of the event. This included information relating to whether the enemy had gone down on fire, exploded in the air, crashed on the ground, gone down seemingly under control but had still ended in a crash, or had landed intact. An as accurate as possible description of the event increased the likelihood of receiving confirmation of the claim by matching it with observations by neutral witnesses (ground troops, airspace observers, balloon observers, anti-aircraft units). Attestations from other eyewitnesses to the event were a prerequisite for the official approval of the air victory. The report of the downing of an enemy aircraft in flames was usually confirmed very quickly, as it was seen by many eyewitnesses – even if they were farther away. If the plane went down "in one piece", meaning that it did not burn and disintegrate in the air, confirmation was sometimes more difficult or even impossible to obtain. This was especially true if the opponent went down behind enemy lines.(13) Accordingly, Manfred von Richthofen's statement - if he indeed ever made it in such words – can also be interpreted more innocently as meaning: If an enemy goes down in flames, it is a safely recognized victory for the pilot, the Staffel, and his Geschwader.

We also know that German fighter pilots were usually pleased when an opponent managed to land unharmed in German territory. The vast majority of interviewed former fighter pilots confirmed that they preferred to "bring the other guy down alive." As a multitude of photos show, the prisoner was in many cases treated as a guest at the German Jagdstaffel that had brought him down.

Manfred von Richthofen, in accordance with his military mission, was primarily concerned with eliminating enemy aircraft while not necessarily killing the crew, even if in most cases this was the inevitable outcome of his attack. One has to keep in mind that the aircraft he was facing were largely built of wood and fabric, offering no protection against bullets, and Entente aircrews never enjoyed the luxury of a parachute, allowing them to escape alive from their aircraft should the need to do so arise. His primary intention of bringing down an opponent is proven by the fact that he ceased firing as soon as it was obvious to him that the enemy was going down on the German side, possibly allowing him to land unscathed. This was true for 10 out of his 80 aerial victories. This fact strongly contradicts the "murderous trait of character" ascribed to him by recent authors who either were ignorant of, or chose to ignore, all the evidence to the contrary.(14)

Carl-August von Schoenebeck wrote about Manfred von Richthofen's behavior towards enemy crews:

"Richthofen was the most fabulous comrade that you could ever imagine. Generous, fun-loving, -- he participated in all our stupidities... and he was also comradely to the English pilots who were shot down alive and taken prisoner by us. 'The matter's over now, -- there's no more reason for us to fight!" he said. And so, we treated them like regimental comrades. Before they were transported to Germany, they could move around completely freely and dined with us in the Officers' Mess. Richthofen himself took their word of honor that they would not dare to try to escape – Someone once refused to give his word of honor. Then [Richthofen] said: 'All right! I still won't lock you up. I'll just have you closely

guarded. But if you run away, you could end up being shot by the guard.' He didn't run away."[15]

On 3 September 1917, Manfred von Richthofen brought down Lt. A F Bird, pilot of a Sopwith "Pup" from 46 Sqn. RFC. Anthony Fokker, who was visiting Jagdsgeschwader I at the time to witness the debut of the first two examples of his Triplane, filmed both men smiling and chatting together before Bird was taken to Jasta 11 as a guest that day.[16] This clearly contradicts the modern accusation that von Richthofen had a sadistic hatred for the enemy.

Von Richthofen constantly worked on improving the operational capability of the German fighter units, both in terms of technical development and their tactical deployment. The degree of influence his experience reports had, and the detailed technical requirements laid down by him for the development of a fighter airplane that could be used successfully at the front, are shown by the fact that they were even forwarded to the Hungarian "MAG-Magyar Általános Gépgyár" (MAG-Hungarian General Machine Factory), which produced aircraft for the Austro-Hungarian Army.[17] A report written by von Richthofen, likely in the spring of 1917, was found in the archive of this company. One may assume that this report was also forwarded to German aircraft manufacturers engaged in the development of single-seat fighter aircraft. It reads as follows:

"*A main prerequisite for a fighter aircraft is that at a higher altitude, the machine does* **not lose** *altitude during several sharp 360° turns while under full power. Gaining altitude would be ideal. It is a major weakness of the Albatros D III that this is not the case.*

<u>*Climb performance:*</u> *What is the climb performance at each 1,000 meter step? It is important that rotary engine performance does not decrease too much at higher altitudes.*

<u>*Sensitivity of the Controls.*</u>

A) *Torsion* [by ailerons in this case]*: Small movements of the rudder and ailerons must have a great effect; so* [they must be] *very sensitive. The English have double ailerons. The Albatros D III's ailerons aren't quite effective enough. Aileron effectiveness is a principal requirement for a fighter airplane.*

B) *Rudder Control: The Albatros D III's rudder control is good.*

C) *Elevator Control: Should not be overly sensitive. With the Fokker Eindecker, this was often unpleasantly noticeable in air combat. It is quite agreeable, however, with the Albatros D.*

<u>*Speed:*</u> *High speed is also required at high altitude.*

<u>*Climbing performance:*</u> *The old Fokker D put too much emphasis on climbing performance at the expense of speed. Being able to throttle down an aircraft to a very low speed is absolutely necessary even with a very fast machine.*

<u>*View:*</u> *View above, below, sideways must be flawless. Albatros D III – good; Alb. D II – particularly bad downwards; Alb. D I – upward view unsuitable in a dogfight. On the Halberstadt D, the upper wing is too close to the eyes and exactly at eye level. A little farther from the eye and a little higher would be better during air combat.*

<u>*Strength:*</u>

<u>*Diving:*</u> *The machine absolutely must be able to withstand a twisting, vertical dive of over 1,000 meters; this is not always the case with Albatros D III. The machine must withstand the overpressure of aileron looping; also incomplete loops and repeated sharp overturns with the engine at full speed. The ability to achieve great speed during a dive is vitally necessary. A slow dive, as in a Halberstadt, is generally pointless.*

<u>*Gravity tank:*</u> *Gravity tank for 20 minutes at full throttle.*
Pressure tank for 1½ hours at full throttle.

<u>*Instruments:*</u> *As few as possible; tachometer, small altimeter, pressure gauge.*

<u>*Throttling on the control stick*</u>*: MG-buttons in a righthand grip for the right hand, throttle for the left hand, both on the control handle.*

<u>*Auxiliary Sight*</u>*: Guns close together in the middle. 500 cartridges per gun. Type of bullet is a matter of taste.*

<u>*Windshield*</u>*: If oil from the engine can be avoided, none. Otherwise as small as possible.*

<u>*Seatbelt device:*</u> *Must be present.*

<u>*Installation of a mirror:*</u> *Greatly encourages laziness and carelessness.*

These are only my private views and those of the gentlemen of my Jagdstaffel.

Signed v. Richthofen.[18]

Von Richthofen led the formation and tactical deployment of the first Jagdgeschwader of the German Luftstreitkräfte In addition, he wrote reports on how the individual Allied aircraft were best engaged and how pilots had to behave in dangerous situations, providing them with valuable insight on how to survive. By the summer of 1917, he demanded the development and deployment of better fighter aircraft. It was also he who vehemently pushed for the introduction of the Fokker D VII to the front in the spring of 1918, even though this brought him into conflict with officers in the Inspectorate of Aviation.[19] His treatise on "Geschwader Flights", written in March/April 1918, formed the basis for the operational deployment of the German Jagdstaffeln and Jagdgeschwader until the end of the war.[20] Thus, his importance in guiding the effective use of German fighter units in the First World War cannot be appreciated highly enough.

Josef Jacobs, holder of the Pour le Mérite and the second-most successful German fighter pilot to survive the war with 47 victories, expressed the following in conversation with the author:

"*In my opinion, it was totally irresponsible to have allowed Richthofen to fly at the front in the spring of [19]18. With his popularity, his personality, his influence, his experience, and his tactical abilities, he was far too valuable to expose him to this danger.*"[21]

In summary, Manfred von Richthofen was neither the superhuman hero as he was stylized in the 1920s and 1930s, nor a psychopathic killer, as he is sometimes portrayed (by some German authors) in recent times. He was a soldier

Right: A relaxed Manfred von Richthofen, seemingly enjoying a moment of respite in his office.

through and through in accordance with the times and as a result of his patriotic upbringing. In addition, he was an outstanding leader who had the ability to form an elite unit and lead in battle with great success.

Last but not least, it should be borne in mind that Manfred von Richthofen was just 24 years old when he took over the responsibilities at Jagdstaffel 11, and 25 years when he became commander of Jagdgeschwader I. For a man of his young age, his accomplishments, contributions, and influence were extraordinary.

First Successes for Jagdstaffel 11, January 1917

Above: Sanke Card of Rittm. Manfred von Richthofen with the order Pour le Mérite.

Above: Hermann Lohmeyer, mechanic in Jagdstaffel 11, during a furlough on the balcony of his parents' house in 1917.

For Manfred von Richthofen, being named the Staffelführer of Jasta 11 meant that he was not only responsible for its operations but for the lives and success of the pilots entrusted to him as well. In Jasta 2, he had been a single individual within a well-rehearsed team whose members had mostly been hand-picked and trained by the great Oswald Boelcke himself. When Richthofen arrived on 20 January 1917, he encountered a number of experienced airmen at Jagdstaffel 11. However, not one of them had achieved a confirmed success in air combat. Conversely, Jasta 11's pilots looked upon Richthofen with a mixture of curiosity and respect since he was from the legendary Jasta 'Boelcke', had already been credited with 16 confirmed victories, and was decorated with the coveted Pour le Mérite.

On the same day that Manfred von Richthofen arrived at La Brayelle, Halberstadt D II D.107/16, followed by Halberstadt D.116/16 on the 21st, were sent to Army Air Park 6 as unfit for further service. They were replaced on the 21st by the Jasta's first examples of the Albatros D III: D.1993/16 and D.1996/16. The next day, another exchange took place: Halberstadt D Vs D.412/16 and D.418/16 were swapped for Albatros D IIIs D.1980/16 and D.1991/16.(1)

The pilots were quite pleased that the Staffel was now receiving the Albatros D III as it offered a significant improvement in performance compared to their previous Halberstadt fighters. The Albatros was armed with two machine guns compared to the single weapon fitted to the Halberstadt, and its engine provided a third more horsepower compared to the older fighter. Perhaps inspired by Richthofen's striking red-painted Albatros D III, which he brought along from Jasta Boelcke, some of his pilots began to paint their new Albatros D IIIs with individually colored markings as soon as they got them over the course of the coming weeks.

Nineteen-year-old Gefr. Hermann Lohmeyer was among those who were permitted to carry out this artistic activity. With regard to the motivation behind this, Lohmeyer said that after an airplane was assigned to and tried out by a specific pilot, it was given a personal marking so that the men could identify one another during frontline operations.(2)

Hermann Lohmeyer was born on 5 July 1898. He was a plumber by trade, which turned out to save him from serving in the trenches. When he was called up into the military, he was ordered to Flieger-Ersatz-Abteilung 5 in Hanover on 21 November 1916 to be trained as an aircraft mechanic. After two months of training, he was transferred to Jasta 11 on 21 January 1917:(3)

"'I arrived at the Staffel at almost the same time as Richthofen, but of course my arrival was hardly noticed. I reported to the Offizier zur besonderen Verwendung [Special Duty Officer] *and the appropriate Werkmeister* [foreman], *a Vizefeldwebel, and that's it. I only set eyes on Richthofen later and he would hardly have noticed me,' said Lohmeyer with a grin. 'The day after my arrival I was assigned to Lt. Kurt Wolff's machine. Thus, I was one of his three mechanics. Each pilot had three mechanics. We were responsible for the repair and maintenance of his machine – except for the machine guns. We had separate people for machine-gun repair and maintenance. The machine guns weren't our job. The pilots had to secure the machine guns themselves after landing, and then the ammunition was removed. We had our own workshop with specially trained people for major engine repairs, but we of course also gave them a hand.'"*(4)

Whereas driving snow had prevailed the week before, the weather began to clear on 21 January 1917. It was very cold, but the snow-covered ground was bathed in sunlight and therefore flight activity began to rise.

Enemy air activity at the front was brisk on 23 January and Jasta 11 was called into action that morning and then again in the afternoon when eight aircraft with Richthofen in the lead took off for the front. The flight was subdivided into two Schwärme (literally "swarms") which were probably comprised of four Albatros D III and four Halberstadt aircraft. This was likely necessary as the two aircraft types had different flying characteristics and performance. Once they had joined up over the airfield, the pilots headed northwest toward Hénin-Liétard (now part of Hénin-Beaumont) and followed the railway line to Lens, which they knew was one of the preferred areas of entry for their opponents. Once they arrived over Lens, the Jasta 11 flight headed south again and spotted a group of No.40 Squadron F.E.8 fighters protecting two-seaters conducting reconnaissance. When the German fighters attacked the formation, the reconnaissance aircraft turned west while the F.E.8s tried to cover their withdrawal.

Manfred von Richthofen submitted this combat report:

Above: 2/Lt. J. Hay in the cockpit of his F.E.8 6388. (C. Huston)

"23 January 1917
4:10 p.m. above trenches south-west of Lens
No details, airplane fell on the enemy side of the lines.
About 4:10 p.m., I together with seven of my airplanes attacked an enemy squadron west of Lens. The airplane I had singled out caught fire after 150 shots discharged from a distance of 50 meters. The airplane fell burning. Occupant fell out of airplane at 500-meter height. Immediately after the airplane had crashed on the ground, I could see a heavy black smoke cloud rising. The airplane burnt for quite a while with frequent flares of fire."(5)

F.E.8 6388, flown by 2/Lt. J. Hay, was the aircraft that was shot down. Royal Flying Corps' Communiqué No. 72 reported this concerning the fight:

"2nd Lt J Hay, No.40 Squadron, engaged a hostile machine over La Bassée. The wings were seen to break away from the German machine before it crashed. 2nd Lt Hay then attacked and destroyed the leading machine of a formation of eight. He was then brought down himself and killed."(6)

With regard to this communiqué, it would appear that two different engagements were mixed into one, because Hay's victory occurred at 10:15 a.m. British time over an FA(A) 240 machine crewed by Offz.Stv. Wilhelm Mohs and

17th Victory.

January 23rd, 1917.
4.10 p.m.
Trenches above south-west of Lens.
No details, plane dropped on the enemy's side.

About 4.10 p.m. I attacked together with seven of my planes enemy squad. west of Lens. The plane I had singled out caught fire after 150 shots discharged from a distance of 50 meters. The plane fell, burning. Occupants fell out of plane in 500 meters height.

Immediately after the plane had crashed on ground, I could see a heavy black smoke cloud arising. The plane burnt for quite a while with frequent flares of flame.

(Sig.) Frhr. v. Richthofen.

Acknowledged.

Above: English translation of the original air combat report by Manfred von Richthofen, 23 January 1917. (via Alex Imrie)

Lt. Wilhelm Riehl that came down near Provin, east of La Bassée. This event was in no way connected to the afternoon engagement with Jasta 11, where 2/Lt. Hay was shot down in F.E.8 6388. The second victory attributed to him during the struggle with Jasta 11's eight machines was in error because Jasta 11 suffered no losses that day. Perhaps his comrades mistook his burning airplane for that of his alleged victim – something that could happen all too easily during the hectic action of a dogfight.

40 Sqn RFC's report stated the following regarding its clash with Jasta 11 aircraft that afternoon:

Lt. E.L. Benbow of 40 Sqn. RFC, flying F.E.8 7627, was a witness of the airfight and gave the following report:

"I saw 8 hostile aircrafts west of Lens about 2,000 ft. below me. I saw F.E. 8 attack leader and I attacked another hostile aircraft. My gun stopped, apparently owing to cold, I turned round to remedy this when he was attacked by ***a red machine*** *from the front. I dived and got away and when I got the gun working the hostile aircraft had flown east."*(7)

Jagdstaffel 11, after three and a half months of existence and over 200 individual fighter sorties to the front, had finally achieved its first success in air combat. The "new chief", with his 17th confirmed victory during his first flight with his new Staffel, had gotten them off to a promising, new start.

In response to reports coming from the front, the Staffel took off again under Richthofen's command in good weather at noon the next day. Their route took them once again over Hénin-Liétard toward Lens. Shortly after their flight over the village, Richthofen detected a group of No. 25 Squadron F.E.2b aircraft to the southwest near Rouvroy that were busy photographing German positions, and proceeded to attack the formation. He went after F.E.2b 6937 with Capt. Oscar Greig as its pilot and Lt. John E. MacLennan as his observer, and later reported:(8)

"24 January 1917
12:45 p.m. west of Vimy
Fixed Motor
Plane No. 6937 Motor No. 748

Accompanied by Sgt. Howe, I attacked the command airplane of an enemy formation at about 12.15 p.m. After a long fight, I forced my adversary to land near Vimy. The occupants burned the airplane after landing. I myself had to land as one wing had cracked at 300 meters. I was flying an Albatros D III.

According to the English occupants, my red plane was not unknown to them, as when being asked who had brought them down, they answered: 'Le petit Rouge'.

Two machine guns have been seized by my Staffel. The airplane was not worth being removed, as it was completely burned."(9)

One photo show that following his emergency landing, Richthofen visited the downed airplane that had already been incinerated. He cut the serial number from its undamaged

10.

18th Victory.

January 24th, 1917.
12.15 p.m.
West of Vimy.
Fixed Motor
Plane No. 6937.
Motor No. 748.

Occupants: Pilot - Capt. Craig.
Lieut. McLennan.

Accompanied by Sergt. Howe I attacked about 12.15 p.m. the commanding plane of an enemy formation. After a long fight I forced adversary to land near Vimy. The inmates burnt plane after landing. I myself had to land, as one wing had been cracked in 300 meters. I was flying Albatross D.III.

According to the English inmates my red painted plane was not unknown to them, as when being asked who had brought them down they answered: "Le petit Rouge".

Two machine-guns have been seized by my staffel. The plane was not worth being removed, as it was completely burned.

(Sig.) Frhr. v. Richthofen.

Was acknowledged.

Above: English translation of the original air combat report by Manfred von Richthofen, 24. January 1917. While the report sates the airframe and engine serial numbers of the British plane in the upper left corner, no such information relating to his own airplane was included in the report. Here, the serial number of the British aircraft has been wrongly noted as being 6937, while photographic evidence proves that it actually was 6997. (via Alex Imrie)

rudder and took it with him as a souvenir. By that time, the two British crewmen had already been taken away, however. Neither Capt. Greig nor Lt. MacLennan ever mentioned having met Richthofen. According to them, they only found out the identity of their adversary much later, as MacLennan wrote to Floyd Gibbons:

"As regards the red machine, we had previously seen it, but we did not know who it was. I am glad to hear that he had to land, as I did not know this."(10)

Accordingly, Richthofen must have heard about the identification of his aircraft as "Le petit Rouge" from a third party.

25 Sqn RFC's report on this dogfight was short and concise:

A photographic patrol of No. 25 Squadron was continually interfered with by hostile scouts... 2/Lt. J.L. Leith with 2/Lt. W.D. Matheson attacked a scout which had forced one of our machines to land in German territory. The German aeroplane was driven down in a spinning nose-dive. It was not seen to crash but is believed to have done so."(11)

As usual, the captured British crewmen were interrogated by a German intelligence officer as seen in these excerpts from his full report:

"1. Personnel:
Capt. Greig, 26 years old, born in London, career officer.
Lt. MacLennan, 20 years old, also a career officer who formerly belonged to the Scottish Rifles (8th Division) and has been with the RFC for about one year...
3. Orders:
Reconnaissance flight with special task of taking photographs. Accompanied by six other aircraft which for some reason did not fly over the German lines.
6. Air Combat
The observer (MacLennan) started taking photographs and did not notice our [German] *airplane's approach because it approached the enemy in a very clever manner. Intense fire immediately destroyed the fuel line and container. Greig was shot in the right ankle. He* [MacLennan] *couldn't fire because his*

Above: Manfred von Richthofen's emergency landing with "le petit rouge" due to a broken wing following his fight with FE.2b 6997. This photo is also clear evidence that this aircraft was not the one with the Albatros company logo on the rudder left out during the painting process, that aircraft having previously been erroneously identified as "le petit rouge". (R. Kastner)

machine gun failed to work. Capt. Greig set the apparatus on fire after landing. [MacLennan later said that he set the airplane on fire because Greig was wounded]

9. In general

MacLennan: Described the maneuvering of our airplane in front of him shortly before landing as careless because he certainly could have fired at him if his machine gun had been in order.[12]

It could be that Richthofen had stopped firing and moved into a dangerous flying position because he was confident that the British crew had surrendered. On the other hand, he might have gotten into such a position while trying to avoid more stress on his Albatros D III's cracked wing. In any event, he was lucky that the British aircraft's machine gun had failed.

After the war, John E. MacLennan gave a much more detailed report of this air combat that generally corresponded to the statements made to the German interrogation officer:

"Greig and myself were sent up to take mosaic photographs of the Vimy Ridge. This entails great accuracy in the actual exposures, a stopwatch being used so as to entail a correct overlap of each exposure. It is further necessary to keep at an exact altitude and to fly in a direct line, otherwise results obtained are unsatisfactory. Thus, the taking of photographs fully employs the attention of both the pilot and observer…I was actually taking photos when a burst of machine gun fire from behind notified the attack. I looked round and perceived a red enemy machine diving away. In the first attack he shot through both oil and petrol tanks and splintered the propeller; Captain Greig was also shot through both legs. As a result, the machine could only glide and the maneuverability was greatly impaired through lack of any engine power. Richthofen attacked all the way down till the machine was but a few hundred feet from the ground. He attacked each time from below and behind, in which position we were unable to return fire. Only once did he get in front when we managed to get off two bursts from the front gun, but at longish range and without result."[13]

Capt. Oscar Greig maintained a scrapbook along with his own written accounts of his adventures that he later donated to the Imperial War Museum. It was published in 2002 and read in part:

"The observer was looking through the camera sights and just beginning to take exposures. There had been a complete absence of AA shells, but I thought this was because they were waiting for us to get further from our lines. At this moment I heard a machine-gun firing and saw several holes appear in the left wings. We were taken completely by surprise. I turned sharply to the right, banking steeply, hoping to get the enemy in front of me, but on completion of half a circle the enemy fired another burst from my right putting the engine out of action and hitting me in the right ankle. I continued in circles endeavoring to get a sight of the enemy but he succeeded in keeping below and behind my machine, making it impossible for either of us to fire at him while he continued to pour bullets into my machine. I distinctly saw several tracer bullets pass through the instrument board between

Above: German soldiers seize the opportunity to sneak into the picture alongside the burnt-out skeleton of FE.2b 6997. (T. Henshaw)

Above: Richthofen appears on the scene, only to be quickly surrounded by curious German soldiers. (R. Kastner)

my observer and myself.

I had no pain in my left foot, only an unpleasant sensation of warmth, and I could still use my rudder.

I was losing height very rapidly. At about 4000 feet altitude by the aneroid, which was probably really about 2000 feet owing to the descent being too rapid for the aneroid to register correctly, the firing stopped, and I straightened out making for our lines which were then only some three miles distant. My observer pointed behind us, indicating another attack, and tried unsuccessfully to raise the back mounting. As he could not get the gun in action he sat down again. A second after a very fast small scarlet biplane passed over us and went away to the right front.

I tried everything I could to get the engine going but it would only splutter feebly. Seeing that we could not cross Vimy Ridge even if we threw the guns and everything possible overboard, I turned east and landed on a flat plane between Vimy and Fresnoy, gliding through a batch of about a dozen field telephone wires breaking them with the undercarriage, and rolling through some more before the machine stopped. There were several small shell holes about but luckily I did not hit any.

When we stopped, I stood up and asked if McLennan if he was hit. He said, 'No, are you?' I told him I had got a 'blighty'

Left: The burnt-out F.E.2b 6997 is put under guard by a German soldier. The rudder fabric carrying the serial number was cut out by Manfred von Richthofen and kept by him as a souvenir. (Bayerisches Hauptstaatsarchiv, Abt 4 München)

one in the foot and was damned sorry to bring him here; at which he asked if we were in Hunland and was not pleased to find we were. I noticed that he had several bullet holes in his coat, as I had also.

By this time there were a number of Germans running towards us from all directions. I said we had better burn the machine, and Mac lit the lighter carried for that purpose, handed it to me, and jumped out. I got out onto the wing and put the lighter against the celluloid of my map case to make sure of burning the maps and got down. Whilst doing this I noticed one of the front center section bracing wires was shot in two. Also the floor of the observer's part of the nacelle was covered in chips of wood made by the bullets. Directly I got down I saw that there were quantities of oil, petrol, and water, streaming out onto the ground and I wished I had put the lighter in this, but it immediately caught fire making a tremendous blaze and a huge cloud of black smoke.

We walked a few yards away to get clear of the Very lights and Lewis gun ammunition that were being set off by the fire and sat down in the snow at the edge of a shell hole.[14]

Greig and MacLennan decided to set fire to the airplane before German soldiers appeared. Norman Macmillan, a former pilot of 45 Sqn, interviewed Oscar Greig and wrote the following report:

"MacLennan lit the lighter carried for the purpose, handed it to Greig, and jumped out. Greig climbed on the wing. He did so he saw that one of the front cross bracing wires was shot in two and the floor of the observers cockpit was littered with wood chips made by the bullets. He put the lighter against the celluloid of his map case to make certain of burning his maps. He climbed down to the ground.

Oil, petrol, and water streamed from the ruptured engine and tanks. For a moment he wished he had waited and thrust the lighter into this. Even as he wished, he saw it catch fire. A big flame bleached, followed by the black cloud of petroleum burning in the open air. The tyres, saturated with dripping petrol, caught fire and burst. Greig knew they were lucky to have come down without catching fire in the air.

Very lights and ammunition exploded amid the flames. The two men walked clear of the dangerous fireworks and sat down on the snow-covered edge of a shell-hole. The nearest German was brandishing a naked bayonet over his head. When about 50 yards away he shouts in German Greig and MacLennan presumed it was a demand to know if they surrender.

They stood up an held up their hands. The soldier resheathed his bayonet. Greig set down again, for his foot was hurting now. The soldier approached in a friendly manner and stood beside them, waiting for an approaching officer. When the latter arrived Greig stood up and saluted. The German returned the salute. Greig sat down. The officer asked if he was wounded. Without thinking, Greig replied in French. The German officer asked if they were French. Greig replied in English and thought the German's change of manner showed a preference for the British.[15]

Greig's foot was bandaged. They were then taken to a small farmhouse upon a little wagon. Presently a doctor entered, examined Greig's foot, and rebandaged it. Another officer brought cigarettes, a bottle of wine, and two glasses, but he apologized for not being able to offer food. Later they were picked up by an officer in a car and taken to a building in Douai, where they were separated. At 17:00 a Leutnant of the German Air Service waited in a car for him outside. Greig later tried to identify him from the famous Sanke Card portrait of Jasta 11 pilots and believed him to be Kurt Wolff. However, he never told Greig his name, and Greig reported that he had a large scar on the side of his jaw (which Wolff did not). He asked if Greig would give him his Sam Browne belt as a souvenir, which Greig did. Greig was taken to the Institution Sainte Clotilde in Douai, which had been turned into a temporary hospital.[16]

Kurt Wolff, in his diary, had this to say about the event:

"The first flight under the new reign was on the 22nd, and Richthofen immediately shot down an Englishman (No.17); No.18 followed the next day. While doing so, Richthofen had a wing breakage with the Albatros D.III, which was grounded as a result. Then he had to fly a Halberstadt, much to his great annoyance."[17]

Above: Albatros D III D.2006/16 was likely the next Albatros flown by Manfred von Richthofen. It carried a far simpler personal marking, being marked with just the red cap ribbon of the Ulanen-Regiment Kaiser Alexander III. von Russland (Westpreußisches) Nr.1 as personal marking in the form of a fuselage band. As an additional detail, the spinner was also painted red. Perhaps it was not expected to enjoy a long frontline service, after the experience gained with the previous "le petit rouge"? (L. Bronnenkant)

In a 27 January 1917 letter to his mother that was published in *Der rote Kampfflieger*, Richthofen wrote:

"I was lucky. On my first day I shot no.17 down and on my second day no.18. As I was shooting my 18th down, one of my wings broke in two at 300 meters' altitude during the air combat. It was a miracle that I reached the ground without going to pieces. Three airplanes from Jasta Boelcke fell the same day, including nice little Imelmann [Lt. Hans Imelmann] – *how wretched! It's possible that the same thing happened to them as happened to me."*[18]

Clearly, Richthofen indicated that it had been a structural failure that had forced him to land. And the Inspektion der Fliegertruppe (Idflieg) responded a few days later to his incident and others like it by temporarily grounding the new Albatros D III until an investigation could determine the reasons for the failures and how to possibly remedy them.

On 24 January, the same day as Richthofens's emergency landing, Halberstadt D V D 420/16 was exchanged for Albatros D III 2006/16.[19] The Staffelführer then claimed this aircraft for himself as a temporary replacement for his damaged Albatros D III *le petit rouge*. As usual, it was eventually marked with a personal identifier, in this case a red band around the fuselage. But as mentioned above, the Albatros D III had been grounded – an order that was not rescinded until 19 February 1917. In the meantime, Richthofen had to make do with a Halberstadt fighter.

As if these structural failures were not concerning enough, Richthofen was dissatisfied with Jasta 11's performance as expressed to his mother in the 27 January letter reproduced in *Der rote Kampfflieger*:

"You are no doubt wondering why I haven't written. So much has happened in the interim that I don't know what to write down first. I have become the leader of Staffel 11 in Douai. I departed my dear old Jasta Boelcke only with much reluctance. All my resistance was of no use. Staffel 11 has been in existence for as long as my old one, only it hasn't shot anyone down yet; so, the work here is bringing me very little joy at present."[20]

The last sentence conveyed his disappointment that the pilots of Jasta 11 were not meeting his expectations, as Carl Allmenröder mentioned weeks later during a visit home. He described how Richthofen, during one of his first flights with his men to the front, had attacked a British formation in his usual manner, believing that his Staffel would follow close behind. But when he reached the British aircraft and began to attack, he realized that his pilots had remained at a respectful distance – that is, they had opened fire at the enemy from so far away that hitting anything was highly improbable. His machine, however, suffered a number of hits and he had to break off his attack and spin away to avoid being shot down. When he landed his machine at the airfield, the other pilots had already done so and were sheepishly hanging around in front of the aircraft hangar. Richthofen climbed out of his machine and, without any words of reproach, passed by the waiting men and went to get some sleep. Later, Carl Allmenröder and Kurt Wolff reported to

Above: Shooting on targets on the ground was one of the regular training exercises for Jasta 11 pilots to improve their marksmanship. (L. Bronnenkant)

him, wanting to apologize. Instead of bombarding them with reproach, as was expected, Richthofen calmly asked them to take a seat and explain their point of view. The ensuing conversation revealed that the men's reluctance in attacking was by no means due to a lack of courage. They simply lacked the knowledge and experience of how to carry out an attack in Staffel formation at close range. Richthofen calmly listened to this admission and concluded: "*I should have thought of that, but we'll change that right now.*"(21)

The very next day, Richthofen began a rigorous training program where attacking an enemy formation in unison and firing machine guns at targets on the ground were practiced. Moreover, Richthofen lectured them on the flight characteristics of various British aircraft. This training program, combined with their frontline operations, pushed the men to the very edge of their abilities and each night they went to bed exhausted.

Left: As Staffelführer of Jagdstaffel 10, Lt. Erich Loewenhardt strictly followed Manfred Richthofen's air combat instructions and enforced this on his pilots. This was one reason for the successes of Jagdstaffel 10 under his leadership.

January 1917

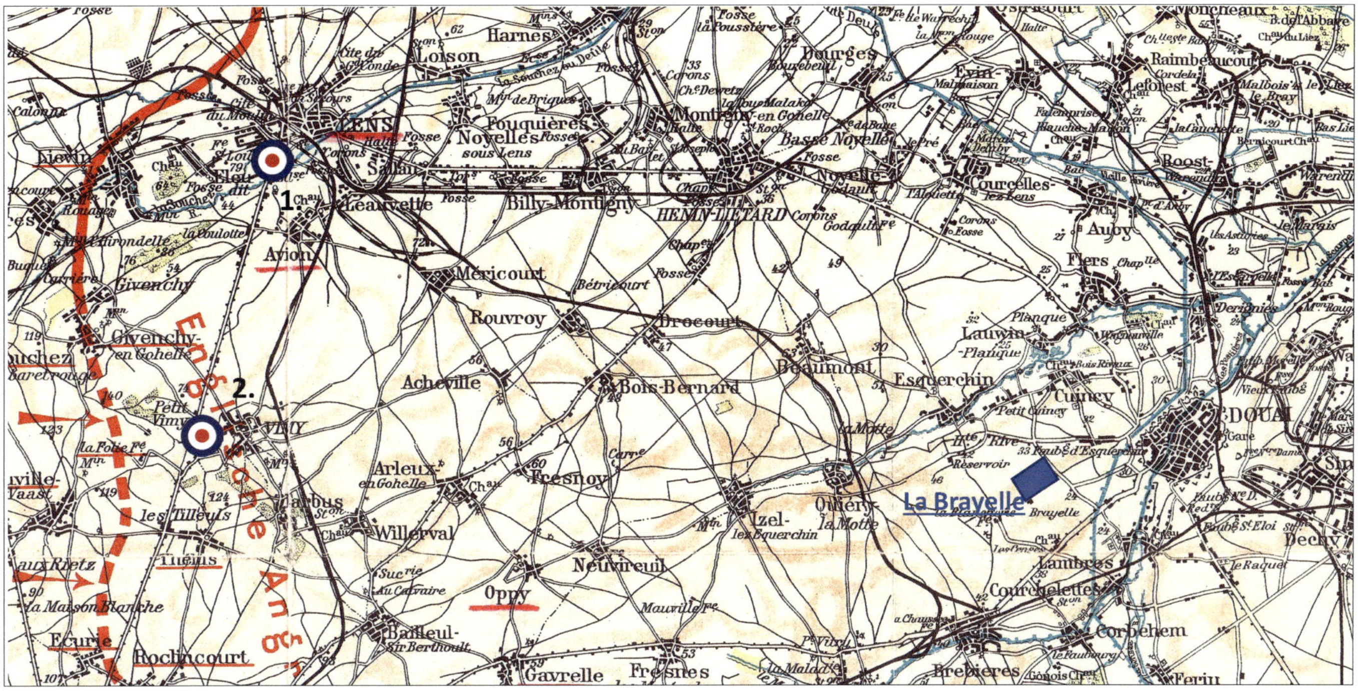

Above & Below: Map and details of the aerial victories of Jagdstaffel 11 in January 1917.

Nb	Date		Hour	Pilot	Location	Aircraft	Line
1.	23.01.17		16.10	Lt. Manfred von Richthofen	s.w. Lens	F.E.8 6388, 40. Sqn.	Allied
2.	24.01.17		12:15	Lt. Manfred von Richthofen	w. Vimy	F.E. 2b 6937, 25. Sqn.	German

Carl Allmenröder wrote to his fiancée, Helene Kortenbach:

"My life consists solely of flying, practicing, and learning – sometimes I'm almost too tired for dinner…"(22)

In a letter to his aunt dated 8 February 1917, Kurt Wolff confirmed just how demanding the training and missions under von Richthofen were:

"*Dear Pos!*

Suddenly, while enjoying one of the lovely cigars you sent me for my birthday, I began to doubt whether I really wrote the thank you letter I intended. I know for sure that I sat down at my desk with that intention, with a cigar in my face, but I cannot remember exactly whether I actually carried out my resolution or whether I wrote to another address out of habit. If my fears turn out to be true, I apologize and thank you very much. You will probably be surprised that I am so dopey at such a young age, but that is only because I am tired. From January 20th up to today, a full four weeks, we have flown every single day without exception and gone up at least twice, sometimes even three times a day. You climb up to around 4000 m on every flight and have enough dogfights, so naturally you are dead tired in the evening and don't feel like doing much more. Today, after a long time, the sky is completely overcast and it's raining in places, i.e. up to now at about 3 o'clock, but I don't know whether the evening will be nice. We've done well, and in the last 4 weeks 9 Englishmen have definitely been shot down by the Staffel; but nothing is known about others who fell far behind the front. You've probably read about my Staffel leader in the newspaper. After my success on the 5th, when I and a non-commissioned officer shot down a Lord who fell burning into the enemy lines, I've unfortunately only had unsuccessful combats."(23)

Carl-August von Schoenebeck also reported that "Schulunterricht" ("school lessons") was an integral part of the established training and educational programs at Jasta 11 and the Jagdgeschwader. Manfred von Richthofen compiled all information on the flight characteristics of enemy aircraft. Those characteristics were discussed during "school time" and precise instructions were given on how to attack each type. Richthofen demanded that his pilots should be able to recite the flight characteristics of enemy aircraft "in their sleep." Schoenebeck related how Richthofen regularly had fun during the evening meal and social gatherings – even after a few beers or some wine – by surprising one of his pilots with questions about the enemy's aircraft:

"So how would you attack an S.E.5 when he is in a superior position above you?'

He expected the answer to be fired back immediately. Nothing was more embarrassing than when the person who was asked couldn't answer. So, we 'crammed' on the flight characteristics of enemy aircraft so that we wouldn't embarrass ourselves. We soon understood that this acknowledgement would not only be the key to success, but it would also save our lives."[24]

Just how carefully Richthofen compiled his experiences in air combat in order to pass his knowledge on to other pilots can be seen in this 16 February 1917 report, sent to the 6th Army's Kommandeur der Flieger (Kofl). Author Floyd Gibbons came across it at the Reichsarchiv (Imperial Archive) in Potsdam while doing research for *The Red Knight of Germany*:

"The adversary often slips downward over one wing or lets himself fall like a dead leaf in order to shake off an attack. In order to stick to one's adversary, one must on no account follow his tactics, as one has no control over the machine when falling like a dead leaf.

Should the adversary, however, attempt to evade attack by such tricks, one must dash down without losing sight of the enemy plane.

When falling like a dead leaf, or intentionally falling wing over wing, the best pilot loses control of his machine for a second or two, therefore, it is a maneuver to be avoided.

Looping the loop is worse than worthless in air fighting. Each loop is a great mistake. If one has approached an adversary too close, a loop only offers a big advantage to the adversary. Change of speed should be relied on to maintain the desired position, and this is best accomplished by giving more or less gas.

The best method of flying against the enemy is as follows: The officer commanding the group, no matter how large, should fly lowest, and should keep all machines under observation by turning and curving.

No machine should be allowed either to advance or to keep back. More or less, the whole squadron should advance curving. Flying straight on above the front is dangerous, as even machines of the same type of plane develop different speeds. Surprises can be avoided only when flying in close order. The commanding officer is responsible that neither he nor any of his pilots are surprised by the enemy. If he cannot see to that, he is no good as a leader."[25]

Richthofen's teachings were also deeply rooted in Jagdgeschwader I, as Kurt Delang later testified to the author. His Jasta 10 Staffelführer, Lt. Erich Loewenhardt, always made sure that no pilot left the formation and that enemy groups were only attacked in close formation. Once, when Delang 'flew off' to attack an opponent, he was chewed out by Erich Loewenhardt in front of the assembled men after landing:

"You are completely crazy to leave the formation! Don't you realize how dangerous that is? My orders are clear and unambiguous: Only attack in closed formation! Do you really want to be shot down? If you disobey my order again, I'll kick you out! Do you understand me?"[26]

Richthofen's increasing demands on his pilots can also be seen in his Kofl's 21–28 January 1917 weekly report. During this period, Jasta 11 conducted 75 "Kriegsflüge", totaling 62 hours.[27]

On 27 January 1917, the same day Richthofen shared his disappointment with Jasta 11's performance with his mother, the unit had a rather cushy assignment. That day they had to "Sperrefliegen," or carry out a barrier flight, during the Kaiser Parade, which meant that they flew back and forth along an assigned sector behind the lines to protect the Kaiser during the ceremony.

Two days later, on the morning of 29 January 1917, the Staffel's Halberstadt single seaters took off and once again made their way toward Lens. Just east of the city, they had a dogfight with British F.E.2b aircraft that again belonged to 25 Sqn:

"Five F.E.2bs, No. 25 Squadron, accompanied by three F.E.8s, No. 40 Squadron, encountered a formation of 12 hostile scouts, all Halberstadts, while on photographic reconnaissance. 2nd Lt A V Blenkiron, 25 Squadron, was hit in the thigh at the beginning of the encounter but managed to stand up and fire at a hostile airplane at point-blank range. This machine went down out of control with columns of smoke and then flames coming out of it. This was confirmed by two other pilots. 2nd Lt J L Leith in another F.E. dived to the help of 2nd Lt Blenkiron and fired a drum at 15 yards range into another hostile aeroplane. This German machine stalled and side-slipped and was seen to fall out of control. Two more Halberstadts were forced to land, and the formation was entirely broken up."[28]

It was Lt. Georg Simon who had been slightly wounded by a grazing shot around 10:50. He had attacked an F.E.2b and wounded its observer, 2/Lt. Blenkiron, with his first bursts of fire. Blenkiron, however, mustered all his strength, got up again, fired back at the attacker, and saw an Albatros go down "out of control". At the same time, a second F.E.2b dove at another Halberstadt to bail out a beleaguered comrade. After a burst of fire, they reported that they saw their opponent sideslip right near them and fall out of control, smoking. Simon, however, fell neither smoking nor burning and had merely spun away when he was in trouble and flown back to Douai where he landed smoothly. After a medic "plastered up" his superficial wound, Simon was fine again and the 6th Army's Kofl mentioned that but reported that no German aircraft had been lost that day.[29]

Even though no German aircraft had been lost, the British formation had succeeded in repelling Jasta 11's attack, as Kurt Wolff recorded in his diary the same evening:

"A formation of six of us encountered 12 Vickers on the 29th. Unfortunately, we did not get any down but were pretty shot up ourselves and Simon lightly wounded."[30]

During his first two weeks as the commander of Jasta 11, Richthofen had laid the foundation for the unit's future successes not only by example with his two air victories but more importantly through his rigorous training program, as would become clear in the months that followed.

Jagdstaffel 11 during Operation "Alberich", February 1917

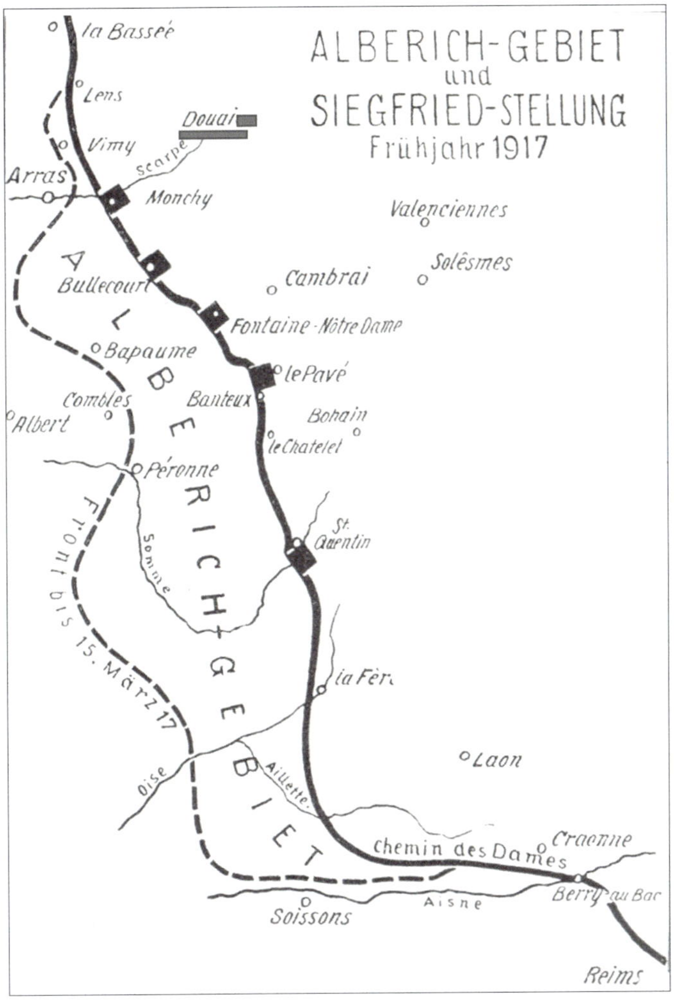

Above: Map of "Operation Alberich", the purpose of which was to bring the German front line back to the heavily developed "Siegfried Line (Hindenburg Line)".

Above: Vzfw. Sebastion Festner sits above the cockpit of his Albatros D III. On 5 February 1917, while flying a Halberstadt fighter, he had become the first Jasta 11 pilot other than von Richthofen to score a victory. Then on 16 February, he became the first of Richthofen's men to score while flying a D III. (L. Bronnenkant)

By October 1916, the German Supreme Army Command decided to move its front between Arras and Soissons back to a new, well-developed trench system known as the "Siegfriedstellung" by the Germans and the "Hindenburg Line" by the Allies. The cover name for this operation, inspired by a mythological dwarf featured in a Germanic legend who could make himself invisible by the use of a magic cloak, was "Alberich." The withdrawal, led by the "Kronprinz Rupprecht" Army Group, began on 9 February 1917 and primarily affected the German 1st and 2nd Armies as well as the right wing of the 7th Army.

In the north, the 1st Army evacuated the area north of the Somme and northeast of Bapaume and withdrew to a new front line along Bullecourt-Havrincourt-Banteux-Le Catelet. The 2nd Army moved out from their territory south of the Somme, abandoning the towns of Péronne, Nesle, Noyon, Chauny, and Ham, and retreated to a position along Bellicourt-St. Quentin-the Oise near La Fère. In the south, the 7th Army gave up their line along Carlepont-Autrêches-Nouvron, abandoned Coucy, and repositioned their front along St. Gobain–Brancourt–Vauxaillon–Laffaux to the Aisne River near Missy/Condé.[1] This measure shortened the German front line by about 45 kilometers. Orders were also issued to level all villages, bridges, and roads in the pullback region so that the enemy would not find cover while advancing through it. It was also important during the withdrawal to prevent Allied reconnaissance aircraft from entering German airspace in order to conceal the construction of the new German defenses.

Operation "Alberich" also affected the southern part of the German 6th Army's front from Quéant in the south to the city of Lens in the north, the exact operational area of Jagdstaffel 11. Although this was only one small part of the operation, it was nevertheless a strategically important one

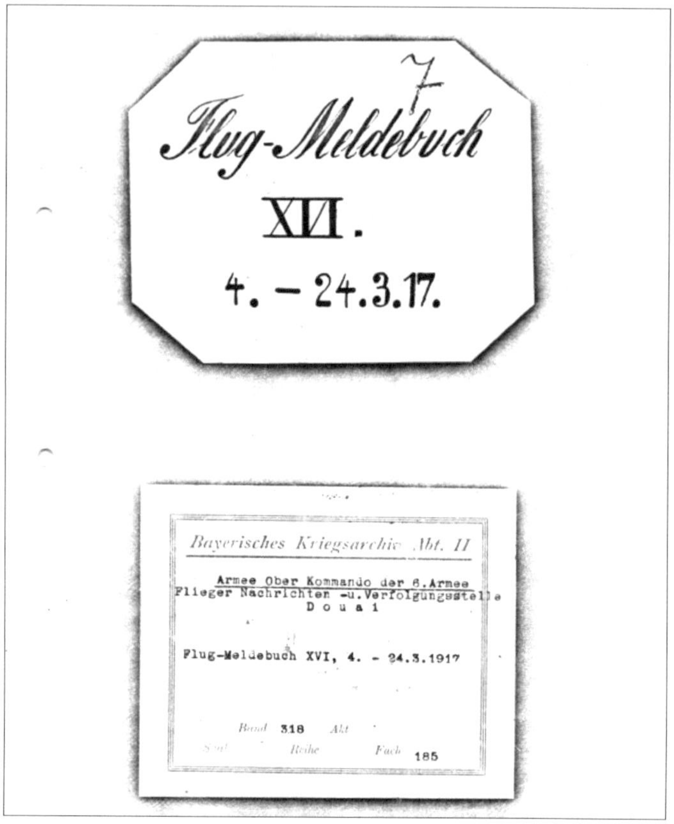
Flug-Meldebuch
7
XVI.
4. – 24.3.17.

Bayerisches Kriegsarchiv Abt. II
Armee Ober Kommando der 6.Armee
Flieger Nachrichten -u.Verfolgungsstelle
Douai
Flug-Meldebuch XVI, 4. - 24.3.1917
Band 318 Akt
Reihe Fach 185

Right: A detailed minute-by-minute report on the daily flight activities within the operational area of the German 6th Army was recorded in the Flugmeldebuch of this Army. (Bayerisches Hauptstaatsarchiv, Abt. 4, München)

because German Intelligence expected a British offensive to be launched there in the spring. It therefore became even more important to block Allied reconnaissance aircraft in Jasta 11's sector which was the only fighter unit deployed there at this time. It did, however, receive some support from Jasta 'Boelcke,' located to the south with the German 1st Army.(2)

The first week of February brought fine weather with almost cloudless skies that resulted in increased flying activity on both sides of the front and particularly in the German 6th Army's northern region.(3)

Following their morning operations on 1 February 1917, a Kette from Jasta 11 took off again at around 16:00 on a flight that took it south of the airfield along the Douai-Arras railway line. When it arrived at the front, the Kette headed north and its leader, Richthofen, soon spotted a British B.E. that was directing artillery fire into German positions. His subsequent combat report gave the following account:

"1 February 1917
B.E. Two-Seater
4:00 p.m. over trenches 1 kilometer southwest of Thelus.
Occupants: Lt. Murray – Lt. McBar, both wounded and died on 2 February.
Airplane No. 6742

About 4 p.m., while flying with Lt. Allmenröder at an altitude of

1,800 meters, I spotted an artillery flyer. I managed to approach, apparently unnoticed, within 50 meters of him in my Halberstadt airplane.

From this distance, up to only the length of an airplane, I fired 150 shots. The enemy airplane then went down in large, uncontrolled right-hand curves, pursued by Allmenröder and myself. The airplane crashed into the barbed wire of our first lines. The occupants were both wounded and were made prisoners by the infantry. It is impossible to remove the airplane."(4)

According to British records, three British and four German aircraft were involved in this aerial battle.(5) The injured British crew in No. 16 Squadron RFC's B.E.2d 6742, Lt. P. W. Murray und Lt. D J McRae, were rescued by German soldiers but succumbed to their serious injuries the next day. Their airplane was destroyed by British artillery fire about 20 minutes after the emergency landing to prevent it from falling into German hands.

Following this, his 19th air victory, Manfred von Richthofen went on his first leave from the front since becoming a Jagdstaffel pilot. It is unclear from available documents who led the Staffel in his absence. What we do know, however, is that Richthofen headed to Berlin to

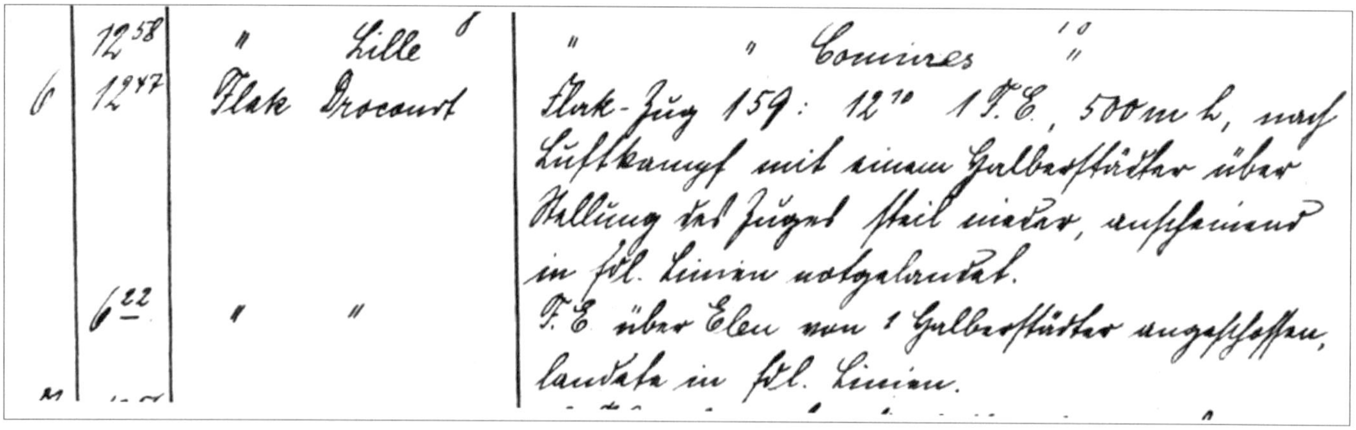
	12⁵⁸	" Lille 6	" " Comines 10 "
6	12⁴⁷	Flak Drocourt	Flak-Zug 159: 12¹⁰ 1 F.E., 500 m h, nach Luftkampf mit einem Halberstädter über Stellung das Flugzeug steil nieder, anscheinend in feindl. Linien notgelandet.
	6²²	" "	F.E. über Elen von 1 Halberstädter angeschossen, landete in feindl. Linien.

Above: Copy of the original Flugmeldebuch entry relating to the Flak Drocourt report dated 5 February 1917 detailing Vzfw. Festner's first aerial victory. (Bayerisches Hauptstaatsarchiv, Abt. 4 München)

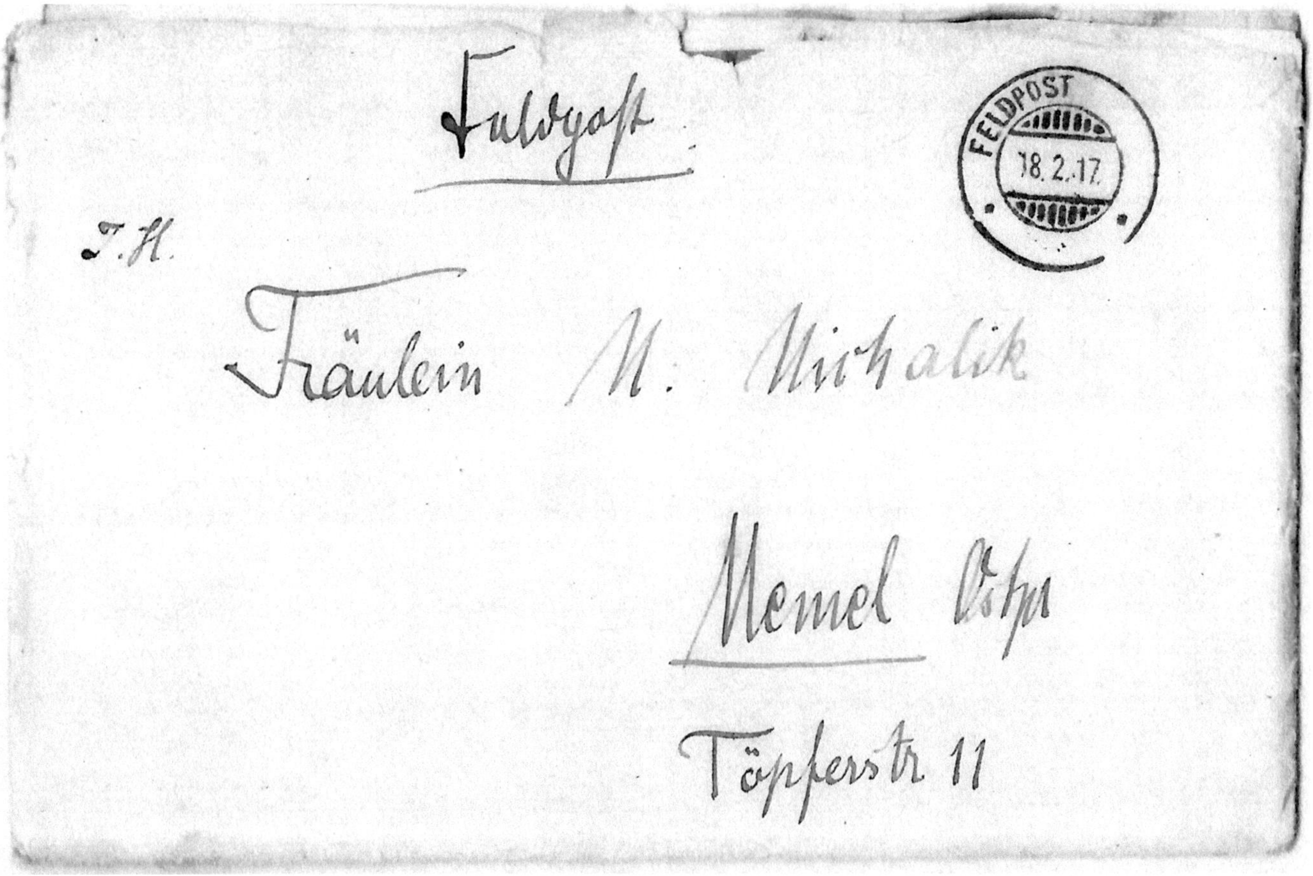
Feldpost
FELDPOST
18. 2. 17.
I.H.
Fräulein M. Michalik
Memel Ostpr
Töpferstr. 11

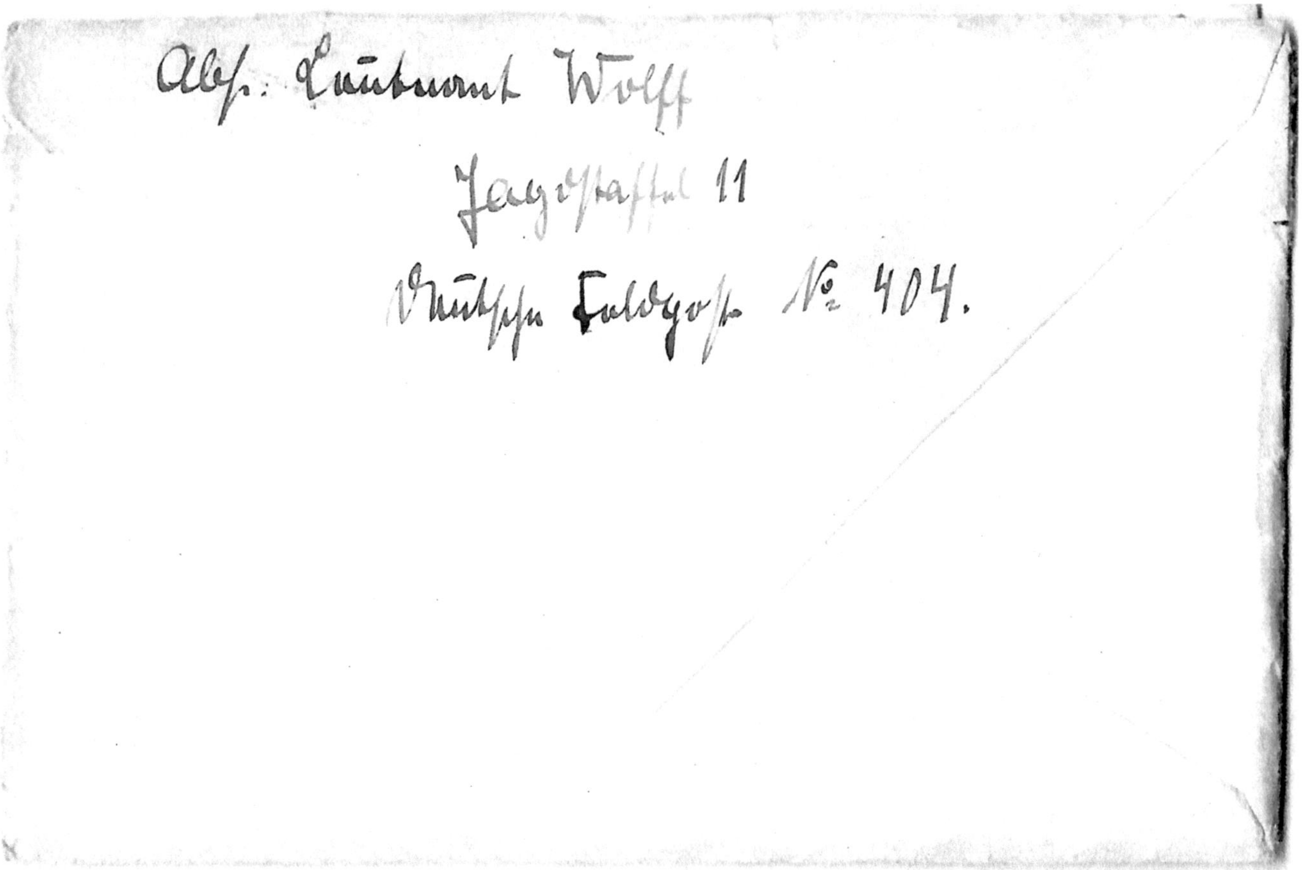
Abs.: Leutnant Wolff
Jagdstaffel 11
Deutsche Feldpost No 404.

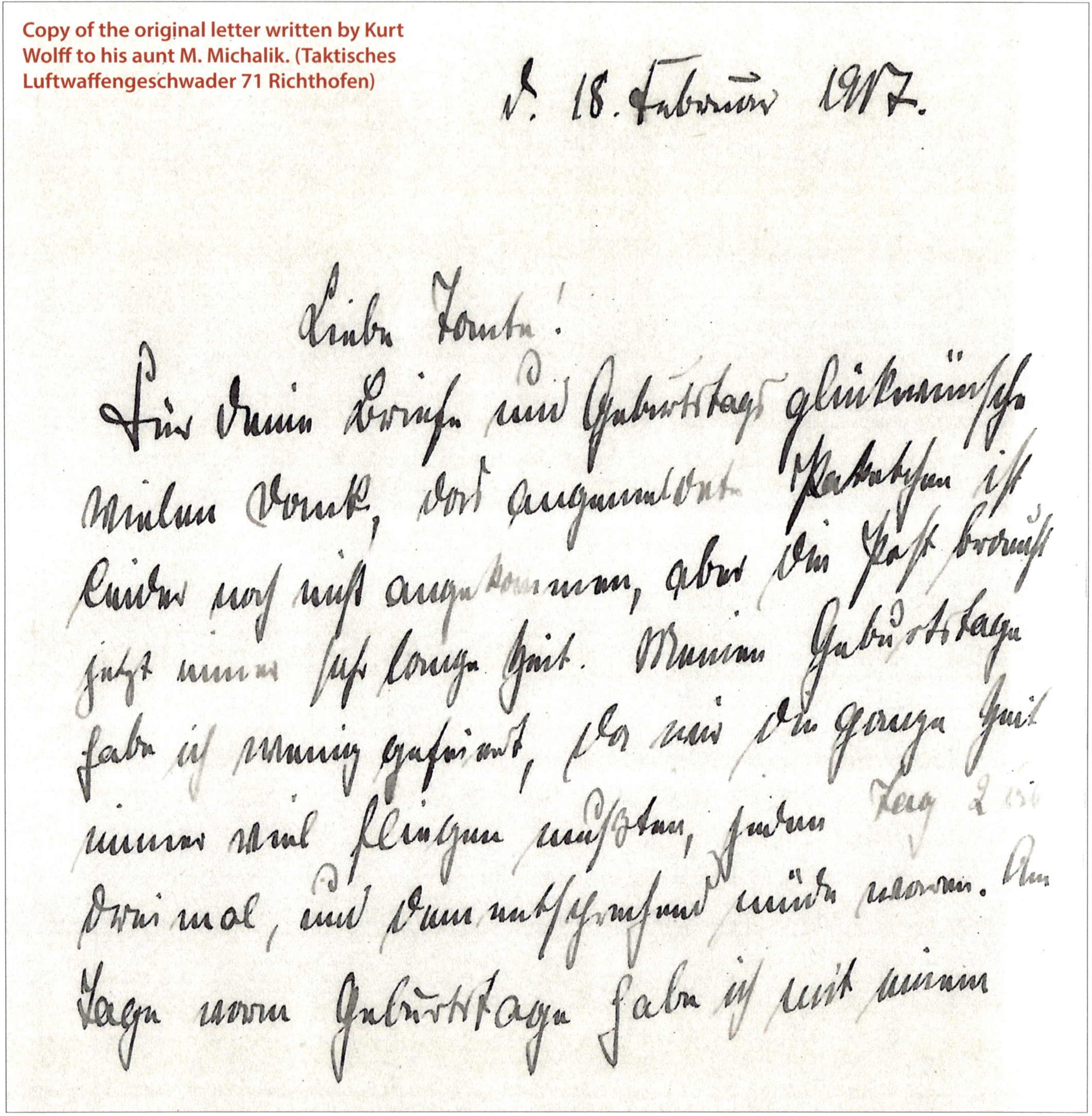

Copy of the original letter written by Kurt Wolff to his aunt M. Michalik. (Taktisches Luftwaffengeschwader 71 Richthofen)

d. 18. Februar 1917.

Liebe Tante!

Für deinen Brief und Geburtstagsglückwünsche vielen Dank, das angemeldete Päckchen ist leider noch nicht angekommen, aber die Post braucht jetzt immer sehr lange Zeit. Meinen Geburtstag habe ich wenig gefeiert, da wir die ganze Zeit immer viel fliegen mussten, jeden Tag 2 bis dreimal, und dann entsprechend müde waren. Am Tage vom Geburtstage habe ich mit meinem

participate with other Staffelführers in a conference focused on solving the Albatros D III's wing fracture problems; and before he returned to his Staffel, he detoured to Schweidnitz to visit his mother.

The 5th of February began with fog and cloudy skies that only gradually dissipated, leaving the pilots to stand idly on their airfield. This morning of low flying activity only began to change in the late afternoon. Following reports from the front, the Staffel's Halberstadts took off in the direction of Vimy/Farbus where British aircraft were directing artillery fire on German positions. Flak clouds from German shells pointed the way for Jasta 11. At 17:02, just as they were arriving on the scene, Flak Platoon 31 managed to shoot down a British artillery airplane from No.16 Squadron RFC, which crashed on the other side of the lines.[6] Shortly afterwards, two Jasta 11 machines reached the enemy artillery flyers and attacked. Kurt Wolff wrote in his diary:

"I flew with Vzfw. Festner on the evening of the 5th and met up with a B.E. at the same place. I shot 180 rounds into his hide with both machine guns and then backed off the enemy somewhat, whereupon Festner tucked himself behind it and shot another 150 rounds at him until he crashed in flames. We had followed the

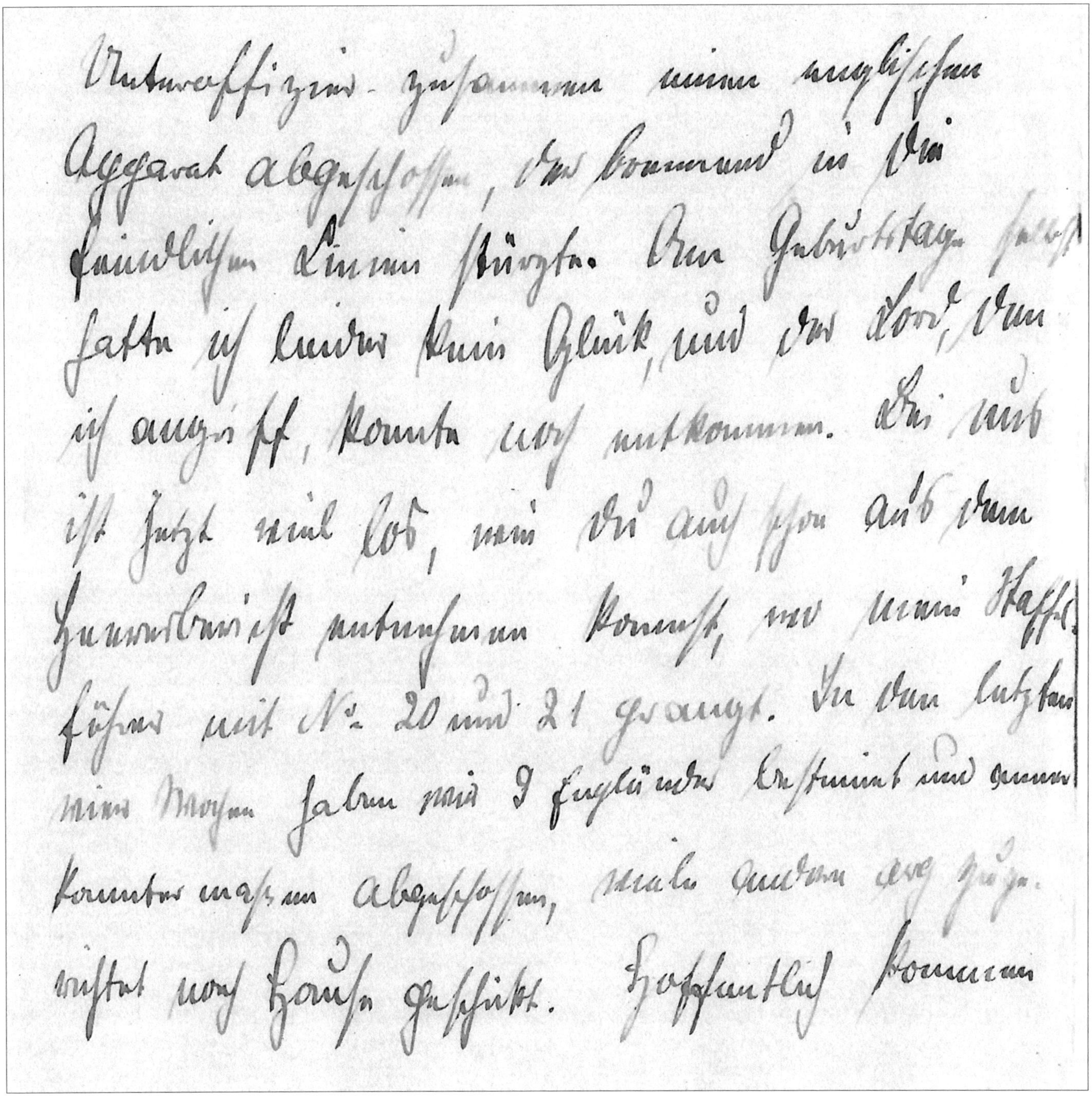
Unteroffizier zusammen einen englischen
Apparat abgeschossen, der brennend in die
feindlichen Linien stürzte. Den Geburtstag selbst
hatte ich leider kein Glück, und der Lord, den
ich angriff, konnte noch entkommen. [illegible]
ist jetzt viel los, wie Du auch schon aus dem
Heeresbericht entnehmen konntest, wo meine Staffel-
führer mit Nr. 20 und 21 genannt. In den letzten
vier Wochen haben wir 9 Engländer bestimmt und einen
[illegible] abgeschossen, [illegible]
[illegible] nach Hause geschickt. Hoffentlich kommen

Englishman from 2,000 to 200 meters and the return flight over the English positions was unpleasant. The apparatus had gone down far over them at 300 meters height where all the rifles and machine guns shot at us; however, I had only two infantry hits to the fuselage – one of them through the longeron – and I also caught a shot through the propeller."[7]

The B.E. ultimately brought down by *Vzfw.* Sebastian Festner had smashed into the ground south of Neuville-St. Vaast at 17:15. Its crash was witnessed by Flak Platoon Drocourt which at 18:14 reported:

"17:02 A B.E. shot down by Flak Platoon 31 fell into enemy lines. At the same time, a second English aircraft opposite the Arnulf Sector crashed following air combat and went down in enemy lines."[8]

"The Arnulf Sector" referred to the section of the front occupied by the Royal Bavarian 12th Infantry Regiment 'Prince Arnulf' between Vimy and Arras.

Some days later Kurt Wolff wrote to his aunt:

"Dear Aunt!

Thank you very much for your letters and birthday wishes, unfortunately the announced small package has not yet arrived, but the mail always takes a very long time now. I didn't celebrate my birthday much, because we always had to fly a lot all the time, two to three times every day, and I am tired as a result. The day

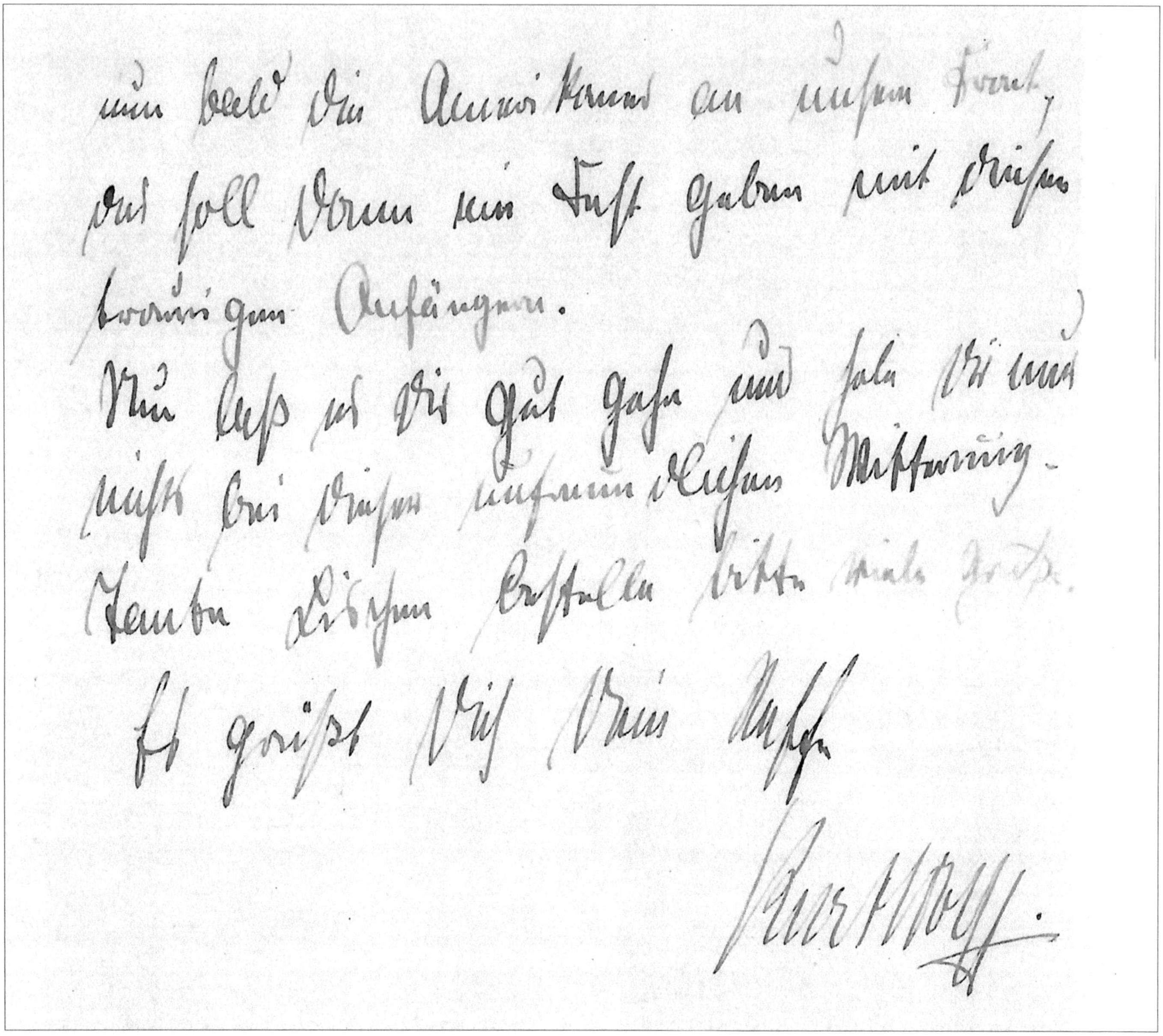

before my birthday, I shot down an English apparatus together with a non-commissioned officer [Vzfw. Festner], *which fell burning into the enemy lines. Unfortunately, I had no luck on the birthday itself, and the Lord I attacked was still able to escape. There is a lot going on with us now, as you can already see from the army report, where my Staffel leader is emblazoned with numbers 20 and 21.*

In the last four weeks, we have brought down 9 Englishmen, and the claims were recognized and confirmed, while many others were sent home badly dressed. Hopefully the Americans will come to our front soon, which should be reason for us to celebrate with these sad beginners.

Now have a good time and don't catch anything in this unfriendly weather. Please forward many greetings to Aunt Lieschens.

Greetings from your nephew Kurt Wolff."[9]

Upon his return to the front, which appears to have occurred between 7 and 9 February, Manfred von Richthofen could be satisfied by the fact that his strict training program had borne its first fruit. His Staffel had been able to carry out a successful air fight without his personal guidance. This first success by one of his pilots clearly showed that they were gaining self-confidence as a result of their constant exercises. The methods driven into them were proving to be correct and therefore only needed to be continued rigorously.

That "one swallow does not a summer make" became clear over the next few days, however, as demonstrated by Kurt Wolff's diary:

"On the 6th, I attacked a B.E. at the same location with 50 rounds close up but it didn't go down. I shot it out with two Sopwiths (40 rounds) on the 10th, but Richthofen, who came along, couldn't haul them down either. On the 11th, a fight with an F.E. single-seater, but I didn't get in a shot."[10]

Above: Carl Allmenröder relaxes between missions. The shoes he is wearing in the photo were certainly not standard military issue.

According to the activity report recorded by Kofl 6th Army, Jasta 11 conducted 95 combat flights totaling 97 hours during the period 29 January to 8 February 1917.[11]

On 9 February 1917, the Staffel received another Albatros D III with military serial number D.1985/16. But due to the type's ongoing flying ban, it was relegated to only standing around in a tent.[12]

On 14 February, Richthofen flew to Jasta 2 'Boelcke''s airfield at Pronville, about 20 kilometers south of La Brayelle. Following his meeting there, he took off shortly before noon to return to his Staffel but chose not to take a direct route. Instead, he flew westwards toward the front to follow it north to Arras before turning east to Douai. Shortly before his turning point, however, he saw an enemy aircraft outside of Loos that was directing artillery fire onto German positions:

"14 February 1917
B.E. Two-seater
12:00
Lens-Hulluch road west of Loos
Occupants: one killed, the other heavily wounded.
Name of pilot: Lt. Bonnet (died)

After flying back from a conference with Staffel Boelcke I spotted an enemy artillery flyer at a height of 2000 meters west of Loos. I attacked the enemy and approached him unnoticed to within 50 meters. After several hundred shots, the plane dashed down, falling into our trenches. The pilot was killed in the air, observer seriously injured when landing."[13]

The British airplane crashed in the front lines south of Loos, as reported by the anti-aircraft units stationed near Drocourt and Fachoda:

"A B.E. was shot down by an Albatros south of Loos. The downed airplane lies in grid square 5008/10. One occupant dead, the other wounded, [secured] *by Infantry Regiment 153. The wounded crewman, an officer, was taken to the 11th Medical Company at Épinoy. Apparatus badly damaged, covered by heavy* [enemy] *machine-gun fire. As a result, it isn't possible to determine if the dead crewman was an officer. The wounded man threw away all his identifying badges and had no papers. The wounded Englishman is unfit for questioning, has sustained a skull injury and concussion, and had been brought senseless to Field Hospital 39 at Carvin."*[14]

Pilot 2/Lt. C.D. Bennet was the man who suffered skull damage upon impact. His observer, 2/Lt. H.A. Croft had been killed in combat and could not be recovered by German forces due to heavy fire from the British. In addition, contrary to what the Flak units reported, Richthofen was flying a Halberstadt fighter.

2/Lt. Bennet was eventually interrogated following his recovery and this is an excerpt from the intelligence officer's report:

*"**1. Personnel:***
Lt. C.D. Bennet, RFC, 19 years old, son of a Moscow brewery director. Joined the army in England on 16 May and trained in a military aviation school; at the front with the 2nd Squadron since the end of August....

3. Orders:
Directing fire, without airplane escort, on a battery of four guns in the vicinity of Hulluch. The pilot himself observed this with the aid of a photograph described as quite good, whose whereabouts the prisoner was pressed about. The endeavor's goal was to circle as directly as possible above the object to be fired upon."[15]

A medical report from Field Hospital 39 where 2/Lt. Bennet was treated stated that he no longer had any memory of the dogfight as a result of his head injury.[16]

Enemy flight activity continued to increase that afternoon, and six Jasta 11 Halberstadts, including Richthofen's, headed for the front around 4:30 p.m. They did so by turning west at a point north of Douai and flying along their well-known route over Hénin-Liétard towards Lens. Richthofen reported:

"14 February 1917
B.E. Two-seater
4:45 p.m.
Station, 1,500 meters southwest of Mazingarbe
No details, as plane landed on enemy side

About 4:45 p.m., I and my Staffel of five airplanes attacked

Above: On 16 February 1917, Carl Allmenröder forced B.E. 2 4179 to land on the German side of the lines. The captured aircraft attracted the curiosity of German soldiers who happened to be near the landing site.

Below: B.E. 4179, crewed by 2/Lt. Lindley and 2/Lt. Munn, photographed by Carl Allmenröder, on the road between Fampoux und Roeux.

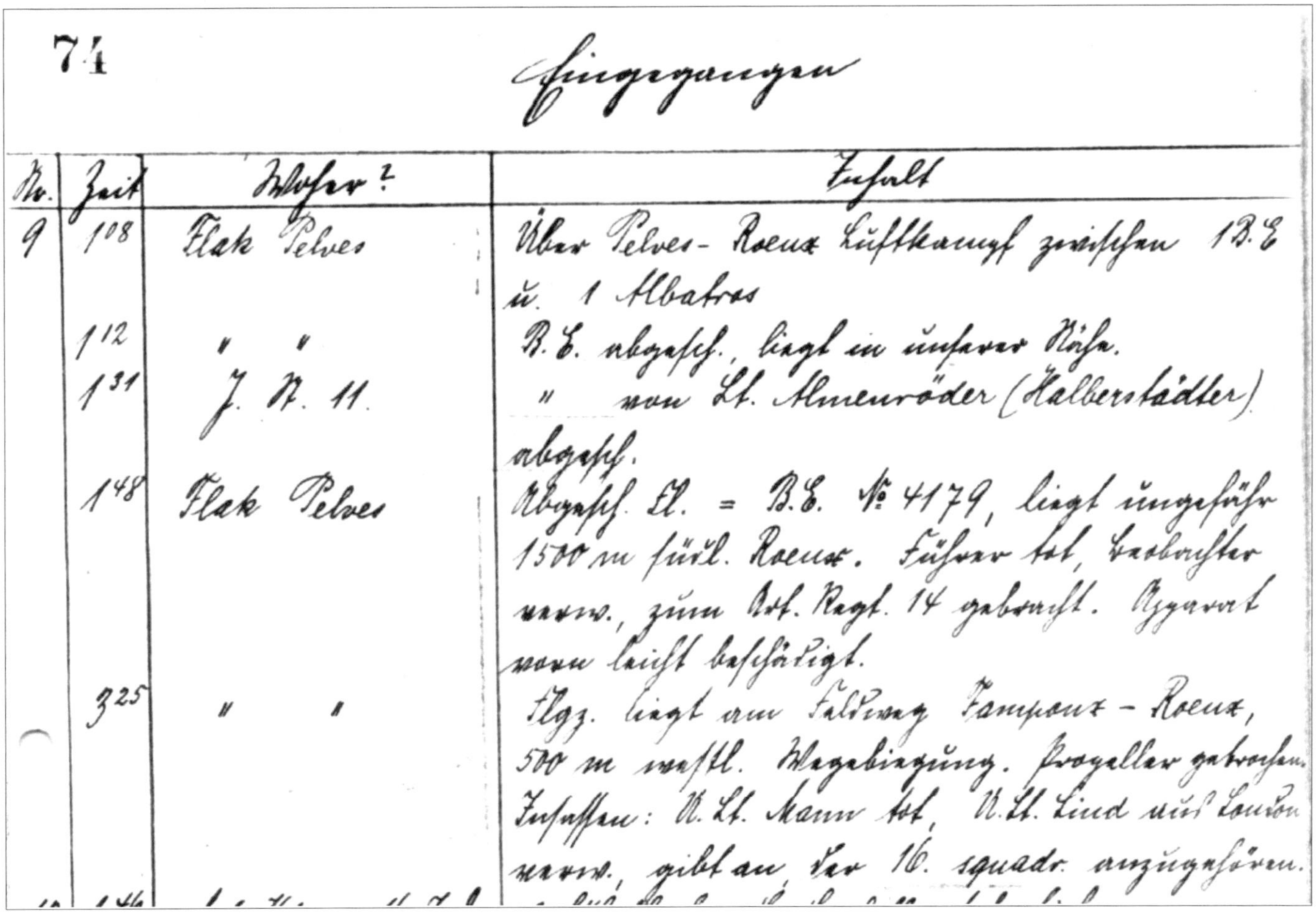

74 Eingegangen

Nr.	Zeit	Woher?	Inhalt
9	1^08	Flak Pelves	Über Pelves-Roeux Luftkampf zwischen B.E. u. 1 Albatros
	1^12	" "	B.E. abgeschoss., liegt in unserer Nähe.
	1^31	J. St. 11.	" von Lt. Almenröder (Halberstädter) abgeschoss.
	1^48	Flak Pelves	Abgeschoss. Fl. = B.E. No 4179, liegt ungefähr 1500 m südl. Roeux. Führer tot, Beobachter verw., zum Inf. Regt. 14 gebracht. Apparat nur leicht beschädigt.
	3^25	" "	Flgz. liegt am Feldweg Fampoux – Roeux, 500 m westl. Wegebiegung. Propeller gebrochen. Insassen: U.Lt. Mann tot, U.Lt. Lind aus London verw., gibt an der 16. squadr. anzugehören.

Above: Extract from the Flugmeldebuch: Reports by Flak Pelves and Jagdstaffel 11 relating to the downing of B.E. 4179 by Lt. Allmenröder. (Bayerisches Hauptstaatsarchiv, Abt. 4 München)

artillery flyers at low altitude near Lens. While my gentlemen attacked a second B.E., I attacked the one flying nearest to me. After the first 100 shots, the observer stopped firing. The plane began to smoke and twist in uncontrolled curves to the right. As this result was not satisfactory to me, especially over the enemy's lines, I shot at the falling plane until the left part of the wings came off. As the wind was blowing at a velocity of twenty meters a second, I had drifted far over the enemy side and therefore could observe that the enemy plane touched the ground southwest of Mazingarbe. I could see a heavy cloud of smoke arising from the place where the plane was lying in the snow.

As it was foggy and already rather dark, I have no witnesses either from the air or from the ground.[17]

Richthofen's victim was B.E.2c 2543, another aircraft from No. 2 Squadron RFC. The pilot, Capt. G. C. Bailey, was wounded, but his observer, 2/Lt. G.W. Hampton, was uninjured. But it turns out that they did not crash and Bailey was able to return the damaged airplane to his airfield, as reported by 2 Sqn's records:

"During reconnaissance B.E. 2d 6250 flew just over us from East, pursued by three enemy machines flying in formation. Pilot Capt. G.C. Bailey turned machine to go to assistance – observer manned rear gun. Enemy fired machine gun wounding pilot in knee. Pilot instructed observer to man front gun. Pilot made spiral descent under fire. Rear gun mounting was hit, and gun fell onto pilot and overboard. Enemy machine retired in north-east direction. No opportunity of firing at them during fight. Machine then returned to aerodrome, fuselage and wings and pilot's seat being hit.[18]

The next day, 15 February, brought fine, clear weather that led to lively Allied and German air activity and air battles during which a Flieger-Abteilung 5 (LB) pilot and a Flieger-Abteilung 13 observer were wounded.[19]

After receiving a request from the front, a Jasta 11 Kette set off north around 12:00 p.m to the Lens area where they encountered enemy "kleine Gitterschwänze" ("small lattice tails"). Vzfw. Sebastian Festner attacked F.E.8 7635 from 40 Squadron, RFC on the German side of the lines near Lens. After a long fight and pursuit, the British machine made it over its lines at a height of 500 meters and crash-landed between Liévin and Grenay.[20]

Flak Platoon 159 at Drocourt witnessed this and reported:

"12:10 after air combat with a Halberstadt over the platoon's

position one F.E. at 500 meters altitude fell steeply and then made an emergency landing in enemy lines."[21]

Vzfw. Sebastian Festner's opponent was 2/Lt. L.B. Blaxland, who escaped uninjured.

The next morning, 16 February, began with ground fog but soon gave way to clear weather. The Staffel took off shortly before 13:00 and its Halberstadt fighters followed the river Scarpe towards Arras. Lt. Carl Allmenröder then spotted a B.E. over the La Scarpe river valley and attacked it. He managed to get behind the British machine and opened fire. After several volleys, the observer's machine gun in B.E.2c 4179 from No. 16 Squadron RFC fell silent and the airplane began to lose height, whereupon Allmenröder stopped firing. He then followed the airplane that finally landed about 500 meters west of the bend in the dirt road between Fampoux and Roeux.

Allmenröder briefly circled above the landing site before flying the short distance back to La Brayelle. Once there, he grabbed his camera and a squadron car and drove with a few comrades via Vitry to the landing site near Roeux – a distance of about 12 kilometers. By the time he arrived, the seriously wounded pilot had already been taken to a hospital. Allmenröder learned from some soldiers that the observer had been killed by his first volley. He then took two photos of the downed B.E., which was undamaged apart from one propeller blade. He removed the fin of its vertical stabilizer that carried its 4179 factory number, stowed it in the car, and brought it back to La Brayelle as a souvenir.

The observer who had been killed outright was 2/Lt. L.V. Munn. His pilot, 2/Lt. E.W. Lindley, died of his wounds in a German hospital two days later.

Flak Platoon Pelves issued the following report:

"Air combat between one B.E. and one Albatros over Pelves-Roeux. B.E. was shot down near us."

It was later supplemented by this entry:

"B.E. shot down by Lt. Allmenröder (Halberstadt). Downed aircraft: B.E. No. 4179 lies approximately 1,500 meters south of Roeux... Propeller broken. Occupants Lt. Munn dead, Lt. Lind [sic] *from London who belong to 16 Squadron.*"[22]

This demonstrates that the identification of German aircraft types by Flak units could be unreliable. Misidentifications were indeed very common.

With Carl Allmenröder's success, a second pilot from the Staffel other than Richthofen had finally scored an air victory.

On 17 February, the Staffel received an entirely new type of aircraft for testing: Siemens-Schuckert (SSW) D I 3507/16. This essentially was a copy of the French Nieuport 17 made by the Siemens-Schuckert company. The aircraft was not well received by pilots of Jagdstaffel 11 and the type received mixed reviews from others who had an opportunity to try it. Kurt Student, Staffelführer of Jasta 9, *"praised it as having the best flying qualities of any fighter at the front at that time"*, whereas Rudolf Nebel, *Jasta* 5, described its flight characteristics in an interview as *"extremely toxic"*.

Above: Following his 16 February 1917 victory, Carl Allmenröder flew in this Halberstadt D V to Lille to visit his brother Wilhelm at Fliegerabteilung (A) 227.

This statement was surely influenced by the fact that his Staffelführer, Hptm. Hans von Hünerbein, had crashed fatally during a test flight of the type.[23]

Flying activity along the front declined sharply due to heavy cloud cover, rain, and fog, and largely came to a standstill between 17 and 22 February. Jasta 11 was only able to undertake 35 front flights for a total flight duration of 43 hours during this period.[24]

February 1917 also saw more personnel changes at Jagdstaffel 11. Sgt. Erwin Howe left Douai and was transferred to Armee-Flug-Park 6 on the 17th, whereas two new pilots were brought in: Lt. Hans-Georg Lübbert from Fliegerabteilung 18 on the 18th and Lt. Carl-Emil Schaefer from Kampfgeschwader II's Kampfstaffel 8 on the 21st.

Hans Georg (Henry) Eduard Lübbert, called "Eddy" by his friends and relatives, was born on 29 December 1893 in Hamburg as the eldest son of Hans Julius Lübbert and his wife, Eleonore Valentine del Banco. His father's family was of Jewish descent but had converted to the Protestant faith in the 1820s and had shortened their surname from Oppenheimer-Lübbert to just Lübbert.[25] Eddy's mother also came from a Jewish family and had become a Protestant following her wedding. She was the older sister of the well-known artist, Alma Aline Henriette del Banco (1862–1943).[26]

Hans Julius Lübbert was a successful, respected, and wealthy businessman and co-owner of Cuxhavener Fischerei AG that later became known as Nordsee Deutsche Hochseefischerei Bremen-Cuxhaven AG.[27] Accordingly, Eddy and his younger brother, Friedrich-Wilhelm, enjoyed a carefree and untroubled childhood. The family was characterized as liberal, cosmopolitan, generous, and patriotic. Eddy had been named after his step-grandfather, Major General Eduard Lübbert, whom he greatly admired. Following his example, Eddy chose a military career and enlisted in Dragoner-Regiment König Karl von Rumänien (1. Hannoversches) Nr. 9, based in Metz.[28] Friedrich-Wilhelm,

Above: Carl Allmenröder climbs out of the cockpit of his Halberstadt after landing at the airfield of Fliegerabteilung (A) 227. His aircraft had a black cannonball as personal marking. (G. VanWyngarden)

five years younger than Eddy, also joined Dragoner-Regiment Nr.9 but as a war volunteer in the late summer of 1914.(29)

Eddy was proud to be a member of the Dragoons. Nicknamed the "Caesars of Hohenfriedberg", who in their cornflower blue tunics had "put the fear of God" into their opponents on the battlefields of the Seven Years' and Napoleonic Wars, had enjoyed a legendary reputation since the days of Frederick the Great. It was considered a privilege to serve with them; and on 12 August 1914 Eddy went to war with them.

On 19 August 1914, Eddy, on behalf of himself and his brother, wrote a letter home on the occasion of their father's 44th birthday that reflected the family's patriotic spirit:

Warmest congratulations on your birthday, a holiday among the series of serious days of action. We thank you on this day for protecting and nurturing us for so long that we can now, strong and healthy, fight for the Fatherland. May our grateful wishes for the one who did everything for us and made everything possible accompany you forever, and we hope at this time that you too may take part in this fight in a position that allows you to achieve the very best. And that you, once we have triumphed, can continue to achieve great things as you have in recent years.(30)

At the beginning of November 1914, Eddy's unit was sent to the Russian Front where it was tasked with preventing Cossack patrols from entering German territory. It went straight to the front lines and Lübbert wrote home:

"On a country road in the middle of Poland. The war has overwhelmed us all. It's much larger and more terrible than you think at home. The thunder of cannons accompanies this letter; the one who's writing it lies on a flat field under a threatening winter sky, gray clouds that presage snow. And the troops are following the retreating Russians from all sides, silent legions passing over the hilly landscape all around, incredibly inspired by the thought: 'Forward triumphantly!'"(31)

On 12 November 1914, he learned firsthand how there was no longer a place for the proud cavalry in this war. His regiment attacked the Russian positions at Niemjowo

Above: Members of Fliegerabteilung (A) 227 surround the visiting single-seater from Jagdstaffel 11. (G. VanWyngarden)

Left: Carl Allmenröder´s quarters at Roucourt Castle in April 1917. The rudder of B.E. 4179 rests against the wall.

4.) Eigene Flugtaetigkeit

Abteilung:	Flüge:	Stdn:	Bildmeldgn:	Luftkämpfe
5	3	6	16	1
18	4	9	38	-
13	1	2	10	1
204	1	3	12	-
227	9	8	22	-
292 bay.	8	10	23	-
240	4	4	12	-
235	10	16	32	2
202	9	8	34	-
11. J. St.	35	43	-	8
30				
	84	109	199	12

Above: Report by Kofl 6 on the flight activities of the flying units in the area of the German 6th Army from 17 to 22 February 1917. Jasta 11 had made 35 frontline flights for a total flight duration of 43 hours, while Jasta 30 had not yet undertaken any front flights. It was the last report in which the operational flights were broken down by unit. (Bayerisches Hauptstaatsarchiv, Abt. 4, München)

and rode into the enemy's machine-gun fire. It ended in a bloodbath. Only meager remnants of the proud regiment were left after the attack. A shaken Lübbert wrote home:

"They are still lying around and bleeding everywhere... Of our four squadrons, only one-and-a-half still exist with hardly any horses. We ride by the field, this field of the dead, silent and shattered. The living bury their comrades."[32]

Thereafter, the proud 9th Dragoons were mainly used as infantry. In December, Lübbert fell ill with dysentery. Following his recovery, he was deployed to Courland where he was wounded by a shot to the shoulder in May 1915. On 20 September he was promoted to Offizier-Stellvertreter and on 26 October he was raised again to Leutnant. After another convalescent period and additional time on the Eastern Front, he reported to the Fliegertruppe in the spring of 1916 and was transferred to Flieger-Ersatz-Abteilung 11 on 10 May 1916.[33]

On 21 May 1916, he wrote to his parents: "*Just now commanded to Flieger-Ersatz-Abteilung 2 Breslau.*" He was going to Fliegerschule Hundsfeld.[34]

In another letter to his parents six days later, he provided this insight into a flight school's daily routine:

"...I am awakened at 4:00 a.m. and enjoy the morning hour...at 4:50 a.m. the auto that takes me to the airfield is at the door. The flying starts right away outside. You train on machines that have dual controls for instructor and student, so that your hand is always guided. I have an excellent instructor, a calm, determined man who is quite capable. Lessons are then from 8:00 to 9:00 a.m. It's back home at 9:00 a.m. to have a good sleep. It all starts again at 4:30 p.m. and lasts until 8:00."[35]

Eddy Lübbert, now a pilot, was transferred via Armee-Flug-Park 6 to Feldflieger-Abteilung 18 on 17 November 1916.[36] One of his observers there was cuirassier officer Oblt. Hans-Helmut von Boddien, who would also later serve as a pilot in Jagdstaffel 11.

Above: Lt. Hans-Georg Eduard (Eddy) Lübbert wearing the uniform of Dragoner-Regiment König Karl von Rumänien (1. Hannoversches) Nr. 9. (L. Bronnenkant)

On 10 January 1917, Lübbert was mentioned in the Kofl

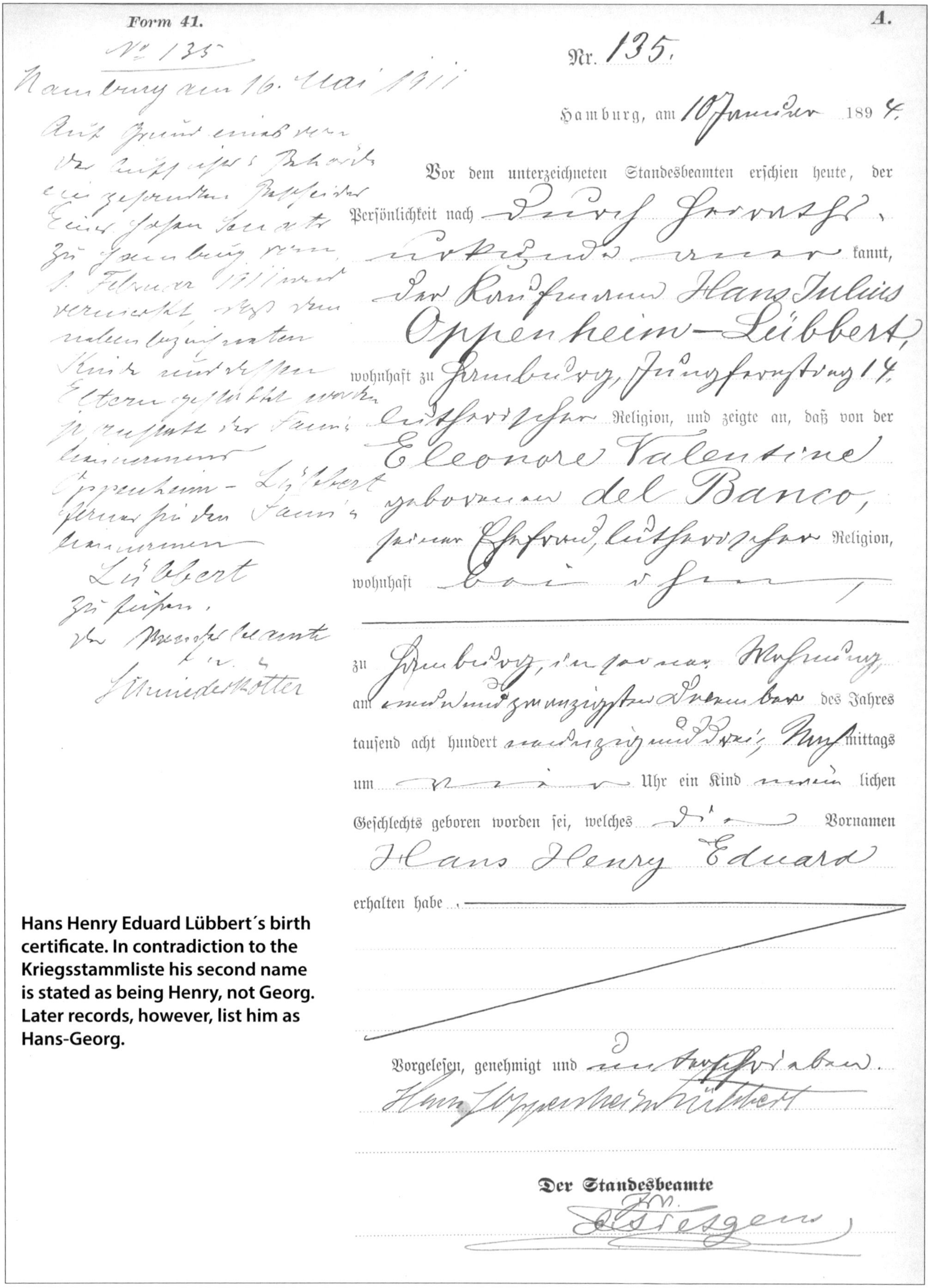
Form 41. A.

Nr. 135.

Hamburg, am 10 Januar 1894.

Vor dem unterzeichneten Standesbeamten erschien heute, der Persönlichkeit nach [illegible] kannt, der Kaufmann Hans Julius Oppenheim-Lübbert, wohnhaft zu Hamburg, Jungfernstieg 14, lutherischer Religion, und zeigte an, daß von der Eleonore Valentine geborenen del Banco, seiner Ehefrau, lutherischer Religion, wohnhaft bei ihm, zu Hamburg, in seiner Wohnung, am einundzwanzigsten Dezember des Jahres tausend acht hundert neunzig und drei, Nachmittags um vier Uhr ein Kind männlichen Geschlechts geboren worden sei, welches die Vornamen Hans Henry Eduard erhalten habe.

Vorgelesen, genehmigt und unterschrieben.
Hans Oppenheim-Lübbert

Der Standesbeamte
[illegible]

No 135
Hamburg am 16. Mai 1911
Auf Grund einer von der Aufsichtsbehörde eingesandten Abschrift des Senats zu Hamburg vom 1. Februar 1911 wird vermerkt, daß dem nebenbezeichneten Kinde und dessen Eltern gestattet worden ist, anstatt des Familiennamens Oppenheim-Lübbert fernerhin den Familiennamen Lübbert zu führen.
Der Standesbeamte
i. V.
[illegible]

Hans Henry Eduard Lübbert´s birth certificate. In contradiction to the Kriegsstammliste his second name is stated as being Henry, not Georg. Later records, however, list him as Hans-Georg.

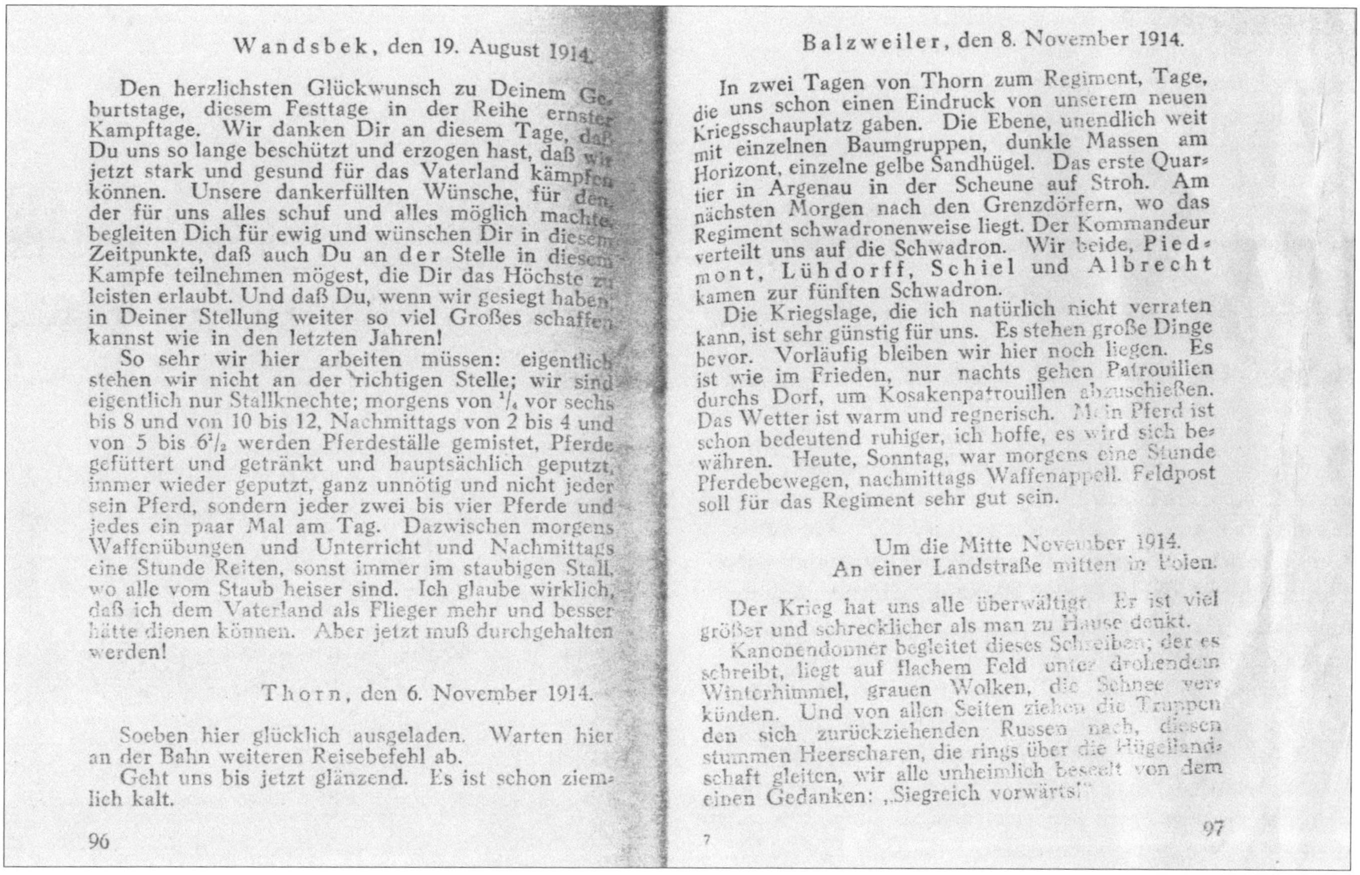

Wandsbek, den 19. August 1914.

Den herzlichsten Glückwunsch zu Deinem Geburtstage, diesem Festtage in der Reihe ernster Kampftage. Wir danken Dir an diesem Tage, daß Du uns so lange beschützt und erzogen hast, daß wir jetzt stark und gesund für das Vaterland kämpfen können. Unsere dankerfüllten Wünsche, für den, der für uns alles schuf und alles möglich machte, begleiten Dich für ewig und wünschen Dir in diesem Zeitpunkte, daß auch Du an der Stelle in diesem Kampfe teilnehmen mögest, die Dir das Höchste zu leisten erlaubt. Und daß Du, wenn wir gesiegt haben, in Deiner Stellung weiter so viel Großes schaffen kannst wie in den letzten Jahren!

So sehr wir hier arbeiten müssen: eigentlich stehen wir nicht an der richtigen Stelle; wir sind eigentlich nur Stallknechte; morgens von ¼ vor sechs bis 8 und von 10 bis 12, Nachmittags von 2 bis 4 und von 5 bis 6½ werden Pferdeställe gemistet, Pferde gefüttert und getränkt und hauptsächlich geputzt, immer wieder geputzt, ganz unnötig und nicht jeder sein Pferd, sondern jeder zwei bis vier Pferde und jedes ein paar Mal am Tag. Dazwischen morgens Waffenübungen und Unterricht und Nachmittags eine Stunde Reiten, sonst immer im staubigen Stall, wo alle vom Staub heiser sind. Ich glaube wirklich, daß ich dem Vaterland als Flieger mehr und besser hätte dienen können. Aber jetzt muß durchgehalten werden!

Thorn, den 6. November 1914.

Soeben hier glücklich ausgeladen. Warten hier an der Bahn weiteren Reisebefehl ab.

Geht uns bis jetzt glänzend. Es ist schon ziemlich kalt.

96

Balzweiler, den 8. November 1914.

In zwei Tagen von Thorn zum Regiment, Tage, die uns schon einen Eindruck von unserem neuen Kriegsschauplatz gaben. Die Ebene, unendlich weit mit einzelnen Baumgruppen, dunkle Massen am Horizont, einzelne gelbe Sandhügel. Das erste Quartier in Argenau in der Scheune auf Stroh. Am nächsten Morgen nach den Grenzdörfern, wo das Regiment schwadronenweise liegt. Der Kommandeur verteilt uns auf die Schwadron. Wir beide, Piedmont, Lühdorff, Schiel und Albrecht kamen zur fünften Schwadron.

Die Kriegslage, die ich natürlich nicht verraten kann, ist sehr günstig für uns. Es stehen große Dinge bevor. Vorläufig bleiben wir hier noch liegen. Es ist wie im Frieden, nur nachts gehen Patrouillen durchs Dorf, um Kosakenpatrouillen abzuschießen. Das Wetter ist warm und regnerisch. Mein Pferd ist schon bedeutend ruhiger, ich hoffe, es wird sich bewähren. Heute, Sonntag, war morgens eine Stunde Pferdebewegen, nachmittags Waffenappell. Feldpost soll für das Regiment sehr gut sein.

Um die Mitte November 1914.
An einer Landstraße mitten in Polen.

Der Krieg hat uns alle überwältigt. Er ist viel größer und schrecklicher als man zu Hause denkt.

Kanonendonner begleitet dieses Schreiben; der es schreibt, liegt auf flachem Feld unter drohendem Winterhimmel, grauen Wolken, die Schnee verkünden. Und von allen Seiten ziehen die Truppen den sich zurückziehenden Russen nach, diesen stummen Heerscharen, die rings über die Hügellandschaft gleiten, wir alle unheimlich beseelt von dem einen Gedanken: „Siegreich vorwärts!"

7 97

Above: After the end of the war, Friedrich Wilhelm Lübbert had the letters of his deceased brother printed as a book for the surviving family members. On the bottom of page 97, the letter "Around the middle of November 1914…" can be found.

6th Army's weekly report for the first time:

"On 7 January 1917, aircraft crew Lt. Lübbert and Lt.[d.] *R. Zühl (Flieger 18) attacked an English formation of five F.E. biplanes in a commendable way and after a long, spirited air combat, forced an English F.E. biplane to make a prompt emergency landing on the other side of the lines (witnessed by Flak)."*(37)

A little over two weeks later, on 28 January 1917, he was again mentioned in the army report:

"An airplane from Fl. Abt. 18, pilot Lt. Lübbert, observer Oblt. von Boddien, attacked the bridge over the Authie river, 20 kilometers south of Étaples. The plan to blow up the bridge after landing failed because of the strong garrison there. The aircraft dropped explosive devices from a height of 30 meters instead and scored hits on the roadbed just south of the bridge. It also successfully machine gunned two trains north of the bridge, one of which caught fire. I express my appreciation to the crew for this particularly gratifying deed on His Majesty's birthday.

gez. Von Falkenhausen" (38)

Eddy Lübbert was awarded the Iron Cross, 1st Class for this achievement. But the work of a Flieger-Abteilung pilot eventually did not suit his temperament or his search for adventure. Therefore, he pursued Jagdfliegerei (fighter aviation), about which he wrote home:

"…hard battles against the English five-fold superiority in the air! In a few days, I'll switch over from a two-seater to a single-seater and will only be a Jagdflieger; that's where one is needed most now."(39)

His wish was further fulfilled when he was posted to Jasta 11 on 18 February 1917.(40)

Carl-Emil Schaefer was born on 17 December 1891 as the oldest child and only son of a wealthy silk goods manufacturer in Krefeld-Bockum. Two sisters followed after him. He attended the town's Realgymnasium (secondary school) and then joined his father's firm. He completed his military service in 1911 - 12 as a one-year volunteer with Hannoversches Jäger-Bataillon Nr.10 in Goslar. In order to gain some international experience, he then worked as a businessman in London and Paris with the aim of eventually taking over the family business.

Emil Schaefer returned to Germany from Paris on 1 August 1914 and at the outbreak of the war was called up as an Unteroffizier with Westfälisches Jäger-Bataillon Nr. 7 – the legendary "Bückeburger-Jäger." He earned both the Iron Cross 2nd Class and a promotion to Vizefeldwebel on 18 September 1914 but on the 26th of that month was seriously wounded in the left thigh during an assault. As a result, he later walked with a slight limp. Following six months of recuperation, he returned to his battalion at the front, and on 18 June 1916, he was raised to Leutnant.[41]

Half a year later he reported to the Fliegertruppe and was sent on 1 January 1916 to Flieger-Ersatz-Abteilung 8 to be trained as a pilot. At the end of the month, on the 30th, he was commanded to the Militärfliegerschule (Military Aviation School) at Köslin.[42] He wrote about this to his mother:

"I've been successful, despite my shortened leg, in becoming a duly appointed apprentice aviator at the Köslin Military Aviation School, where I can look forward to 4 - 5 months of peaceful pilot training time…"[43]

After completing his training, he was transferred on 22 June 1916 to Kampfgeschwader II, deployed at Kovel on the Eastern Front, and subsequently assigned to the unit's Kampfstaffel 8 on 29 July. On 18 November he was switched over to Kagohl II's Kampfstaffel 11.[44] He trained as a single-seat pilot during the winter of 1916 and became a founding member of Kagohl II's fighter contingent when it was formed on 20 December after Kagohl II had been recalled to the Western Front.

In January 1917, Kagohl II moved to the sector of the Western Front near Verdun. In contrast to the Eastern Front, where enemy flight activity was rather low, the Kampfgeschwader encountered fierce enemy resistance at Verdun. To protect their two-seater aircraft, the single-seaters of the squadron had been assembled into Kampfstaffel 11, as attested to in Schaefer's letter home: *"Our Geschwader has recently formed a Jagdstaffel: von Hausen, Albert, Scheele, Hermann, Stern, and myself."*[45]

Schaefer was the first of them to score when he managed to shoot down a Voisin near Fey-en-Haye, thereby earning his first official victory. He wrote this description of his success in his book, "*Vom Jäger zum Flieger*":

"Scheele and I buzzed off on a little tour of the front. We reached the lines at St. Mihiel at 4,000 meters altitude, flew along them, crossed the Moselle at Pont à Mousson, and saw the Seille River with Nomeny right ahead of us. Two French Voisin biplanes then emerged behind and below us from the right. We descended in a wide arc toward France, so that we pushed over the front and plunged down after them 'out of the sun.' They obviously did not notice us until we began firing at close range. Unfortunately, we came roaring at them with much too much speed; we were only able to get off a few shots before we were by them and had to turn again. As we flew past each other – at about 30 meters distance – I saw 'my' Frenchman's observer give me a salvo; at that crazy speed it naturally went wide of us. The Frenchman twisted and turned in a steep gliding flight and tried to escape over the front. We had dropped from 2,500 to 800 meters when I had a gun jam in my left gun, and during the short time that I tried to fix it, he took advantage of the situation and went right past me toward France. I immediately got behind him and let him have it from there with my right gun. I saw the trenches whizz by under me at 400 meters altitude. To be on the safe side, I then tried the motor because we had been in an [unpowered] *glide all the while. Just my luck – it didn't start. I turned around and headed towards home in a very shallow glide. I saw the French trenches under me at 300 meters altitude; at 200 meters, the foremost German lines moved slowly, slowly underneath me. Thank God! I wouldn't have to land among 'Joffre.'* [i.e., among the French], *after all. Everything was open fields up to the second lines, but if I landed there I'd be shot to pieces immediately. Just behind the lines, however, was a ravine, and if I reached it I'd be saved. I pulled out no more than five meters above the ravine's first row of trees; too bad, it was solid forest, there was no possibility of landing, that was it for my poor 511* [Albatros D II D.511/16]. *I let it float, float, float until it had no more speed; then came the crown of a tree that I couldn't avoid. I 'bowed' before it and then pulled the machine up at an angle into the tangle of branches; wild splintering and cracking, a strong thud on the nose of the fuselage, and my D.511 slowly slid from the tree down on its left wings, which folded up like an accordion. I only had to unbuckle, stick out my left arm,*

Above: Lt. Emil Schaefer as a member of the famous "Bückeburger Jäger" before he joined the Fliegertruppe.

Above: Emil Schaefer during a different kind of hunting expedition. He was, like Manfred von Richthofen, a dedicated huntsman.

then go out upside down. After some thrashing about in the thick overgrowth, I stood upright in a thicket of bushes deeply covered in snow."(46)

After returning to his Kampfstaffel, Schaefer learned that five eyewitnesses in the forward lines had reported the crash of the French airplane.

In mid-February 1917, Schaefer sent a telegram to Manfred von Richthofen, whom he had gotten to know when both men were serving in Kagohl II: *"Do you have any use for me?"*

Richthofen replied: "*You've already been requested."*(47)

The official transfer to Jagdstaffel 11 via Armee-Flug-Park 6 was issued on 21 February 1917.(48)

Rittm. Georg von Ompteda, a journalist for *Die Woche* magazine, visited Jasta 11 in the spring of 1917 and described Schaefer's appearance in these terms:

"The man has a good, large appearance. A small, shaved head with light blond hair sits on a tall, slim body that exceeds the normal measure of men. His face is round, his teeth – visible when he speaks and laughs – good, and his eyes clear. Despite his young years, he is at his best form, exceedingly confident, as one imagines a man who has looked death in the eye countless times..."(49)

On 19 February 1917, the Albatros D III flight ban was lifted after technical improvements had been tested and approved. In addition to the Albatros D IIIs it already had on hand, Jasta 11 received six new ones between 25 and 28 February. These were Albatros D III D.2014/16, D.2015/16, D.2016/16, D.2026/16, D.2036/16 and D.2062/16. Thus, according to a 28 February 1917 report, the Jasta's complete

Above: Emil Schäfer, pictured on the right, as a pilot of Kampfstaffel 11, Kampfgeschwader II, winter 1916/1917. Rudolf Windisch is on the left

Above: Emil Schäfer crashed with Albatros D II D.511/16 in Priesterwald in late January 1917. At the time he was still a member of Kampfstaffel 11.

inventory consisted of twelve Albatros D IIIs and five Halberstadt aircraft.[50] But now that the unit had a full complement of Albatros D IIIs, its Halberstadts fell into disuse and were ultimately returned to the Army Flugpark.

On 23 February, Jagdstaffel 4, under the command of Oblt. Kurt-Bertram von Doering, moved from its airfield at Le Chatelet in the German 3rd Army sector to Douai airfield.[51] Its aircraft were housed in hangars situated right next to those used by Jagdstaffel 11. Its pilots were quartered in Douai whereas its ground personnel were housed in shacks at the airfield. That evening, Manfred von Richthofen and his pilots hosted a small welcoming party for their new comrades.

In the days that followed, von Richthofen and Doering began to coordinate their units' operations. Both got along excellently from the onset, especially since Oblt. von Doering was also a cavalry officer who had served in 1. Großherzoglich Mecklenburgische Dragoner-Regiment Nr. 17. But this was not the only connection that the two men shared. Both had a common view of the way Jagdstaffeln should be employed, and both were officers who knew how to present their opinions to higher officials. However, it is worth keeping in mind that, at this time, Doering was yet to be credited with his first confirmed victory, an event that would only occur later on 14 April.

Kurt-Bertram von Doering was the older and, as an Oberleutnant, more senior officer; in exchange, the younger and lower-ranking Richthofen had the prestige of being a Knight of the Orden Pour le Mérite as well as a highly successful combat pilot. In line with his upbringing, von Richthofen deferred to von Doering and treated him with courtesy and respect.

As Doering's widow shared with the author: *"My husband always spoke in the very best way about Manfred von Richthofen as an officer and comrade."*[52]

Jasta leaders and some pilots were brought together on 25–26 February 1917 at Jagdstaffelschule I to observe a demonstration of new phosphorus ammunition for captive balloon attacks. The meeting also led to a comprehensive exchange of views among these men, who gathered together for an interesting group photo at its conclusion.

As a final note, Hans Bethge's Jasta 30, stationed at Phalempin, about 18 kilometers north of Douai, announced on 28 February that it too was now ready for action within the 6th Army sector.[53] Thus, the German 6th Army finally had three operational Jagdstaffeln at its disposal.

Manfred von Richthofen, Kurt-Bertram von Doering, and Hans Bethge used the time in February 1917 to speak with Hptm. Maximilian Sorg (Kofl 6th Army) and other members of the Army High Command about their views on the best use of fighter units and to dissuade them from ordering what they considered to be nonsensical "Luftsperre" ("air barrier" flights). Bethge, highly regarded by Richthofen, shared Doering's and Richthofen's views, so the trio mutually supported one another and took advantage of their combined

Above: Emil Schaefer stands in front of Albatros D II 1724/16, a fighter he used while serving in Kagohl 2's Kasta 11. (L. Bronnenkant)

Above: The Albatros D II 1724/16 flown by Lt. Schäfer, Kampfstaffel 11.

prestige – Richthofen as a Knight of the Pour le Mérite Order and Doering and Bethge as senior Oberleutnants -- during these discussions. Hptm. Sorg, who had previously worked out aerial tactics with Max Immelmann, was open and positive towards the views and suggestions of the three Jasta leaders. Together, they succeeded in convincing the 6th Army's senior staff to abandon the idea of "Luftsperre". In addition, airspace observation and the telephone networks to the Jagdstaffeln were expanded. The success in the coming weeks and months more than proved that the three Staffelführers were right.

As a result of their discussions, Richthofen, Doering, and Bethge now largely had a free hand in how their Jagdstaffeln were deployed in the period that followed and they met regularly to coordinate their efforts.[54]

Flying activity was low during the week of 23 February through 1 March 1917 because of poor weather conditions. The result was that the 6th Army's three Jagdstaffeln could only mount 39 combat flights totaling 27 flight hours.[55]

Compared to January 1917, the flight activity and corresponding air combats along the 6th Army's front increased significantly in February. While only two victories had been registered in January, the next month saw 18

Right: Oblt. Kurt-Bertram von Doering, commander of Jagdstaffel 4, was highly regarded by Manfred von Richthofen. (L. Bronnenkant)

Above: Group photo of the various Staffelführers and fighter pilots who attended the meeting at Valenciennes on 26 February 1917. (W. Bock)

Meeting of Staffelführer and Pilots on the 26 February 1917

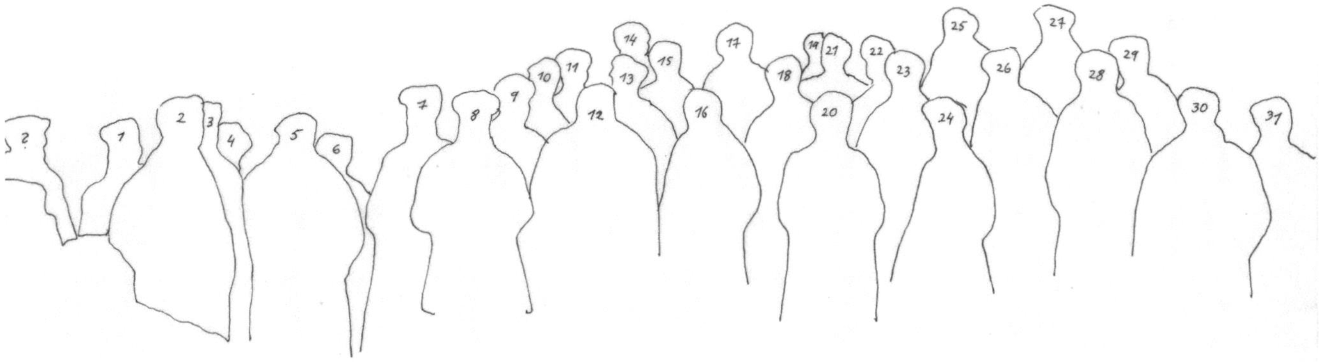

1. Lt. Dostler (6), 2. Oblt. Hoenemanns (22), Oblt. von Esebeck (17), Oblt. Heising (20), Lt. von Richthofen (11), 6.?, 7. Hptm. Schwarzenberg (Idflieg), 8. Oblt. Rummelpacher (10), 9. Oblt. Hahn (19), 10. Lt. von Lersner, 11. ?, 12. Hptm. Walz (B), 13. Oblt. Zander (Jastaschule), 14. Rittm. Wulf (Ober-Ost), 15. ?), 16. ?, 17. ?, 18. ?, 19. Oblt. Lang (29), 20. ?, 21. Oblt. Grieffenhagen (18), 22. Oblt. Berthold (14), 23. Hptm. Stenzel (8), 24. Oblt. von Döring (4), 25. Lt. Bolle (28?), 26. Oblt. Kohze (3), 27. ?, 28. Oblt. von Brederlow (17), 29. ? 30. Lt. Mohr (3), 31. Lt. Schlenker (3).

Above: Dr. Gustav Bock´s identification of the persons seen in the photo.

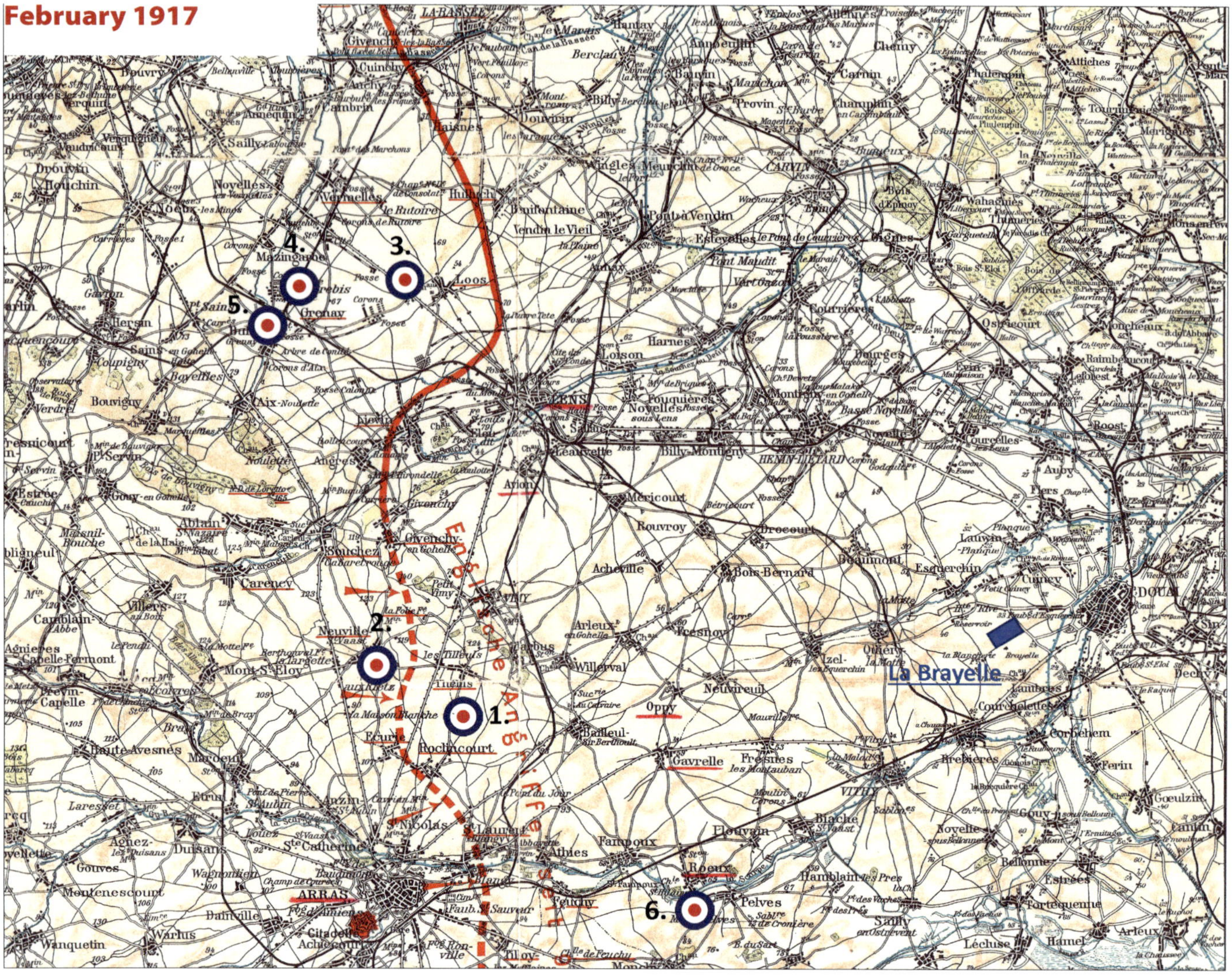

Above & Below: Map and details of the aerial victories of Jagdstaffel 11 in February 1917.

Nb.	Date		Hour	Pilot	Location	Aircraft	Line
1.	01.02.17		16:00	Lt. Manfred von Richthofen	s.w. Thelus	BE 2d 6742, 16. Sqn.	German
2.	05.02.17		17:00	Vzfw. Sebastian Festner	Neuville-St. Vaast	BE	Allied
3.	14.02.17		12:00	Lt. Manfred von Richthofen	w. Loos	BE 2d 6231, 2. Sqn.	Allied
4.	14.02.17		16:45	Lt. Manfred von Richthofen	s. Mazingarbe	BE 2c 2543, 2. Sqn.	German
5.	15.02.17		12:10	Vzfw. Sebastian Festner	w. Grenay	FE 2b	Allied
6.	16.02.17		13:00	Lt. Carl Allmenröder	s. Roeux	BE 2c 4179, 16. Sqn.	German

confirmed successes. Of those, Jasta 11 accounted for six, Jasta 'Boelcke' (operating from the 1st Army sector) had nine, two were attributed to Flak Platoons 31 and 29, and one was credited to a machine-gun unit on the ground.(56)

Above: Manfred von Richthofen (left) in conversation with Hans Bethge, Staffelführer of Jagdstaffel 30. Incidentially, this unit received two or three Halberstadt D V fighters that were formerly flown by Jagdstaffel 11 in February and March 1917.

Dienststelle : Jagdstaffel 11.

(Standort) : Flughafen Douai , den 28. Februar 19 17.

This Page & Two Following Pages:
Personnel roster, Jagdstaffel 11, dated 28 February 1917.
(Bayerisches Hauptstaatsarchiv, Abt. 4, München)

Personal-Bestandsnachweisung.

Veränderungen zur Kriegsrangliste. (Nur für Jdflieg. auszufüllen.)
Abgang : (wohin ?), Beförderung, Patentverleihung, Umschulung, Prüfung.
Bei Zugang : (woher ?) Kriegsrangliste auf besonderem Bogen beifügen.

eingetr. !

I. V.

(Unterschrift und Dienstgrad : Hartmann

Leutnant und Staffelführer.

Flugzeugführer :

Lfde. Nr.	Dienstgrad	Name	Geburtsdatum	Patent	Truppenteil	Zugehœrig oder kommandiert	Abzeichen seit	Bei der Dienststelle seit	Woher zur Dienststelle gekommen	Bemerkungen
1	Leutn.	Frh. v. Richthofen	2.5.92	19.11. 1912	1.Ul.-Regt.	zugeh.	3.6. 1916	19.1. 1917	Jagdst. Boelcke	Führer
2	„	Hartmann	31.5. 1892	18.8. 1912	JR.50	„	11.3. 1916	30.1. 1917	A.Fl.P. 6.	Offz.z. b.V.
3	„	Kleinhenz	28.8. 1892	1. 8. 1914	b.5.FAR	„	25.7. 1916	11.10. 1916	KEK. III	
4	„	Allmenröder	3.5. 1896	30.3. 1915	FAR.62	„	8.1. 1917	21.11. 1916	F.Fl. A.18	
5	„	Wolff	6.2. 1895	18.4. 1915	E.R.4.	„	Juni 1916	5.11. 1916	A.Fl.P. 6.	
6	„	Schaefer	17.12. 1891	18.6. 1915	1.E./W. Jäg.7	„	15.8. 1916	24.2. 1917	Kgohl.2	
7	Lt.d. Res.	Österreicher	1.2. 1890	28.9. 1915	Fl.Btl. Schl.	„	Aug.14	11.10. 1916	KEK. III.	
8	„	Krefft	18.1. 1892	28.11. 1915	Fea. Schl.	„	Jan. 1915	11.10. 1916	„	
9	„	Simon	4.1. 1885	29.6. 1916	Fea.5.	„	29.2. 1916	7.11. 1916	A.Fl.P. 6.	
10	„	Anslinger	1.5. 1887	25.8. 1916	Fea.9.	„	28.2. 1916	16.10. 1916	„	
11	Vzf. d.R.	Baierlein	16.11. 1886	.	Fea. Schl.	„	10.8. 1914	26.11. 1916	„	[illegible] 5 [illegible] [illegible] 5225 [illegible] 6648
		Beobachter :								
12	Vzf.	Festner	30.6. 1894	.	Fea. Schl.	„	15.2. 1917	11.10. 1916	KEK.III.	
~~13~~	~~Vzf.~~	~~Braumüller~~	~~4.12. 1889~~	.	~~Fea.2.~~	„	~~27.2. 1916~~	~~16.1. 1917~~	~~A.Fl.P. 6.~~	gem. Kofl. 6 Nr. 23648 v. 27.2.17 zum A.Fl.P.6 versetzt
~~14~~	~~Vzf.~~	~~Prehn~~	~~20.4. 1894~~	.	~~Fea.2~~	„	~~25.12. 1915~~	~~16.10. 1916~~	~~Jagdst. 10~~	

Bemerkungen über kranke, zukommandierte bzw. abkommandierte Flugzeugführer und Beobachter :
(bei Kranken, ob Stelle für sie offen bleibt)

~~Vizef. Prehn gem. T.Bef.Kofl.6 v.18.11.16 Z.1.) zur Flugstation Tondern kommandiert. Kommando bis 28.2.17 verlängert. (Kogenluft Nr.57554).~~

Leutn. Kleinhenz gem. Kofl.6/ 23453 vom 24.2.17 auf die Dauer von 4 Wochen zur Jagdstaffelschule Valenciennes kommandiert.

Sonstige Offiziere und Beamte :

Lfde. Nr.	Dienstgrad	Name	Geburtsdatum	Patent	Truppenteil	Zugehœrig oder kommandiert	Abzeichen seit	Bei der Dienststelle seit	Woher zur Dienststelle gekommen	Verwendung
1	Uzhlm.	Schenk	11.5. 1893	.	1.Pi.Btl.	zugehörig	.	17.10. 1916	E.F.K. 5 6.Armee	Zahlm.

Dienststelle : Jagdstaffel 11

(Standort) : Flughafen Douai , den 28.Februar 1917..

This Page & Following Page:
List of the aircraft complement of Jagdstaffel 11 dated 28 February 1917. (Bayerisches Hauptstaatsarchiv Abt. 4, München)

Flugzeuge.

(Unterschrift, Dienstgrad) :

Leutnant und Staffelführer.

	Stand am ~~1~~ 28.2. 1917		Abgang vom 1. Januar 1917 bis ~~3~~ 28. Februar 1917.			
	Fabrik	Nummer	Fabrik	Nummer	Grund	Verbleib
R						
G						
D	Halb.D.V	410	Ha D V	415		am 4.1.17 A.Fl.P.6
			„ D II	107		„ 20.1. „ „
	„	416	„ D III	116		„ 21.1. „ „
			„ D V	412	kriegsun-	„ 22.1. „ „
	„	417	„ D V	418	brauchbar;	„ 22.1. „ „
			„ D V	420	umgetauscht	„ 24.1. „ „
	Halb. D	802	„ D V	413	gegen Alb.	„ 18.2. „ „
			„ D V	1118	D III.	„ 25.2. „ „
	„ D.V.	1099	„ D V	421		„ 26.2. „ „
			„ D	801		„ 26.2. „ „
	Alb.D III	1980	„ D	805		„ 26.2. „ „
			„ D V	1100		„ 27.2. „ „
	„ „	1985	„ D V	414		„ 27.2. „ „
	„ „	1991				
	„ „	1993				
	„„ „	1996				
	„ „	2006				
	„ „	2062				
	„ „	2015				
	„ „	2016				
	„ „	2036				
	„ „	2014				
	„ „	2026				

The Birth of the Richthofen Myth, March 1917

Enemy flight activity, which had been increasing steadily in February 1917, reached a new high in March 1917. In January, only a few enemy aircraft had flown across the lines, but now large formations were routinely entering German airspace. As a result, Jagdstaffel 11's operational activities increased as well to meet the challenge and in March 1917, weather permitting, the unit normally carried out four to five missions a day. When aircraft returned from the front, their black-garbed ground crewmen had to 'turn the crates over' as quickly as possible – i.e., quickly refuel and rearm them to prepare them for the next mission. In-depth servicing and repairs occurred only at night, and sometimes even into the early morning hours, not long before the aircraft would take off on their next mission.

"We had the feeling that something much bigger was going on at the front," recalled former mechanic Hermann Lohmeyer.(1)

His sense was correct because something really was "in the works". A situational assessment based on reconnaissance results and reproduced in Kofl's weekly activity report made it clear that the German 6th army's commanding generals right up to Supreme Command were aware of the redeployment of British units:

"The assembling and preparation of large enemy forces in front of the army's southern flank appears to be nearing completion. The zone behind the enemy front in the Lens-Arras sector has been prepared for greater battle conditions in terms of quartering capabilities and the expansion of the small and full-size rail network. The fourfold increase in enemy air activity in the Lens-Arras sector compared to the prior week also indicates the beginning of an enemy offensive against the army's southern flank."(2)

The plan of what is now known as the "Nivelle Offensive" was that it would begin with the French 3rd Army attacking near St. Quentin while the British 1st, 3rd, and 5th Armies assaulted the Arras region to capture high ground and divert German reserves from the Aisne and Champagne fronts. This would be followed by a major French offensive across the Aisne River and deep into German territory. Then, the British and French forces would join up in a pincer movement, capturing considerable territory, enemy forces, and supplies. It was a very ambitious undertaking and would be launched in early April 1917.

The area of attack for the British forces against the German 6th Army ran north and south of the Scarpe River over a 35-kilometer front from Loos-en-Gohelle in the north to Fontaine-lès-Croisilles in the south. It would therefore take place within Jasta 11's theater of operations.

The Royal Flying Corps was assigned an important role in the preparations leading up to this major offensive. The assembling and preparation of British troops and the delivery of huge amounts of artillery, weapons, rations, and munitions were to be concealed from German reconnaissance by the constant presence of British aircraft at the front. They were issued orders to prevent German aircraft from entering their airspace. In addition, they were tasked with scouting out German infantry and artillery positions and their staging areas. Ammunition and supply depots needed to be pinpointed so that they could be destroyed by artillery fire and bombing attacks. Photo reconnaissance would play a key role in this since it promised to obtain the most accurate insight into the situation behind German lines. As a consequence, shooting down the enemy's reconnaissance aircraft became the most important job for the German Jagdstaffeln.

1st March 1917 – 6th March 1917

On 1 March, dense fog initially hampered that morning's flight activity, but it soon cleared, and the activity increased. Reports from the front lines came to Jagdstaffel 11, whose pilots climbed into their Albatros D IIIs, secured their seatbelts, and took off in response. It was Emil Schaefer's first flight to the front with his new unit and he wrote about the difficulties he encountered:

"We conducted a pursuit flight on 1 March that was a complete failure because we didn't fly in close formation and therefore came up against superior [force of] *opponents individually. I grappled briefly with four Sopwiths but couldn't for very long because others were always on my back. We flew again in the afternoon and this time the formation flight went smoothly; but there weren't any Englishmen around for it."*(1)

That same day, Vzfw. Georg Braunmüller was transferred out to Armeeflugpark 6 to become a flight instructor with Flieger-Ersatz-Abteilung 4 at Posen.

The next day, the order to transfer Lt. Anslinger via Armeeflugpark 6 to Jagdstaffel 35 arrived.(2)

Flight operations virtually came to a standstill over the next two days due to rain and fog, which gave Kurt Wolff the time to write a letter to his sister Elisabeth:

"Dear Elisabeth!

… Now I'm coming to bother you again. Little Hans Georg's birthday is on 13 March. Maybe when you walk through Wertheim or Tietz [large department stores in Berlin] *or anywhere else, you'll find an airplane model or something like that, because the little guy is so eagerly into flying. Well, you catch my drift, and please don't be angry that I bother you again. Concerning the money, you don't have to be afraid. I paid my tailor the last debt the day before yesterday and I am now saving for the next war bond, because I don't need to save for vacation for the time being. First of all, many more are in line for it before*

me, and then I don't feel like it yet. Things are really getting started now and I feel like I'm up to my best performance now, I'd rather go on leave later when I really need it. Now it's a nice hunt here and the English have also appeared a little more numerous. Yesterday, the Staffel shot down no less than five English apparatuses, two of which crashed on fire, one burst in the air and two others smashed to the ground. The Lords must have been amazed. Richthofen has now arrived at number 23. Hartmann is now a good support for us, with the beautiful successes and the increase in work going along with it. The fact that he came to us is of course a push from me and him, but we are only now satisfied with it.

Your brother Kurt[3]

The last sentence mentioned in the letter implies that the O.z.b.V., Lt. von Hartmann, with whom Wolff had flown as "Franz and Emil" at Kampfstaffel 26 (Kagohl V), had been transferred to Jagdstaffel 11 in the meantime, as a result of a request on behalf of Kurt Wolff to Manfred von Richthofen.

On 4 March, an early light ground fog quickly dissipated and yielded a beautiful clear day. This led to a considerable increase in flight activity on both sides of the front as well as numerous aerial fights.[4]

Following a morning flight that went without incident, Lt. Emil Schaefer led Schwarm I (a "swarm" was half of the Staffel) on another mission to the front around 11:30 a.m. in response to a request from there.[5] This certainly speaks volumes concerning Richthofen's confidence in Schaefer's leadership capabilities in that he entrusted half the Staffel to him just a few days after his arrival.

The unit's Albatros D IIIs, fighting against gusting winds, flew along the straight road to Henin-Liétard, passed Harnes, and headed west in the direction of the city of Lens where they saw British airplanes directing artillery fire on German targets in the area of the "Loosbogen," or "Loos Bend" – an eastward indentation in the German front lines. Their opponents were Sopwith two-seaters from 43 Squadron that were being protected by F.E.8 fighters from 40 Squadron.

Emil Schäfer wrote:

"Our Schwarm II had been out in the morning without accomplishing anything. Swarm I got itself ready. Richthofen wanted to come along but had to stay home because of a minor defect. As a result, I led it. We had just arrived at the front when an English formation appeared that flew into the Loos Bend. We attacked it from Lens at the same time that three Germans (from Jasta Boelcke, as we later learned) did. My first opponent evaded me by diving. Before I could follow him, I saw Allmenröder being harassed by two Englishmen and gave him some room. As a result, a Vickers single-seater got on my neck. I did a half loop and let myself go into a spin. Two comrades who saw this thought I'd been shot down, as did the Vickers, who then left me alone. I pulled myself somewhat clear of the mix-up to get a quiet look and calmly took on a Sopwith two-seater which, after about 100 rounds, began to burn, sideslipped, overturned, and fluttered earthwards in burning bits and pieces – whereupon I couldn't refrain from a loud 'Hurrah!' But since one of my tension wires had been shot through, I took the quickest way home."[6]

Schaefer's opponent, Sopwith Strutter A1109 from No. 43 Squadron, crashed near the village of Haisnes (about 8 km north of Lens). Its crew, 2/Lt. P.L. Wood and 2/Lt. A.H. Fenton, did not survive.

The "Vickers single-seater" that attacked Schaefer was probably flown by 2/Lt. R.E. Neve, as indicated by this 40 Squadron report:

"2/Lt. R. E. Neve sighted six hostile aircraft at about 12,000 feet between Vimy Ridge and Lens. He was in company with a patrol of seven Sopwith. Hostile aircraft attacked formation… He was engaged with various hostile aircraft for about 10 minutes and many shots were seen to enter fuselage of hostile aircraft. One hostile aircraft was seen to dive steeply, apparently out of control."[7]

Schaefer righted his "out of control" Albatros D III and returned to his airfield where he made an emergency landing without any injury to himself.[8]

Kurt Wolff attacked another F.E.8 during the same fight and later wrote:

"There was intense activity on 4 March. We encountered a formation of Sopwiths and F.E. single-seaters near La Bassée, where Schäfer shot a Sopwith down in flames. My F.E. single-seater flew on despite the 180 [rounds] *I had fired into him.*[9]

40 Squadron's 2/Lt. W.B. Hills was involved in a dogfight with another Jasta 11 Albatros:

"I saw a Sopwith being attacked by a fast scout, **painted blue**… *I dived from 10,000 to 9,000 feet and fired nearly a large drum into hostile aircraft at about 30 yards range. Hostile aircraft went into a vertical nosedive, apparently out of control, but was unable to see if hostile aircraft crashed, owing to being attacked by two hostile aircraft. Not having any ammunition left…"*[10]

The German machine described as "blue" might have been the half-yellow, half-blue Albatros normally piloted by Eddy Lübbert, who appears to have spun away, like Schaefer, to escape Hills' attack.

In summary, Jagdstaffel 11 had shot down one British airplane whereas all of its fighters returned to La Brayelle. Schaefer's Albatros, however, had been damaged as a result of his emergency landing.

The defect on Richthofen's Albatros was remedied and he set out around 12:30 p.m. toward Loos to try to join his men. He reported:

"4 March 1917
B.E two-seater
12:50 pm 1 km north of Loos
Details unknown, airplane dropped on enemy's side.

I had taken off alone and was just looking for my Staffel when I spotted a single B.E. My first attack was apparently a failure as my adversary tried to escape by curves and dives. After having forced my adversary downwards from 2,800 to 1,200 meters, he imagined himself safe and once again flew straight on. I took advantage of this, put myself behind him, and fired some 500

Above: Lt. von Hartmann, the new special duty Officer of Jagdstaffel 11.

rounds at him. My adversary dived, but in such a steep way that I could not follow.

According to our infantry observations, the airplane crashed to the ground in front of our trenches."(11)

The machine most likely was B.E.2d 5785 from 2 Squadron, manned by Lt. J.B.E. Crosbee and Sgt. J.E. Prance. Prance received a serious hip wound, and the machine was hit by countless bullets during Richthofen's onslaught. Nevertheless, Lt. Crosbee was able to make an emergency landing in Allied territory and the two men were recovered by British troops.

Lt. Crosbee stated:

"Hostile machine attacked our machine, when on photography [mission] *north of Lens. Neither pilot nor observer saw the machine, as it came right out of the sun which was very powerful. Hostile machine fired burst when diving from the side towards the nose of our machine, the fourth or fifth shot hitting the observer, Sgt. J.E. Prance who was already at the gun. Hostile machine than swung towards our tail and fired another burst before diving below us and swooping above us again, when observer fired one or two bursts apparently well aimed. Hostile machine then shot away left-hand ammunition box. Observer attempted to load one or two remaining drums, which came off, hostile machine again hitting us in rear petrol tank. The observer weakening, a slow spiral was done, following by a steep one, during which hostile machine fired continuously, with a few hits. When our machine was 3,000 feet over trenches, hostile machine flattened out 800 feet above. Observer then fired another well-placed burst, after which hostile machine made off low towards Hénin-Lietard and was lost in the mist. Hostile machine: Halberstadt single seater, time 11:45."*(12)

As was seen previously, Albatros aircraft were commonly confused with Halberstadt fighters at this time in British reports. Richthofen had actually been flying an Albatros D III, as confirmed by Flak Platoon Faschoda around 12:47 p.m.:

"Dogfight between Albatros and B.E. airplane Loos, B.E. apparently damaged went westward."

Half an hour later, Infantry Regiment 93 reported from the forwardmost lines:

"Shot down B.E. crashed near Loos."(13)

Due to the heavy enemy air activity in the German 6th Army's southern sector between Loos and Arras, Schaefer and a "Schwarm" consisting of six Albatros D IIIs took off again. This time, however, they were led by Richthofen. According to reports from the front, they followed a route toward Lens. Just as they reached the area over Billy-Montigny, they came across an English unit that had penetrated German airspace. Richthofen's report on the ensuing fight reads:

"4 March 1917
Sopwith Two-seater
4:20 pm
Acheville
Occupants: Lt. W. Reid and Lt. H. Green, both killed, buried by local command at Bois Bernard.

Accompanied by five of my airplanes, I attacked an enemy squadron above Acheville. The Sopwith I had singled out flew for quite a while in my fire. After 400 shots, the plane lost a wing while trying to make a sudden turn. It plunged downward.

Above: Wreckage of Sopwith 1½ Strutter A/1109, downed by Emil Schaefer in the morning of 4 March at 10:55 As seen before, whenever such a crash occurred out of range of enemy artillery, a huge number of spectators soon arrived at the crash site. (R. Absmeier)

It is not worthwhile to have the plane taken back, as parts of it are scattered all over Acheville and the surrounding countryside. Two machine guns were seized by my Staffel (one Lewis gun No. 20024 and one Maxim gun L.7,500).[14]

The machine was Sopwith Strutter A1108 from No. 43 Squadron RFC.

Emil Schaefer summarized in a letter home:

"We took off again with Richthofen that afternoon. Like before, we immediately got hold of a formation; next to me, Richthofen attacked a Sopwith that went into a turn. I immediately shot at it. I also lost sight of him in the turn. Before I could catch hold of him again, he broke in two under Richthofen's fire and tipped over out of control."[15]

After the combat described above, Richthofen reassembled his unit north of Lens. While he and two others turned south toward Arras, Lts. Kurt Wolff, Franz Anslinger, and Emil Schaefer sighted a British formation flying west which they followed and attacked. The English formation consisted of nine Sopwith Strutters from No. 43 Squadron with an escort of F.E.8s from No. 40 Squadron, about which Emil Schaefer reported:

"I, together with Wolff and Anslinger, trailed after a formation flying west and attacked a Vickers single-seater which, after a few shots, first began to smoke, then to burn, and then went down out of control."[16]

While Schaefer watched the British airplane falling far over the lines at Béthune, a second "Vickers single-seater" shot at him from above. He immediately began to bank. After a minute-long dogfight, he finally got into firing position and opened up:

"He smoked, stopped shooting, and tried to escape by falling down. Each time he righted himself, I sat on his neck and shot. I pushed him down to the ground this way. He wanted to land, couldn't manage the site, and smashed his crate into the trees of a country road. I, however, rushed home at 200 - 500 meters' altitude. I was so deep in France that it took me six minutes to reach the front. During that time, the good English expended a fortune in military ammunition on me without my crate suffering a scratch."[17]

Schaefer's opponents may have been Lt. C.O. Usborne and 2/Lt. K.S. Henderson. Both of them managed to land in their own territory and escape injury. Usborne, who was flying F.E.8 A4872, reported:

"Saw hostile aircraft diving on Sopwith and engaged hostile aircraft from below. Fired 30 or 40 rounds. Hostile aircraft then left the Sopwith and returned fire, damaging my rudder, and puncturing my petrol tank. I spun to escape from hostile aircraft and landed near Bethune, crashing, having no rudder control. I saw another F.E. 8 (2/Lt. K. S. Henderson) attacking at the critical moment."[18]

Lt. Kurt Wolff, who was also involved in the fight, wrote in his diary:

March 4, 1917

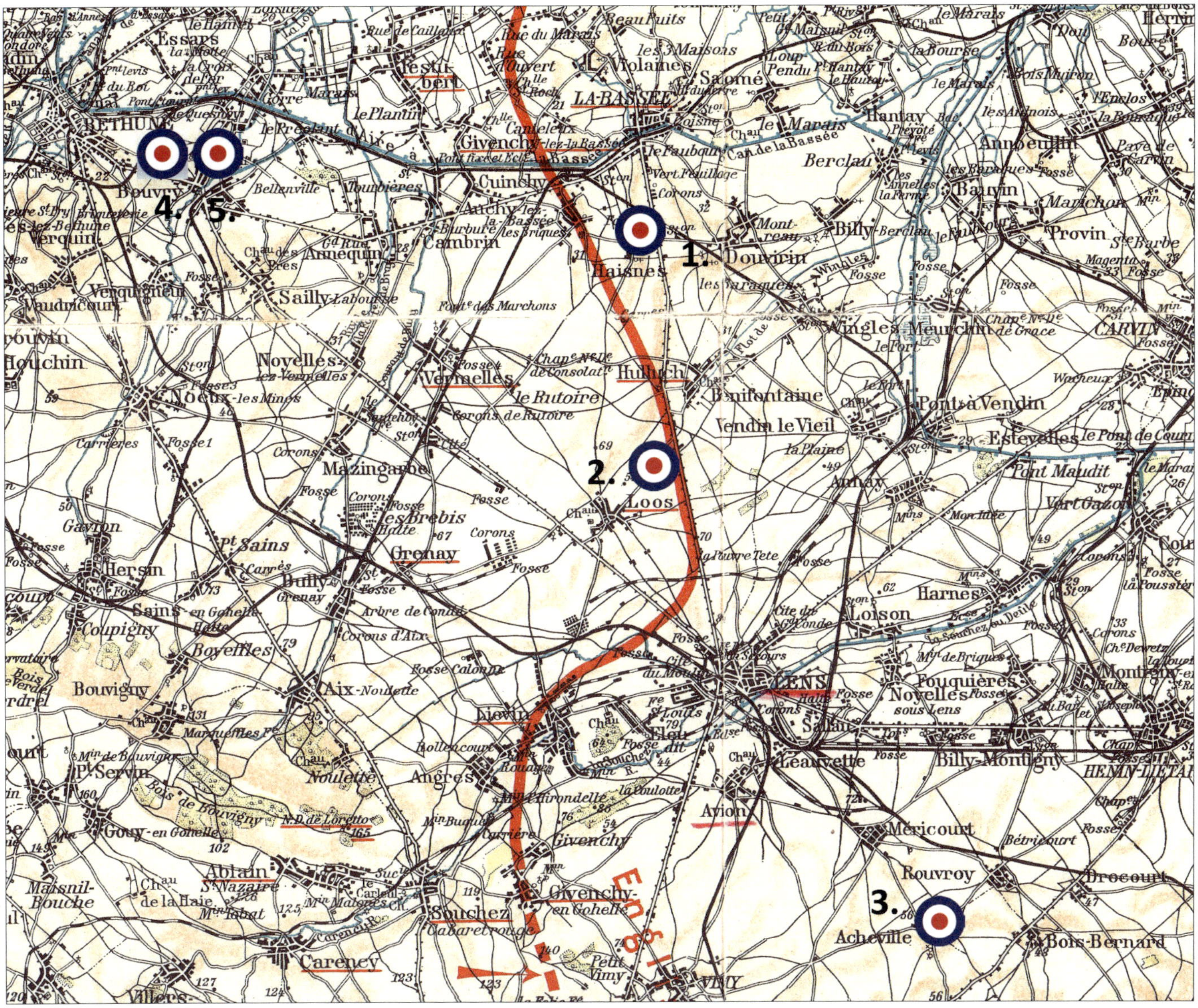

Above & Below: Map and details of the aerial victories of Jagdstaffel 11 on March 4, 1917.

Nb.		Hour	Pilot	Location	Aircraft	Line
1.		11:15	Lt. Emil Schäfer	Haisnes	Sopwith Strutter A1109 43. Sqn.	German
2.		12:45	Lt. Manfred von Richthofen	North-east of Loos	B.E.2d 5785, 2. Sqn.	Allied
3.		16:20	Lt. Manfred von Richthofen	Acheville	Sopwith Strutter A1108, 43. Sqn.	German
4.		17:20	Lt. Emil Schäfer	Bethune	F.E.8 A4872, 40. Sqn.	Allied
5.		17:20	Lt. Emil Schäfer	Bethune	F.E.8, 40 Sqn.	Allied

Above: The two claims by Lt. Schäfer at 17:20 on 4 March 1917 were not confirmed.

"In the afternoon, Richthofen shot down a Sopwith from out of a large formation that fell apart in the air; then Schäfer finished off two F.E. single-seaters far behind the front that had just risen up. I had no success. The Staffel group result was five kills; 18 downings (3 wrongly) in all were reported this day."[19]

Flak 46 received the following confirmation request at 5:16 p.m.:

"Jagdstaffel 11: English airplane shot down by Jagdstaffel near

Neuve Chapelle."

Flak 46 responded to the request at 8:00 p.m.:

"1 Albatros in air combat with 2 Englishmen, both enemy aircraft shot down near Neuve Chapelle-Richebourg."[20]

The reply submitted by Flak 46 was not enough, however, to confirm Schaefer's claims and neither of the aircraft forced down by him was credited as a victory – a fact that reflects the strict application of regulations following requests for air victory confirmations by single-seater pilots.

The next day, 5 March, brought a short break for Jasta 11's pilots. Heavy snowfall prevented flight activities on both sides of the front.

While the pilots enjoyed a day off, their black-garbed ground crewmen remained just as busy as they always were. For the mechanics like Hermann Lohmeyer, the Staffel's working day normally began in the early morning hours. In order to make their machines operational, the mechanics had to be in the hangars a good two hours before daybreak – the earliest possible time for takeoff. The engines were test run and all aircraft functions were checked. Once this was completed, the machines were armed with ammunition – even those that were not slated for the first flight to the front. As many aircraft as possible had to be ready for action by sunrise. When the pilots arrived at the airfield, the airplanes and their mechanics stood ready for takeoff in front of their hangars.

After the brightly-painted Albatros D IIIs had taken off and disappeared from the ground staff's view, Hermann Lohmeyer and the other mechanics could take a short break for a good snack. Then it was back to work because there was always something else to be done. While this work was going on, the men constantly kept one ear open for the sound of approaching aircraft. Lohmeyer and the others had learned to recognize different aircraft types by the sound of their engines because the buzz of approaching aircraft engines could be either from the Staffel returning home or from British airplanes. Approaching British aircraft posed no danger as long as they flew at higher altitudes. But if they were flying low, it could mean a British bombing raid on the airfield.

When the Staffel's aircraft were expected to return, considering their takeoff time and normal flight duration, their mechanics would emerge from the hangars and gather on the airfield. Richthofen's dog, Moritz, was usually the first to notice the approaching airplanes and would run across the airfield, barking. Then one (or two) of Richthofen's mechanics would have to put the dog on a leash to prevent him from running into one of the landing machines. It was hard work dealing with the strength of a full-grown mastiff.

As the planes appeared as small dots in the sky, a count would be taken to determine whether everyone had come home. Once it was realized that a "Schwarm" was complete, a sigh of relief would run through the ranks. If a machine was missing, however, there would be worried faces. When the airplanes were close enough to spot their markings, each mechanic would look for "his" airplane and run toward it as it rolled in. Once the airplane came to a stop, the mechanics would help their pilot out of the cockpit. While Kurt Wolff was known to quickly jump out of his machine, others would sit in their airplane awhile to bring their thoughts back to earth before getting out.

Right: Lt. Wolff's mechanic Hermann Lohmeyer.

As Hermann Lohmeyer later recalled, air combats in January and February 1917 took place farther away at the front. From March 1917 on, there was increased fighting nearer the airfield and sometimes, he could see a dogfight with the naked eye. When that happened, the mechanics would empty out of their hangars. Everyone, standing spellbound and with hearts racing, would watch what was going on in the air. Each feared for his pilot and would cheer each enemy machine that came down.[21]

The snowfall stopped on the morning of 6 March and the skies were essentially clear apart from a few clouds. There was little hostile air activity at first, but it increased as the day progressed and became heavy in the afternoon.

Schwarm I, led by Richthofen, took off at around 10:00 a.m., and as Schaefer later wrote:

"It was pretty hazy that morning. Geschwader II [another name for Schwarm II] *conducted a long flight without seeing a single Englishman. Schwarm I took off at 10:00 and saw only a few tethered balloons standing peacefully at a low altitude. But far, far over the lines, there was a lot of activity above the British airfields. The "Lords," individually and in small groups, could be seen circling at low height, and none was coming forward. We all expected that Richthofen would resolutely wait, perhaps for hours and up to our last drop of fuel, for them to approach because the weather had turned beautiful and the business would surely have to start now. So there was great surprise when our lead airplane suddenly made a sharp turn towards home. Perhaps he had engine trouble or had not seen the English way over there. So I and two others stayed put. But Richthofen was there again, circling us and unmistakably signaling with a wave: home. He called us together at the airfield: 'Gentlemen, we'll take ten minutes to refuel. There will be bloody business at the front in less than half an hour and I want to deploy a greater number of assembled aircraft. Please make sure everything is ready when we set off.'*

It soon became clear how right he was. Telephone reports started coming in about artillery aircraft flying alone near Arras and Vimy, near Loos and Suchez. And then came the anticipated

Above: Manfred von Richthofen and his Great Dane "Moritz", who had to be put on a leash by one or two mechanics before the planes landed, so that he would not run into the landing planes. (E. Wischnowski)

news. A greater English formation had crossed the front at the Loos Bend. Half an hour later, we were all roaring off in our aircraft.

The flak clouds showed us the way. The formations encountered each other about six to eight kilometers in German territory – 15 Englishmen at an altitude of 2,000 meters, 9 of us at 1,500 meters. Peacefully, apparently not concerning ourselves with them, we pulled under them toward the west. When we were at the front, then at about the same height as the enemy, we turned around. If the enemy wanted to go home now, he would have to force his way through us. We had blocked him from getting away without a fight. We could now placidly look at the gentlemen: there were six large Vickers, big lattice-tailed photography airplanes, with nine Sopwith two-seaters, light combat airplanes, above them, which had to be attacked first."

We slowly climbed above them and each of us singled out our opponent. This wasn't made easy for us because the Englishmen didn't fly straight on for even a second. They swirled around one another, constantly turning and circling. It was virtually impossible to go for one in peace. One of them went astray some 500 meters from the formation and Richthofen immediately had him by the scruff of the neck.

I tackled the last one in a group of three, got behind him,

Above: Members of Jagdstaffel 11 observe an air combat. (E. Wischnowski)

Above: Carl Allmenröder in front of the tail of his new Albatros D III. The first digit of the military number is visible in the photo, confirming that it was a machine from the series with the military number 2xxx/16. Seven aircraft within this serial range were delivered to Jagdstaffel 11. This was very likely the aircraft that was later painted "field gray."

Above: Manfred von Richthofen's red Albatros D III heads a line-up of Jagdstaffel 11´s Albatros D III fighters at LaBrayelle, most likely in March 1917. The third aircraft from right is Eddy Lübbert´s blue-yellow Albatros, but the poor quality of the photo prevents further identification of aircraft. (L. Bronnenkant).

Above: Lt. Manfred von Richthofen takes off in his red-painted Albatros D III from La Brayelle airfield in March 1917. In the background is the half-blue-half yellow Albatros D III of Lt. Lübbert can be seen. (L. Bronnenkant)

and gave him a full dose. He twisted and turned but I hung onto his tail from close behind and refused to be shaken off. A second [one] *flew alongside at a short distance and I saw the observer standing upright, firing at me; but his tracer bullets all whizzed past me. At our great speeds, you have to keep a good distance when firing from the side. I was sitting so close to the one Englishman that the other would have had to aim at his comrade to hit me. Since he didn't dare do that, his shots went behind me. So I let him shoot and calmly continued to deal with my original opponent, who finally caught on fire and crashed.*

That was the moment I had waited for to quickly spoil the second one's shooting. I went at him in a righthand turn in which my machine-gun burst raked across his machine diagonally. Then I was behind him with a quick left turn. His rudder came off with the first volley and I saw him slide away. At the same moment, there was a rattling behind me; I turned around and saw within reach a large rotary engine over which red muzzle flashes were sending angry, long, thin threads of smoke at me. The first impacts were already hitting my machine. The only thing to do was to leave. I turned the crate upside down with a jerk and it went down vertically into the depths. When I looked around again in peace and safety, I saw one of the large Vickers fire off a white flare, which meant in German: 'the undertaking is being abandoned.' I was able to fly home contentedly.

When I reported: 'One Sopwith on fire, shot off the rudder of a second but don't know what happened to him,' infernal derisive laughter arose. 'The observer must have thrown his map at you, the observer must have cried over your poor shooting and his handkerchief flew away' being the most harmless of the pestering I had to endure. I was about to give up when there was no confirmation of the downing by the afternoon, but towards the evening the Englishman was found. Infantry Regiment X sent a machine gun, remnants of a photographic apparatus, a piece of wing, and other curiosities along with the report that an airplane from the morning's airfight had spun down out of control and broken up over the trench positions. The fuselage and one wing lay in the English lines while the rest had rained down on our side."[22]

It turns out that Schaefer confused the aircraft types in his account because British records state that their formation consisted of six Sopwith Strutters from 43 Squadron under the protection of six F.E.8 fighters from 40 Squadron.

The airplane Schaefer had downed in flames was Sopwith Strutter A978 with the Canadian crew of Lt. S.J. Pepler (pilot) and Capt. J.D. Stuart (observer). Pepler was killed in action and Stuart died of his severe injuries the next day.

The second Sopwith, A1097, that was manned by Capt. H.H. Balfour (pilot) and 2/Lt. A. Roberts. 2/Lt. Roberts gave the following report:

"I was observing with Capt. Balfour on patrol La Basse [sic] *to Bailleul with 5 other Sopwiths and 6 F.E. 8s. 8 Halberstadts tried to encircle our Sopwiths and attacked us. 3 tried to get on our tail and opened fire at us. I opened fire at one of the machines and he went down out of control. I watched him down at 2,000 feet. I was too busy engaging other 2 machines to observe him strike the ground. The other 2 machines I failed to hit, and they left us firing at us."*[23]

2/Lt. C.D. Knox and Lt. C.R. O'Brian, the crewmen of Strutter A2407 that was also involved in the fight, described the encounter from their perspective following their return:

"While on patrol with 6 Sopwith and F.E.8 of 40. Squadron, we encountered a hostile patrol. Two hostile aircrafts circled round us from the front and got on our tail. Observer fired ½ drum into one hostile aircraft. Pilot spun Sopwith to get out of direct line of fire from hostile aircraft. Hostile machine still on Sopwith tail after spinning. Observer fired rest of his drum into the hostile aircraft which was seen going down out of control. Same time Sopwith was spun again by pilot who was unaware of German's fate."[24]

Contrary to British reports, Jagdstaffel 11 again suffered no losses.[25] The two Sopwith claims put in by Schaefer were approved as his 3rd and 4th official victories.

In the meantime, Richthofen had spotted a Sopwith that was somewhat detached from the British group and attacked it, just as Schäfer had mentioned in his report. The English observer began to fire from a long distance while Richthofen's machine guns were still silent. When he was within 50 meters of the Sopwith and was about to open fire, he heard a bang in his machine. There was a strong smell of gasoline, and his engine began to lose power. Richthofen immediately shut off the ignition and coasted down.

Der rote Kampfflieger provided this description of the incident:

"I'm probably still a few hundred meters high and have to look around carefully to see where I want to land, because such a landing is usually associated with a crackup. And a crackup like that doesn't always go well – so be careful. I find a meadow, not very big, but it's just enough if one goes about the job somewhat carefully. It's also conveniently located for me, right on the paved road near Hénin-Liétard. That's where I'll land. Everything goes smoothly...Now I have time to examine the damage. There are a few hits in [the airplane], *but the one that caused me to break off the fight was one through both fuel tanks. I don't have one drop of petrol left, and the engine is also shot up. Too bad about that, it ran so well.*

I let my legs dangle out from the machine and must have looked a little silly. A large crowd of soldiers quickly gathered around me. Then comes an officer. He's all out of breath. Very excited!...He rushes toward me, gasps for air and asks: 'I hope nothing has has happened to you? I watched the whole thing and am so flustered! Good God, it looked terrible!' I assured him there was nothing wrong with me, jumped down, and introduced myself. He evidently didn't catch a syllable of my name. But he did ask me to drive in his car with him to nearby Hénin-Liétard where his quarters were. He was a Pionier officer.

We sit in his car and are about to drive off. My host still hasn't calmed down. Suddenly he looks startled and asks: 'Good God, where's your driver then?' At first, I didn't really know what he meant, and I probably looked at him somewhat confused. Then I realized he thought I was the observer of a two-seater airplane and was asking about my pilot. I quickly got hold of myself and said quite dryly: 'I drive alone.' The word 'drive' is frowned upon in the Fliegertruppe. You don't drive, you 'fly'. The fact that I 'drove' alone evidently diminished me in the eyes of the good

Above: Manfred von Richthofen's red Albatros D III after the emergency landing on 6 March 1917. Disconnected aileron control wires, possibly damaged during combat, can be seen. The horizontal control surfaces are also missing from the aircraft. In spite of the damage, the overall appearance of the aircraft is still rather glossy and immaculate, an indication that it had only recently been decorated in its current scheme. (L. Bronnenkant).

Above: Lt. Kurt Wolff snapped while taking off in his "Violet Plum" from La Brayelle airfield. The pointed cap he loved to wear under his leather flying cap resulted in a bulged appearance of his head, clearly visible here. (T. Weber)

Above: Lt. Kurt Wolff stands to the right of his Albatros D.III with the violet fuselage. According to his mechanic Hermann Lohmeyer, the insignia on this aircraft was not painted over.

Above: Lt. Hans-Georg „Eddy" Lübbert's Albatros D III just before take-off. By this time, the aircraft was unarmed and no longer served with Jasta 11. (G. Van Wyngarden)

Above: Lt. Hans-Georg „Eddy" Lübbert´s Albatros D III just before take-off. (G. Van Wyngarden)

Facing Page, Below: Eddy Lübberts "half-blue-half-yellow" Albatros D III pictured from the left side, where the upper half was yellow and the lower part was blue. This gave the fuselage a "lengthwise quartered" appearance. (G VanWyngarden)

gentleman. The conversation became somewhat reserved.

Then we arrive at his quarters. I am still wearing my dirty, oil-stained leather jacket and a thick scarf. Along the way, he had naturally bombarded me with an endless number of questions. In general, the gentleman was considerably more excited than I was. Then he forced me to lie down on a sofa, or wanted to do so, saying that I must still be quite upset from my fight. I assured him that I had already fought several times in the air, but he just couldn't get it into his head. I certainly didn't look very warlike.

After some conversation, he came up with the well-known question: 'Have you ever shot anyone down?' As I said, he hadn't heard my name. 'Oh yes,' I said, 'now and then.' – 'Indeed, so, you've perhaps shot down two?' 'No, 24, however.' He smiled, repeated his question, and said that by 'shot down' he meant one that has fallen down and stays on the ground. I assured him that that was my perception of it too. Now I was totally looked down on because he now thought I was a mighty braggart. He left me sitting there and told me that dinner would be served in an hour and if I wanted to, I could dine with him. Then I took advantage of his offer and slept soundly for one hour. Then we went over to the Officers' Mess. I took off my jacket there and luckily had my Pour le Mérite on me – unfortunately, no uniform tunic underneath but just a vest. I asked him to excuse me for not being

Above: Eddy Lübbert's "half-blue-half-yellow" Albatros D III from the right side. The upper half of the fuselage is blue and the lower half yellow. The photo was taken at Roucourt airfield in April 1917. It is not known who was flying Eddy's aircraft at that time. (A. Imrie)

Above: Manfred von Richthofen in the cockpit of von Eddy Lübberts' Albatros D III, at Roucourt airfield in April 1917, ready to take off. (Taktisches Luftwaffengeschwader 71 Richthofen)

better dressed, and my good chief suddenly discovered the Pour le Mérite on me. He became speechless with astonishment and assured me that he didn't know who I was. I told him my name again. Now it seemed to dawn on him that he had indeed heard of me before. I was now given oysters and champagne to drink and actually had a good time until Schaefer finally came and picked me up in my car."[26]

This account contains several factual errors and inconsistencies. For example, Richthofen certainly did not attend the "official" lunch table of the pioneer unit hosting him, and at most might have had a small snack in its mess hall (to shorten the waiting time for his ordered car). This becomes clear by examining the timeline: the air combat began about 11:45 a.m., the engine hit took place shortly before 12:00, and the emergency landing probably just a few minutes after 12:00. Because of the unavoidable waiting time at the landing site, Richthofen's arrival at the pioneer unit's quarters probably took place around 1:00, whereupon

Above: A B.E.2e shot down by Jagdstaffel 11 from the photo collection of Eddy Lübbert. It could be the B.E.2e A2765 of 16 Sqn, shot down by Manfred von Richthofen on 6 March 1917.[39]

Richthofen certainly would have quickly telephoned his squadron in Douai to tell them what had occurred. This does not occur in the story but was obligatory in aviation circles in the case of unscheduled landings, not only to inform and reassure one's comrades, but to tell them where one should be picked up. Afterwards, Richthofen's nap would have taken place from about 1:30 to 2:30. His visit to the Officers' Mess would then have occurred at about 2:45 p.m., when lunch was probably already long over. From that time to when the squadron car picked him up (around 3:00 to 3:30), he would only have had time for a small snack (perhaps leftovers from lunch) and certainly not for an opulent meal. After his return to Douai at about 4 p.m. or shortly thereafter, he would had to have prepared for his next front flight for which he took off at about 4:45. It was on this flight that he achieved his 24th aerial victory at 5:00 which, according to the above conversation with the pioneer officer, he had supposedly already achieved earlier.

We might therefore suspect that Richthofen's ghostwriter, Erich Salzmann, dramatized and embellished the story for *Der rote Kampfflieger*.

The British airplane that Richthofen had forced to land probably was 43 Squadron's Sopwith Strutter A961, manned by 2/Lt. C.P. Thornton and 2/Lt. H.D. Blackburn, who reported upon their return:

"*While on line patrol this morning, we were attacked by hostile aircraft. One dived on the machine in front of me and I dived after it and fired about 20 rounds into it, at rather under 20 yards range. He then fell underneath me, and I saw it falling out of control. Sgd. C.P. Thornton.*

I saw this machine descending in flames, but did not see it strike the ground, owing to be engaged with another hostile aircraft. Sgd. H .D. Blackburn"[27]

Eddy Lübbert, who took off with Schwarm II, had this to say about the eventful day:

"*The morning of 6 March promised to be the best for fighter aviation. Clear weather, a weak southerly wind, and individual cumulus clouds at altitudes of 3,500 to 4,000 meters – just good enough to hide behind and sneak up on the enemy. We had hardly finished breakfast at 9:00 when the first telephone reports of English B.E. aircraft came in, mostly individual ones close to the front at 1,500 to 2,500 meters' altitude that were directing artillery and probably the easiest prey. So, our first Schwarm took off and my machine, which was in the second Schwarm, was rolled to the takeoff spot. We got ready half an hour later. It was already so warm that just a leather jacket and a pair of fur gloves offered enough protection.*

You stow yourself comfortably in your seat, cock both machine guns, and wait for the Schwarm leader to buzz off. ***Lt. von Richthofen's bright red machine*** *buzzes across the airfield, carefully and calmly rises up and climbs slowly in a right bank.*

Now we can take off too.

Schaefer*, our young Kanone* ['big gun' or 'ace'] *who had shot down eight English aircraft in four days, takes off in a more unruly manner in* ***his yellow bird with a black tail. Lt. Wolff's 'plum violet' and Lt. Allmenröder's field-gray crate*** *and my* ***half-blue, half-yellow*** *machine rush across the ground together and follow after Richthofen. Flying like a swarm of mosquitoes, but constantly behind Richthofen and close above him, we climb to our hunting area which extends from the La Bassée Canal to the midpoint between Lens and Arras in the south.*

We climb up to 4,000 meters, looking everywhere for the small, explosive clouds from our anti-aircraft guns that betray invading opponents to us. But there's nothing to be seen. Only two German two-seaters are flying around 1,000 meters below us at Arras. After 30 minutes of searching, we go back home. The mechanics go right to work on getting the machines ready for takeoff again.

We sit ourselves down in the spring sunshine in front of the

March 6, 1917

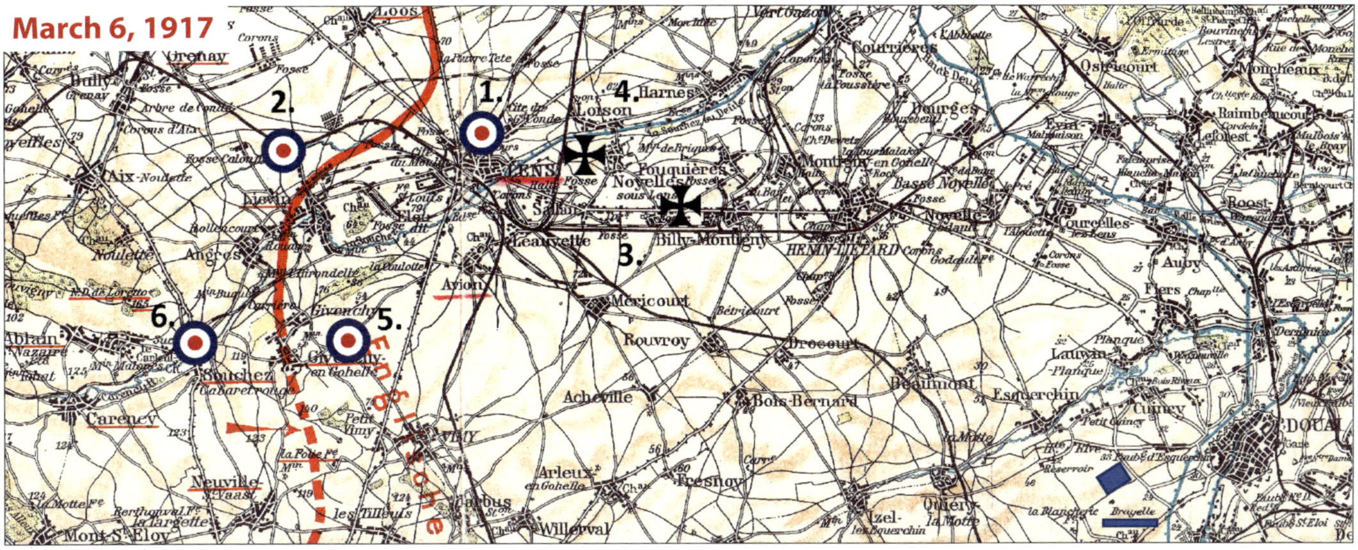

Above & Below: Map and details of the aerial victories of Jagdstaffel 11, March 6, 1917.

Nb.		Hour	Pilot	Location	Aircraft	Line
1.		11:45	Lt. Emil Schäfer	Near of Lens	Sopwith Strutter A978, 43. Sqn.	German
2.		11:45	Lt. Emil Schäfer	w. Lens	Sopwith Strutter A1097, 43. Sqn.	Allied
3.		11:50	Lt. Manfred von Richthofen	Road Lens – Hénin-Liètard	Albatros D III forced to land	German
4.		12:00	Lt. Hans-Georg Lübbert	Lens	Albatros D III, wounded,	German
5.		12:30	Lt. Kurt Wolff	Givenchy	B.E.2d 5856, 16. Sqn.	German
6.		17:00	Lt. Manfred von Richthofen	Bethune	B.E.2e A2765, 16. Sqn.	Allied

little ready house and wait. Very soon, reports come in from the front: 'Eight F.E.s are crossing the front at Lens' and 'Numerous single fliers are beginning to appear.' Right after that we see a formation of six to seven English machines passing by a few kilometers from the airfield. As soon as our apparatuses are filled up, we set off again. We are at the front in a few minutes and begin to climb over a formation of 17 Englishmen trying to cross over the lines between Arras and Lens. The Englishmen fly well, close together: Sopwiths, Vickers, and some single-seat fighters hanging above them.

Richthofen paused for a few minutes, and when the formation got close to the lines, dove onto the last Sopwith who tried to escape through heavy turning, got close behind him, and must have had him well in his sights. I pressed after the second, whose observer was firing at Richthofen. He flew straight ahead, and I was able to drive a lot of shots into him. Suddenly, he shoots off a red flare and makes a sharp, righthand turn. I shoot over him because I'm going too fast; and a moment later a burning Sopwith passes under me to the right. I'm glad that he's finished and attack another Sopwith who at first pushes down, but when I get up close behind him, rears his crate up so that I almost ram into him.

Then three single-seaters plummet down from above and a wild dogfight begins. I take on one of them, get closer and closer to him, and notice how two others – a lattice-tail and small machine with a fuselage – are trying to set themselves behind me. But since I'm catching up pretty well to the one, I don't pay attention to those behind me.

Suddenly, I'm in the field of fire of the one who is behind me to the left, and smack! I get shot in the left breast and right arm.

Crash-dive toward the east!

The Englishman stays behind, high up. I am surprised that the chest shot doesn't weaken me. I notice that I'm bleeding profusely, but I can both breathe and move my arms freely. When I'm at 500 meters' altitude, my fuel pressure goes, and I'm pleased that I can pump [to repressurize it] *so well with my injured right arm.*

Then I looked for a nice place at Hénin-Liétard and landed safely. I switched off the engine, climbed out of the crate, and went up to the soldiers who were streaming in from all sides; but then I went quite limp and sat down by a sign that stood in the middle of the field.

At first, the soldiers – especially an officer who had run up – were disappointed that I was not an Englishman, but soon they touchingly took care of me, undressed me, and put dressing packets on my still heavily-bleeding wounds. I was in the field hospital at Hénin-Liétard a quarter-hour later, where I was bandaged up and put to bed. I got something to eat right away and really

Above: Kurt Wolff, left, and Carl Allmenröder, right, visit the wounded Eddy Lübbert in the military hospital on March 7, 1917.

needed that since I was dying of hunger.

The shot entered at left through my old shoulder wound, broke my collarbone, exited from my left breast without hitting anything else important, and then scraped some skin off my right forearm. So once again, tremendous luck!"(28)

The F.E.8 pilot who attacked Lübbert is believed to have been 40 Squadron's Capt. W.R. Gregory, who gave the following report:

"Type and number of aeroplane: F.E.8 6394.

In company with my patrol met hostile patrol over Givenchy. I fired a short burst at one hostile aircraft at 30 yards range, apparently with effect. I then fired about 25 rounds at under 15 yards range, and bullets were seen to enter fuselage of hostile aircraft which then dived vertically, but I could not follow him owing to the presence of other hostile aircraft."(29)

Maj. A.S.W. Dore and Sgt. W. Cubberly from 43 Squadron witnessed the fight and Cubberly later reported:

"I was in machine on the left of leader when we were attacked by about 8 hostile aircraft. Major Dore fired a few shots at 2 hostile aircraft, which crossed his bow at close range. No more shots could be fired from front gun owing to temporary stoppage. One of these two machines did an Immelmann-turn on our tail. I fired a few shots from back gun and saw them go into the machine near the pilot's seat. Major Dore then stalled to throw him off his tail and saw him go down vertically with an F.E.8 firing on him."(30)

Der rote Kampfflieger gave the following account:

"I learned from [Schaefer] *that Lübbert had once again lived up to his nickname. That is, we called him 'Bullet-Catcher' because his machine suffered badly after each air combat. It once bore 64 hits without he himself being wounded. This time he had a grazing shot to the chest and was already in the hospital. I flew his machine straight back to the airfield. Unfortunately, this outstanding officer – who had what it takes to be a Boelcke – died a hero's death for the Fatherland a few weeks later."*(31)

After the tough dogfight over Lens, Lt. Kurt Wolff assembled the Staffel's remaining Albatros fighters to continue the patrol and headed in an arcing, southerly course along the front. Less than 30 minutes later, probably during their return flight, Wolff spotted a British airplane behind German lines near Lens. He later noted in his diary:

"Intense activity again on the 6th of March. Fourteen Englishmen and seven of our Staffel had a big [fight] *over Lens. Schäfer shot two of them down, one in flames, the other minus its rudder. Lt. Lübbert was wounded in the fight; the shot entered at the shoulder blade and came out under the nipple without injuring any bone. Richthofen's machine was so shot up that he had to make an emergency landing. I had taken off after them because my engine did not start and was only able to observe the big air battle from 500 meters below; but I caught another Englishman a little later. I encountered a B.E. biplane around Lens at about 2,500 meters, took him on, and after just a few shots he went spiraling down. The observer had managed to get off only a few shots, then behaved quite calmly, and I saw him leaning far out of the fuselage. I went down to 650 meters and followed the Englishman to about 100 meters above the ground, where I lost him under my wing. I had shot 190 rounds in the battle and was able to see my bursts of fire precisely with the help of tracer ammunition. The Englishman had crashed at the northeast edge of Givenchy. Both people had clean headshots. The observer had not gotten off 20 shots. The apparatus had the number 5865 and the two Lewis machine guns 1122 and 10961. One 'Lord' had a card with the name 'Semple' on it. Since the apparatus was too far forward, not much of it could be recovered, particularly since it was also shot up by English artillery."*(32)

The aircraft attacked by Wolff was No. 16 Squadron's B.E.2d 5856, manned by 2/Lt. G.M. Underwood and 2/Lt. A.E. Watts, who had been tasked with photographing German positions between Farbus and Bailleul-Sir-Berthoult. Wolff attacked them and drove them toward Givenchy-en-Gohelle where they crashed into the German lines.(33) The loss report from 16 Squadron summarized:

"... it fell at first out of control. When near the ground the pilot apparently gained control, and after gliding safely for some distance landed in Givenchy where it apparently crashed."(34)

The crew of B.E.2e A2795, Lt. A.V. Surlton and 2/Lt. F.H. Baguley, had participated in the fight and witnessed B.E.2d 5856's fall:

A Halberstadt was seen to dive onto a B.E.2c and followed it down. We attacked the hostile aircraft from the side and behind, one drum being fired from the pilot's mounting. The hostile aircraft seemed to go down out of control and must have come to earth at about S.5 . The pilot had to change drums, so lost sight of it; the observer believed he saw it crash in S.5.b. The B.E.2c flew very low towards Givenchy, where it was lost from sight."[35]

The "Halberstadt" reported as "out of control" probably was Wolff's Albatros that had followed 5856 down to about 100 meters from the ground. Wolff landed safely at La Brayelle airfield at around 12:45 p.m. He exited his machine, put his thumb up, peeled off his thick scarf, and called out to his mechanics: *"One shot down, my first!"* The mechanics surrounded him: *"Congratulations Herr Leutnant, congratulations!"* After patting them on their shoulders, he rushed over to the other pilots, who also wanted to congratulate him on his first success. As Hermann Lohmeyer remarked, mechanics were just as proud of a success in air combat as their pilots.[36]

In a letter written to his aunt on 7 March 1917 Kurt Wolff described the combat in more detail:

"It was a very knightly duel, only the two of us alone, he in a double-seater, I in a single-seater. The duel was fought over the trenches. The Lord shot badly and didn't hit at all, then I freed the two guys from their fear and shot them to death. The apparatus is still on our side, close to the trenches. Otherwise, we have always had big fights, squadrons against squadrons. It is too beautiful a sight when suddenly an apparatus rushes from this swarm of bees to the ground burning, or even bursts into pieces up there, or crashes down, tumbling several times.

Yes, flying is pretty and romantic, as long as you fail to remind yourself that there are always people sitting in the apparatus. Yesterday, the Lords wounded a very dashing guy of our Staffel [Lt. Eddy Lübbert], *chest shot, but not severe."*[37]

The events of 6 March did not end upon the return of this Schwarm. After Manfred von Richthofen had been driven back to Douai, he and Lt. Carl Allmenröder took off together around 4:45 p.m. and headed toward Lens and Souchez in response to a report from the front there that well-aimed artillery fire was being guided by enemy aircraft. Once at the front, Richthofen and Allmenröder indeed spotted two enemy artillery aircraft above Souchez, which they dove upon and attacked:

"6 March 1917
B.E. Two-seater
5 p.m.
Souchez

Together with Lt. Allmenröder I attacked 2 enemy artillery flyers at a low altitude over the other side [of the lines]. *The wings of the airplane I attacked came off; it dashed down and smashed on the ground."*[38]

Richthofen's target, 16 Squadron's B.E.2e A2785 and its crew of 2/Lt. G. M. Gosset-Bibby and Lt. G. J. O. Brichta, crashed to earth within German territory at Souchez. Both occupants had perished.

Then the two German pilots went after the same unit's BE.2d 5834, flown by Flt.Sgt. F.H. Quicke and Capt. L.E. Claremont, who gave the following report:

*"Remarks on hostile machine: 2 Halberstadts. One **white machine, one bright red machine.***

Our machine was travelling in the direction of the target over 'no man's land' when I noticed one of our own machines falling in a spinning nose dive. My pilot immediately turned our machine to-wards west. At the same moment I heard the rattle of machine-gun fire. The pilot saw a white machine alone in front of us; a few moments later a red machine which had apparently been directly above the center section & thus out of my field of vision, turned sharply & dived at our tail. On seeing the red machine, I immediately opened fire, firing about 70 rounds & causing it to turn away sharply & fly back over the enemy lines. Left aileron top was put entirely out of action. The outer strut was broken. The pilot side-slipped the machine and landed it safely in a field near Camblain l'Abé."[40]

This report tells us that Richthofen was flying a "red" Albatros and Allmenröder a "white" one, which referred to their fuselage colors.

That evening, Kurt Wolff appeared in the workshop where his three mechanics were still working on his damaged machine. He was carrying a bottle of champagne and four glasses and insisted on toasting his first victory with his mechanics:

"We sat down on some crates and our Leutnant toasted us and thanked us for our work. We were suitably proud", remembered mechanic Lohmeyer.[41]

The next day, as Wolff related in his diary, he and Schaefer visited Eddy Lübbert in the hospital, after which they drove to the site of Schaefer's downed Sopwith Strutter near Liévin. Wolff recalled that the machine was completely incinerated and that only the engine was recognizable.

Eddy Lübbert wrote home the day of their visit in a letter that, despite his wound, was optimistic and reflected his bright, sunny nature:

"At noon, during an air fight in which seven of us fought against sixteen Englishmen, I received two light grazing shots, one on the left chest and one on the right forearm. I was so fortunate, that I will stay in the hospital in Douai for only three or four days and then go back to the Jagdstaffel, where I can fly again in a fortnight. While I was approaching an Englishman, three tried to wrap me up from behind. When I was wounded, I easily ducked out and made a safe landing with our magnificent single-seater."[42]

Eddy Lübbert, his namesake's nephew, added this during an interview with the author:

"My father told me that 'Eddy' had always been a daredevil. As a cavalryman, he rode the wildest cross-country hunts. He was always in a good mood and had a very carefree, funny nature. To him, everything was a great adventure. He probably never thought he'd be shot down and lose his life. That would not have fitted his carefree and cheerful nature."[43]

7th March 1917 – 9th March 1917

The 7th and 8th of March brought a brief respite for the pilots of Jasta 11. Thick fog and intermittent, heavy storms brought flight activities to a halt on both sides. Their ground crews were busy, however, since they could now take the time to thoroughly inspect and repair their aircraft as needed.

As Xaver Leinemüller, a former mechanic with Jasta 10, explained to the author, maintenance of a wooden-fuselaged airplane after a dogfight was considerably more complex and difficult than for one with a fabric-covered, steel-tubed airframe. For aircraft with a wooden fuselage, it was difficult to determine whether or not hits therein had affected its structural integrity. Each hit had to be carefully examined and skillfully repaired. The pilot's seat had to be removed to allow the mechanics to inspect the interior of the fuselage; and sometimes parts had to be replaced at great expense. On the other hand, damage caused by a bullet to a fabric-covered, steel-tubed airframe was quickly and easily determined. If the airframe and the control wires were undamaged, fabric holes were simply patched over, and the aircraft deemed ready for use.(1)

While the mechanics labored away on their aircraft during what airmen called Fliegerwetter, or "flier's weather" – as opposed to Flugwetter, "flying weather", where conditions were good – the pilots could finally sleep in and relax in their quarters and in their Officers' Mess. Some drove to nearby Douai where restaurants, clubs and bars primarily frequented by aviators were at their disposal. On occasion, concerts and even theatrical performances were put on. Manfred von Richthofen gave his pilots a free hand on such days. He himself, however, often used such opportunities to speak with his Kommandeur der Flieger Hptm. Sorg, or write reports for his superiors. If there was no "paper-pushing" to contend with, however, he was known to be happy to accompany his fellow pilots for a good "swallow" or two in Douai. Karl-August von Schoenebeck once recalled with a smile:

"We aviators certainly weren't ascetics or abstainers, and Richthofen could really jump right in and celebrate at dinner. Not one of us ever saw him drunk, however."(2)

The week of 2–8 March was a successful one for German aviation units. Kofl 6 reported 73 aerial fights in their Army sector with eight hostile aircraft shot down and only one aircraft of Fliegerabteilung (A) 292 crashed and two pilots wounded.(3)

The morning of 9 March brought stormy weather with occasional blowing snow, which initially led to moderate enemy air activity.(4) Jasta 11's pilots made themselves comfortable in the Officers' Mess. No one could have guessed that their unit's most successful day to date was in the offing. The weather eventually cleared and aircraft from the Jasta took off for the front. Fortunately, we have several surviving German and British reports that together provide a good picture of the day's events.

Emil Schaefer's "Schwarm" took off around 10:00 in stormy weather and driving snow following a request from the front and headed northwest from their airfield. To their right between wispy clouds, they could see the Canal de la Haute Deûle and followed it. In the vicinity of Pont-à-Vendin, they detected a flight of F.E.8s underneath them that were being heavily shelled by flak. Covered by clouds, they swooped down upon the group. The English pilots, nine in all from No. 40 Squadron RFC, were taken by surprise and scattered in all directions. While a few immediately dropped off and sought the safety of their own lines, the remaining six or seven became engaged in a heavy "Gekurbel" ("wind-up" or "melée").

Kurt Wolff quickly attacked one and forced him to land around 10:20 a.m. about two kilometers south of Pont-à-Vendin near Annay, north of Lens:

"On the 9th, Schäfer, Al[1]menröder, Baierlein, and I flew and encountered a Geschwader of seven F.E. single-seaters east of Loosbogen. Baierlein had a stoppage in both machine guns and could not participate. We three dove on the seven, scrapped around with the fellows and shot down four, three on our side, one on the other which was then shot up by the artillery. I had shot a Lord's fuel tank to pieces, so that he had to go down smoking heavily. He landed near Annay and turned over doing so. The machine had number 6456, the engine – a Gnôme – No. 30456, and the machine gun, a Lewis gun, 18555. Lt. Shepard was the pilot. Schäfer landed next to one of the downed ones and brought the man to our Kasino, where he gave us very interesting information about our recent air combats. Then we took him to the Officers' Prison, where the two other Lords were already. I went to the three downed machines in the evening and ended the day with a meal in a dugout at Lens."(5)

After his return from German captivity, 2/Lt. T.A. Shepard, the pilot of F.E.8 6456, submitted this report of his dogfight with Wolff:

"On offensive patrol with 7 other machines. About Vimy Ridge our formation was attacked by enemy formation (O.C. – Richthofen). I saw five of our machines driven down & myself eventually forced to land owing to my engine being hit and controls shot away. Unwounded."(6)

At almost the same time as Wolff's victory, Emil Schaefer forced another No.40 Squadron F.E.8 – 6397, piloted by 2/Lt. W. B. Hills -- to land at 10:20 a.m by Cité Fachoda near Meurchin. And just two minutes later, Schaefer's gunfire forced 2/Lt. G.F. Haseler in another 40 Squadron "Vickers," F.E.8 A4874, to make an emergency landing near Pont-à Vendin.

Hills gave this account following his return to England after the armistice in December 1918:

"Attacked whilst in formation (7 F.E.8's) by about 30 [sic] *Albatros scouts. Dived to assist machine which was being attacked below.*

Above: German soldiers pose with the F.E.8 6456 of 2/Lt. Shepard, 40th Sqn, RFC, forced to land by Lt. Kurt Wolff. (G. VanWyngaden)

Above: The engine of F.E.8. 6456 damaged by the emergency landing. (L. Bronnenkant)

Above: Crash landed F.E.8 6456 of 2/Lt. Shepard from the front. (L. Bronnenkant)

Above: 2/Lt. Shepard's F.E.8 6456 sometime later after snow had covered the ground. (G. VanWyngarden)

Lt. Kurt Wolff forces 2/Lt. Shepard, 40th Sqn., to land on German territory on March 9, 1917 (Painting by Tom la Padula)

Right: F.E.8 6397 of 2/Lt. W B Hills, brought down by Emil Schäfer, after its emergency landing near Meurchin. (L. Bronnenkant)

Left & Below: As usual, emergency-landed F.E.8 6397 of 2/Lt. W B Hills is surrounded and inspected by German soldiers. (L. Bronnenkant und R. Absmeier)

Above & Below: The F.E.8 A 4874 of 2/Lt. G F Haseler after his emergency landing at Pont-à-Vendin. (L. Bronnenkant)

Engine hit whilst charging drum and disabled. Not wounded."[7]

Haseler, Schaefer's second victim, reported the following after his return to England:

"Whilst on offensive patrol near Loos, 6 enemy machines attacked, piercing petrol tank causing machine to catch fire after diving to 4,000 feet fire went out but engine had also been pierced and was forced to land. – Number of attacking machines increased from 6 to 14. Was not wounded."[8]

40 Squadron's 2/Lt. W. Morrice, who also took part in the fight, reported:

"Hostile aircraft dived on F.E.8 but was driven off by two other F.E.8's. 8 hostile aircraft then appeared from south of Lens 5,000 to 6,000 feet above F.E.8's. 3 hostile aircraft dived on F.E.8. from behind and one from the front. F.E.8 fired a burst at about 30 yards range at hostile aircraft in front, which dived steeply. F.E.8 gun jammed and by spinning and looping F.E.8 shook hostile aircraft off and managed to get across the lines. As jam could not be cleared and lateral control was shot away F.E.8 returned to aerodrome."[9]

At some point during the melée, the machine gun interrupter gear on Schaefer's Albatros malfunctioned and, having nearly shot off one of his propeller blades, he had to make an emergency landing of his own near where his first victim, Hills, had come down. Schaefer's gun malfunction might have occurred during a fight with 2/Lt. H.C. Todd in Squadron's F.E.8 6425. Todd, who afterwards had to land his damaged airplane at 2 Squadron's airfield, claimed an Albatros Scout "out of control" east of Lens, and Schaefer's machine was the only one that would have fit that description.

Finally, Carl Allmenröder pursued another 40 Squadron F.E.8 over the British lines near Hulluch and watched as it

Above: Another photo of F.E.8 4874 of 2/Lt. G F Haseler from the photo album of Carl Allmenröder.

descended for an emergency landing. His probable victim, 2.Lt. R.E. Neve in 6399, who had already crossed swords with the pilots of Jasta 11 on 4 March 1917, disclosed in his combat report:

"F.E.8 in company with 6 or 7 other F.E.8s saw two hostile aircraft east of Annoeulin, out of range. 8 hostile aircraft who appeared to be working with these two hostile aircraft dived from above and attacked formation.

I was engaged with several hostile aircraft for about 5 minutes when gun jammed. During first engagement several shots were seen to enter hostile aircraft's fuselage. One hostile aircraft noticed that my gun was out of action, and fired a burst from right hand rear, and throttled down to F.E.8's speed.

I was then wounded and had great difficulty in breathing – a bullet also grazed my head which was apparently sheared off by the helmet.

Petrol tank was then pierced, and engine stopped. F.E.8 feigned to land in hostile territory. At 500 feet I tried gravity tank, engine restarted, and I managed to cross the lines at 80 m.p.h. Then becoming faint, decided to land as soon as possible and seeing three soldiers endeavored to land near them.

Owing to ailerons being shot away F.E. 8 made a bad landing, turned over and caught fire. I scrambled out with coat and gloves ablaze. I rolled over on the ground, a soldier took off his coat and smothered fire, afterwards carrying me to a place of safety. The machine was then shelled, Time: about 9:25 a.m."[10]

Wolff's and Schaefer's three victims all survived their ordeals and were taken prisoner by German forces. As Schaefer remarked in one of his letters:

"...you prefer to have such a poor creature land near us unharmed, rather than going down in pieces or as a wretched

Above: 2/Lt. Geoffrey Frank Haseler, 40th Sqn. RFC survived the aerial combat unharmed and returned to England from captivity on 19 December 1918. (C. Huston)

Above: The Albatros D III D.2062/16 of Lt. Emil Schäfer after the emergency landing next to the F.E.8 6397. (L. Bronnenkant)

Above: The two fighter planes after the emergency landing. In the foreground the F.E.8 of 2/Lt. Haseler, in the background the damaged Albatros D III of Lt. Schäfer. (G. VanWyngarden)

Left: The serial number of the F.E.8 A 4874 cut out by Emil Schäfer as a souvenir. (L. Bronnenkant)

Above: Damage to the propeller caused by the failure of the machine gun controls forced Emil Schäfer to make an emergency landing. (L. Bronnenkant)

lump of coal and ashes."[11]

It should be noted that one of the captured British pilots, 2/Lt. Hills, was Schaefer's guest at Jasta 11 for an afternoon and evening before he was brought to the prison camp, as Kurt Wolff wrote in a letter on the 12 March 1917:

"We had an English officer at our casino, and he told us very interesting things from our joint air battles. On the same day, Richthofen shot down his 25th, so that 5 victories were credited to the Staffel. On all fronts, only six were brought down, including our five. We can be satisfied with the result. This month we have already hunted down 16 Englishmen on 4 flight days. We have wonderful warm spring weather here; the larks are singing, and we are already sitting with the window open…"[12]

The pilots of Jasta 11 were especially excited to have Hills as their guest since he hailed from South Africa.

Captured British pilots were usually interrogated by an intelligence officer and the following are sections from Shepard's, Haseler's, and Hills' interrogation reports:

"Personnel:

Lt. T. Shepard, architect by profession, son of an active English major, 5 brothers in the field of which 1 killed, two wounded; joined the infantry when the war began…entered R.F.C. in August 1916…with 40 Squadron for three weeks.

Tactical Use of 40 Squadron:

During the three weeks that the prisoner was with the squadron, only offensive and defensive patrol flights and escorting photography airplanes…

Orders on 9 March:

To fly an offensive patrol over the German lines in a flight group of nine airplanes. They were soon attacked by German Albatros single-seaters who had their way with them due to their superiority. Apparently, the prisoner's entire flight was annihilated. The F.E. single-seater is hopelessly inferior to the new Albatros single-seater that has suddenly appeared here."

"Personnel:

2/Lt. G.J. Haseler, who joined the infantry when the war began, has been with 40 Squadron for a year now. The prisoner, who gives the impression of a very educated and thoughtful man, refused [to answer] *other personal and military questions. His behavior while looking at photographs of Aire airfield that were presented to him seemed to confirm 40 Squadron's presence there.*

Aviation Associations:

The German Air Arm's superiority since the appearance of the Albatros fighter is clearly emphasized. The English single-seaters were 9 against the Germans' 4 in his last combat, but the fight became hopeless from the onset."

"Personnel:

2/Lt. W.B. Hills worked on a sugar plantation in South Africa

Above: German soldiers are proudly photographed in front of Emil Schäfer's Albatros D III D.2062/16. (G. VanWyngarden)

Above: Another view of Emil Schäfer's force landed Albatros D III D.2062/16.

(Zululand) before the war, took part in the Southwest Africa Campaign as a volunteer up to the end, was then in the 9th Hampshire Regiment; has been in the field as a pilot for 4 months.

Aviator Organizations:

Aire airfild of 40 Squadron, part of 10 Wing, confirmed. The affiliation of 43 Squadron (Sopwith two-seater) to the same wing confirmed. 43 Squadron's task is photography and reconnaissance for which 40 Squadron will provide protection.

Orders on 9 March:

Two months ago, the squadron could have flown over German lines for hours without being attacked or disrupted. Since the appearance of the Albatros fighters, things have changed so much that on 6 March the pilots of 40 Squadron protested against the order to conduct an offensive patrol over German lines, feeling

Above: 2/Lt. Rupert Ernest Neve, 40th Sqn. RFC, pursued by Lt. Carl Allmenröder, succeeded in making an emergency landing on Allied soil with a badly damaged aircraft. (C. Huston)

Above: Lt. Schäfer with South African 2/Lt. Hills after their 9 March 1917 air combat, on the way to Jasta 11 where Hills was welcomed as a guest.

that their machines were inferior. But since the order had to be followed, things turned out as expected. The F.E. single-seater has completely inadequate climbing ability when compared to its opponent. One is lost once one is below the opponent."(13)

This report underscores how the F.E.8 was completely inferior to the Albatros D III at this point, even in the eyes of the British pilots who had to fly them.

But the day's events were not yet over. After the return of "Schwarm Schaefer", "Schwarm Richthofen" took off for the front. The flight spotted a formation of Airco D.H. 2s from No. 29 Squadron RFC at about 11:50 a.m. over Bailleul-Sir-Berthoult, where they had crossed the German lines to protect some of their own aircraft:

"9 March 1917
Vickers One-seater
11:55 p.m.

Between Roclincourt and Bailleul, this side of the lines. 500 meters behind trenches.

With three of my planes, I attacked several enemy planes. The machine I had singled out soon caught fire and plunged downwards after I had fired one hundred shots.

The plane is lying on our side of the lines but cannot be salvaged because it is nearly completely burned and too near the front."(14)

With that, 29 Squadron's D.H.2 A2571 and its pilot, 2Lt. A.J. Pearson, who was killed, became Richthofen's 25th confirmed victory.

Lt. Pascoe, another member of 29 Squadron, witnessed the dogfight:

"Lt. Pascoe saw 3 hostile aircraft diving on a B.E. With Lt. Pearson he dived in and attacked – opened fire at 100 yards, 2 hostile aircraft dived down, one followed by Lt. Pearson, the other by Lt. Pascoe.

Lt. Pascoe saw the 3rd hostile aircraft above Lt. Pearson so left his hostile aircraft to assist Lt. Pearson. At this moment they all went into a snow cloud and Lt. Pascoe did not see any more of them although he dived down to 3,000 feet below the clouds.

Lt. Pearson had failed to return, and a machine was seen to fall in flames at this time in the locality. Remarks on hostile machine: 3 scouts. Very fast. ***Painted green and brown.****"*(15)

Georg Simon was involved in this fight and told Alex Imrie in an interview that he had been flying an Albatros D III with a green fuselage at the time, which appears to have been mentioned in the report above.

March 9, 1917

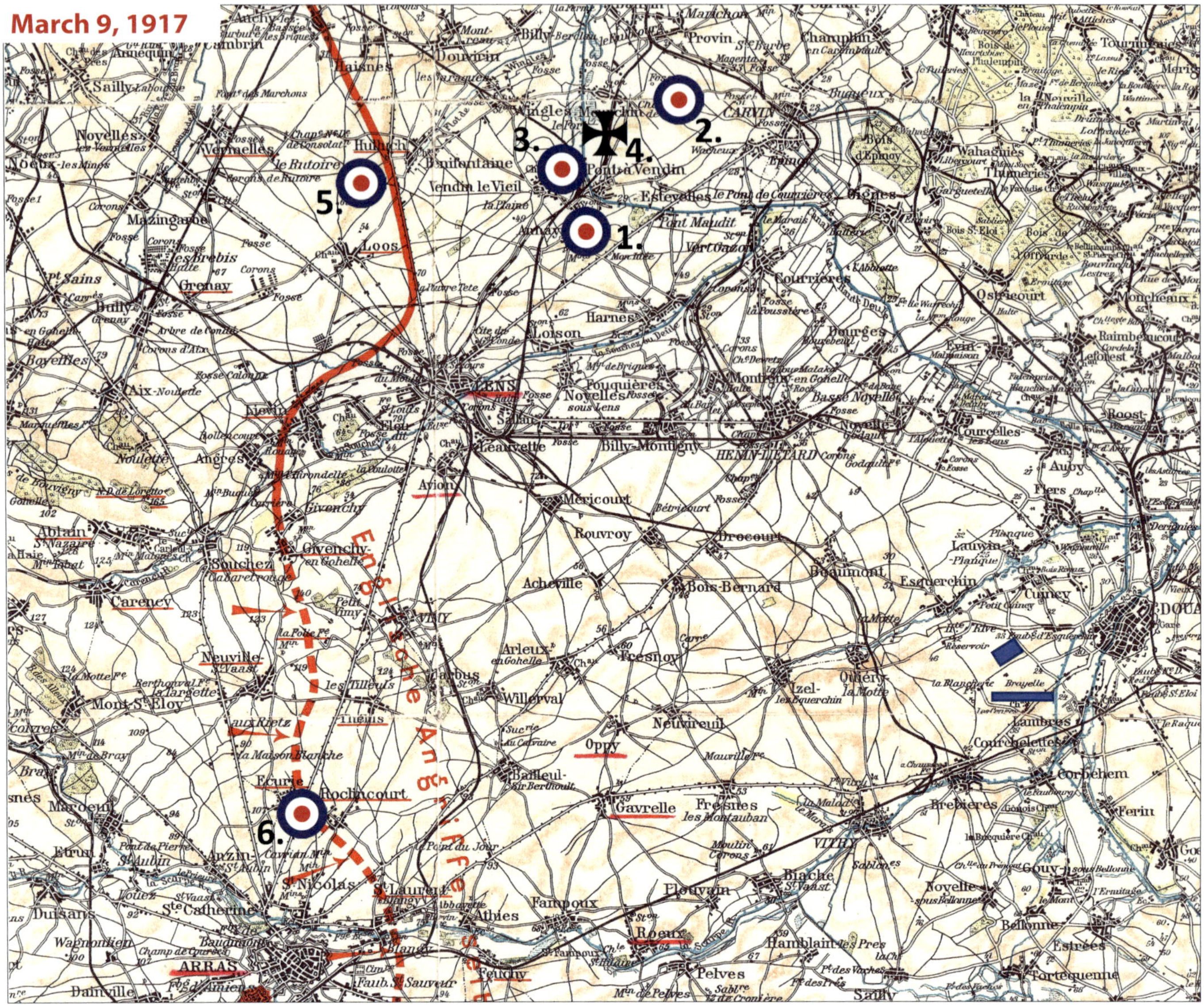

Above & Below: Map and details of the aerial victories of Jagdstaffel 11 on March 9, 1917.

Nb.		Hour	Pilot	Location	Aircraft	Line
1.		10:20	Lt. Kurt Wolff	Anny, north of Lens	F.E.8 6456, 40. Sqn.	German
2.		10:20	Lt. Emil Schäfer	Cite Fachoda near Meuchin	F.E.8, 6397, 40. Sqn.	German
3.		10:22	Lt. Emil Schäfer	Pont à Vendin	F.E.8 A4874, 40 Sqn.	German
4.		10:25	Lt. Emil Schäfer	Near Pont à Vendin	Albatros D III forced to land	German
5.		10:30	Lt. Carl Allmenröder	Hulluch	F.E.8 6399, 40. Sqn.	Allied
6.		11:55	Lt. Manfred von Richthofen	Roclincourt, s.w. Bailleul Sir Berthould	Airco D.H.2 A2571, 29. Sqn.	betw. lines

10th March 1917 – 17th March 1917

On 10 March, there was fog and deep cloud cover and there were no operational flights on either side of the front. The Commander-in-Chief of the German 6th Army, Generaloberst Ludwig von Falkenhausen, therefore took the opportunity to appear at La Brayelle airfield to express his appreciation of Jasta 11's outstanding achievements. At his request, Emil Schäfer, Carl Allmenröder, and Kurt Wolff conducted an air combat exercise over the field, which was observed with much appreciation by the General.

After the distinguished visitor departed, Manfred von Richthofen drove the Staffel car to the Armeeflugpark to personally collect his brother, Lothar. Understandably, the two brothers were very happy to see each other again.

Lothar von Richthofen was born on 27 September 1894 in Breslau. Unlike his older brother, Manfred, he received a civilian education. He did not do well, almost failed a grade, and had not graduated from high school before he took the Officer Candidate's Examination. He took to military life, however, after being accepted into Dragoner-Regiment von Bredow (1. Schlesisches) Nr. 4 and was attending Danzig's War School when the Great War broke out. He fought with his regiment in Belgium and France before being awarded the Iron Cross, 2nd Class, on 21 October 1914.(1)

The 4th Dragoons were transferred to Poland soon afterwards and then to Ukraine, where Lothar fell ill in March 1915 and was sent to the hospital. He did not recover until July and by late autumn had decided to follow his brother by applying to the Fliegertruppe. He was accepted despite his color blindness (red - green).(2) He was sent to FEA 11 in Breslau for observer training on 18 December 1915 and after completing his course was transferred to Kampfgeschwader IV on 12 January 1916 and assigned to its Kampfstaffel 23. On 6 June 1916, he earned the "Beobachterabzeichen" (Observers Badge).(3)

One of Lothar's best friends was Leutnant Joachim von Alt-Stütterheim, who served with him in the Dragoner-Regiment von Bredow (1. Schlesisches) Nr. 4. He wrote about Lothar:

"Lothar was a contemplative youth, slow in his intellectual development, reserved and almost shy, very modest and tactful. In order to speed up his development, his father sent him to the private school of a Herr Butter in Hirschberg for a while and then to a grammar school again."(4)

Manfred von Richthofen wrote this about his younger brother:

"My brother Lothar is a Leutnant in the 4th Dragoons, was at a War School before the war, became an officer right from the start and, like me, began the war as a cavalryman. Whatever heroic deeds he committed then are unknown to me since he never speaks about himself. I was only told the following story: It was the winter of 1914, his regiment was on the Warthe River, the Russians on the other side. Nobody knew whether to stay or go. The riverbanks were partly frozen and therefore difficult to ride over. There weren't any bridges, of course; the Russians had torn them down. My brother swam over, found out where the Russians were, and swam back – all of this in the severe Russian winter at many minus degrees. His clothes froze within minutes, but he claimed he was quite warm underneath them. He rode that way all day until he got to his quarters in the evening. He didn't even catch a cold doing this.

In the winter of 1915, he went into aviation at my urging and, like me, became an observer. Just one year later he was a pilot. Schooling as an observer certainly isn't bad, especially for a fighter pilot."(5)

Pour-le-Mérite pilot Carl Bolle, who had flown many missions with Lothar as his observer in Kampfstaffel 23, characterized him as follows:

"Whoever saw the noble, modest, quiet man with the kind eyes hardly suspected that before him was a marvel of devotion to duty, sacrifice, courage, and tenacity, and that the man was more than just the brother of our great master."(6)

On 5 July 1916, Lothar wrote to his mother from Kagohl 4's new base:

"Mercy-le-Bas, near Landres.

As you can see from the newspaper, there's a lot going on here again in the west. We've moved to the above airfield, somewhat closer to Verdun – where Manfred was previously located. Now he's in the east with the Bugarmee [A.O.K. Bug at the Eastern Front].

I shot down a Frenchman in air combat on 1 July, but it wasn't credited to me because it hadn't been witnessed from below [on the ground]. *It was at Hill 304 at 5:30 a.m. He had just come back from dropping his bombs. A colossal barrage was going on below, so they naturally couldn't see anything. What a shame!"* (7)

Spurred on by his brother's successes, he signed up for pilot training and was sent to Armeeflugpark 1 on 10 November 1916 for that purpose, where he received the Iron Cross, 1st Class, on 11 December 1916.(8) On 10 February 1917, he informed his brother:

"I've just made my first solo flight. Please request me now for your Jagdstaffel so that I don't lose any time. I am getting going on the required flights. Greetings from your brother Lothar."

Lothar's wish was fulfilled one month later when he was ordered to Armeeflugpark 6 on 9 March and sent to Jagdstaffel 11 the next day.

As his mother related in a letter, it was extremely important to Lothar to be transferred to his brother's Staffel:

"Lothar was most attached to Manfred. Both brothers were very different and therefore complemented one another with the utmost harmony. Manfred was always a role model for him, his most loyal friend and comrade in arms."(9)

Wolf-Manfred von Richthofen, Lothar's son, shared what his grandparents – Robert and Margarete Graf von Keyserlingk and Kunigunde von Richthofen – told him about his father:

Above: On 11 March 1917, Generaloberst Ludwig von Falkenhausen paid a visit to Jagdstaffel 11 at their La Brayelle airfield. Here he is pictured in the entrance of the field office hut, together with Manfred von Richthofen, on this occasion.

"Your father was a very friendly and decent human being. Very helpful and kind, as much as possible. In conversation he was not that talkative, more of a listener. He was not shy, but a bit reserved, observing, but like his brother he enjoyed jokes and a good laugh, even though he was not the instigator. He was also very caring and thoughtful with his family; always brought some presents and gifts to family members on his way home on recuperation or leave. Manfred was always the big brother that he looked up to. He wanted to be like him, he was very fond of him."[(10)]

Former Staffel mates described Lothar as a carefree cavalry Leutnant before the war. He thought little about the future and liked to rely on his luck. He was the same in air combat where he took greater risks than his brother. This quality made him a successful fighter pilot. But he was not a born leader like Manfred. Lothar preferred to stay in the background and act as his brother's supporter. If a pilot

Right: Studio portrait of Lothar von Richthofen in the uniform of the Dragoon Regiment von Bredow (1st Silesian) No. 4 with the Order Pour le Mérite added in later. photo taken by J.C. von Dühren, Berlin.

Above: Lothar von Richthofen, at right, and two other flyers, posing in front of L.V.G. B II B.716/15. Most likely, the photo was taken during his observer training. (L. Bronnenkant)

Members of Kampfstaffel 23 of Kampfgeschwader IV. Lothar von Richthofen is seen second from left in the back row.

Left: Pilots and observers of Kampfstaffel 23 gather for a photo on 4 July 1916, on the occasion of the farewell of Oblt. Engwer. Lothar von Richthofen is seen third from right. (M. Schmeelke)

Above: Members of Kampfstaffel 23 celebrate the Kaiser's birthday on 27 January 1917. Lothar von Richthofen is standing second from left. Four female visitors, nurses from a nearby field hospital, can be seen, and apparently there was no shortage of alcohol on this occasion. (L. Bronnenkant)

became depressed due to some criticism by Manfred, Lothar would encourage him. His sense of justice was particularly strong and he could get quite emotional over any injustice. If he had the feeling that someone was being treated unfairly or unjustly, he would immediately stand by his side.

Alex Imrie heard Karl Bodenschatz characterize Lothar in this manner:

"Lothar was always friendly and helpful. When it was someone's birthday, he was the first to congratulate them and sometimes even had a small gift for them. He was always honest and straightforward when asked his opinion and therefore undiplomatic at times. All the aspects of flying were incredibly easy for him; here he was superior to most, but he sometimes had difficulties with other things. He enjoyed the company of others, especially playing cards and billiards, but he also kept quiet times when he would retreat to his room or go for walks alone. He was very athletic. In the service, completely devoted to duty."(11)

Carl-August von Schoenebeck recalled:

"Lothar was always reserved in the presence of his big brother and you could see his admiration for his brother. Although he was rather quiet and restrained, he could definitely be the center of a social gathering due to his personality. He had a lot of charm, an elegant demeanor, and a good sense of humor. He was happy to join in on every joke but never instigated one. He was a daredevil in air combat. I saw this on several occasions and his attacks almost took my breath away. Like his brother, he would immediately come to your aid if you got into trouble."(12)

Hans-Georg von der Osten, during an interview with the author, added:

"Lothar was completely dependable in action and, like his brother, you could rely on him one hundred percent of the time to come to your aid if you got into a difficult situation. I was followed by two Englishmen [Sopwith Camel] *once who were really shooting me up. Lothar saw the mess I was in. He immediately made a tight turn and attacked the Englishmen head on; I thought he was going to ram one of them. The two Englishmen immediately let go of me and dived down. When we got back home, I saw that his machine was pretty shot up too. The motor was ruined. My crate looked pretty worse for wear too. He had, without regard for himself, flown directly into the bullets to get me off the hook.*

Incidentally, he was the better pilot of the two [brothers] *and an excellent marksman. He repeatedly took too many risks in air combat that led to his injuries."*(13)

When Wolf-Manfred, Lothar's son, met with many of the former pilots of the Richthofen Geschwader, he asked about his father and was told:

"At first, Lothar was not as disciplined as a pilot and fighter pilot. He did not analyze his situation and combats as much as Manfred did. In the beginning he made all sorts of mistakes: using up all his ammunition, sometimes not planning the right strategy before an attack, a lack of situational awareness, too much focus on his target, etc. He was very brave, almost reckless in his daring. But this changed with time and he became much more aware and conscious about the various aspects of air combat. He was also a very good comrade and was not the military disciplinarian that his brother was – he had a gentler approach."(14)

Manfred von Richthofen confirmed this in an interview with the journalist Erich von Salzmann in the spring of 1917:

"Now he [Lothar] *can do it quite well... I'm just worried that he's attacking too sharply; when he has the enemy in front of him, he sees red. He mustn't do that."*(15)

Hans Bethge, Staffelführer of Jasta 30 and a friend of Manfred von Richthofen, described Lothar as:

"The more ambitious of the two brothers. This trait made him into a successful fighter pilot; but as a Staffel leader, who is supposed to prudently lead his aircraft into combat, he did not compare with his brother."(16)

Another character trait, according to former Staffel mates, was his ambition. He always strove to do the best and be the best among the best. This is not only how he was in combat, but it was said that he was the same during free time and particularly while playing billiards where he strove to be the winner, but always in a fair and honest way. When he lost, he congratulated the winner in a sincere way, but the winner could be sure to be asked for new game."

Whereas his brother usually went to bed around 10:00 p.m., Lothar would stay up late to party and play card games or billiards. One or more glasses of wine or champagne would be emptied; but the next morning he would throw himself back into action with his usual bravado even if he had hardly slept.

Lothar was tall and this physical presence was underscored by his alpha-male personality and even the riding crop that he always carried. But he also had a lot of charm and empathy and according to his comrades, these combined traits made him quite attractive to the opposite sex with whom he was not averse to flirting and more.(17)

One critical statement about Lothar, however, comes from Hptm. Franz Hailer, Abteilungsführer of Fliegerabteilung (A) 292, as Otto Fuchs remembered:

"Franz Hailer, a Bavarian, had for some reason a personal aversion towards 'Prussians' in general and even more to members of the nobility. I remember that he once either had a disagreement or was witness to disagreement with [Lothar von] *Richthofen. He called him 'an arrogant Prussian Junker'."*(18)

However, Otto Fuchs added:

"I cannot confirm this negative opinion of Hailer's. Lothar von Richthofen was a quiet, modest person and had a very correct, polite, elegant, and comradely demeanor. During meetings with the pilots at Douai, I also saw that he had a tremendous sense of humor. However, I did see that he could also become arrogant if he clashed with someone."(19)

Considering all of these statements together, one forms the impression that Lothar's masculine demeanor and the arrogance he sometimes displayed during conflicts was the self-protective mechanism of a very sensitive person. He was no less intelligent than his brother although he did struggle

Above: Snapshot of Lothar showing his friendly and carefree nature. (J. Tinnerholm)

with some things. On the other hand, he was also superior to others as when it came to flying. Today he might be described as a person with Savant Syndrome. [20]

Despite his ambition and commitment, Lothar was still a rookie fighter pilot at Jagdstaffel 11; and even though he was the Staffel leader's brother, he received no special treatment and had to go through Manfred's tough training program. He completed it with much enthusiasm because he aspired to be not just good but wanted to be one of the very best.

Lothar later had the rear portion of his machine's fuselage and its control surfaces painted yellow. This was a tribute to his parent Dragoon regiment to which he felt closely connected and whose cap bands, collars, and cuffs were yellow.[21] But at first he had to use Manfred's old Albatros D.2006/16.

Lothar made an interesting statement about Jasta 11's tactical deployment after he had joined the unit:

"The Staffel was divided into two groups, which means that half the Staffel always flew together. As Jasta leader, my brother sometimes flew with one group, sometimes with the other. Schäfer led my group and Wolff, Allmenröder, and Lübbert flew with me. At the time, that was the most advantageous number to fly together."[22]

Dividing a Jagdstaffel into two half-Staffels appears to have occurred among several units at this time. For example, we know that Jasta 30 at Phalempin was also tactically split into two "Halbstaffeln" at this point in the war.

While Lothar was happy to finally be at the front as a fighter pilot, Eddy Lübbert was still struggling with his fate in the hospital as he wrote on 10 March:

"My small wound is improving quickly, but not fast enough

Above: Hptm. Franz Hailer, Abteilungsführer of Fliegerabteilung (A) 292 had a critical opinion about Lothar von Richthofen.

for me. I'm longingly waiting for the moment I can fly again. Just imagine, yesterday my comrades shot down the Englishman who wounded me and took him prisoner unharmed. He told us a lot of interesting things."[23]

Lübbert's statement about his conqueror from 40 Squadron having been shot down and taken prisoner on 9 March was incorrect because none of the 40 Squadron men brought down that day – i.e., Haseler, Hills, and Shepard – claimed a victory on the 6th when Lübbert was wounded, and only Hills and Shepard are known to have participated in the 6 March fight. When visiting Lübbert in the hospital, Kurt Wolff and Emil Schaefer likely informed him that three pilots from the unit they had fought with together on 6 March had been brought down on the 9th and Lübbert misinterpreted this in the way he presented in his letter. It is also possible that one or more of the captured British pilots had told a 'little white lie' to their captors.

The weather improved and was reported as changing cloud cover on 11 March. The morning began with only moderate flight activity that increased and at times became quite lively as the day progressed.

Emil Schaefer began his day with a flight over the lines at the Loosbogen with the aim of dropping off various

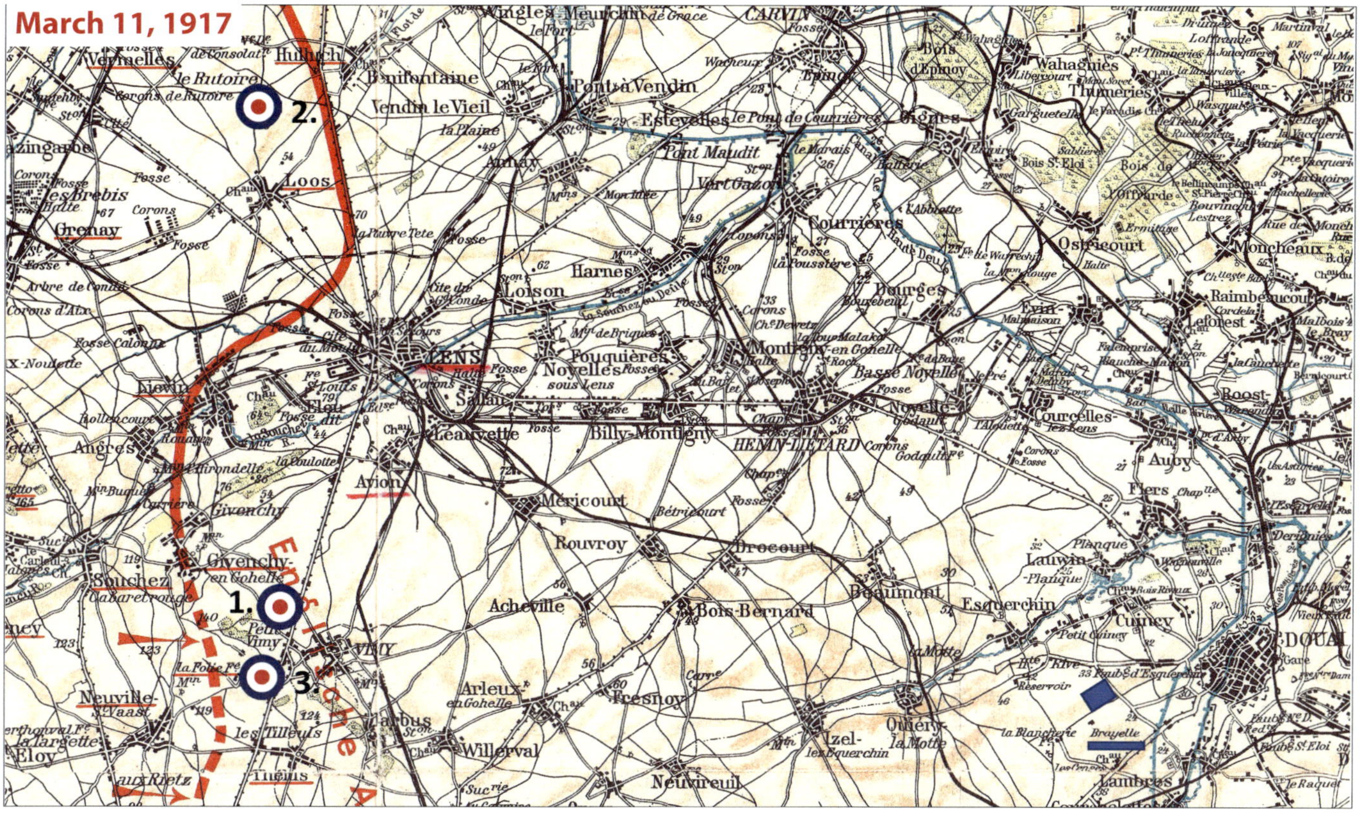

Above & Below: Map and details of the aerial victories of Jagdstaffel 11, March 11 1917

Nb.		Hour	Pilot	Location	Aircraft	Line
1.	◉	11:00	Lt. Konstantin Krefft	Between Givenchy and Vimy	?	Allied
2.	◉	10:20	Lt. Emil Schäfer	South-west of Hulluch	B.E.2c 4541, 2. Sqn.	Allied
3.	◉	10:22	Lt. Manfred von Richthofen	La Folie-Forest south-west of Vimy	B.E.2d 6232, 2. Sqn.	German

messages concerning the fallen and captured British pilots from the action two days earlier:

"*. . . We let them* [the prisoners] *write a few words to their parents and packed everything in a military mail packet. It was weighed down with a few pieces of coal to show them over there that there hadn't been a coal shortage for quite some time, bundled together, and tied up with an Iron Cross ribbon. One fine morning, I flew with this improvised drop sack deep into the Loosbogen from Lens and dropped it far behind the lines. I don't know if it was found, we didn't get an acknowledgement of receipt.*

A light morning fog hung in the air, which is probably why the flak guns' shooting was so God-awful. That was fine with me and I happily – uplifted by having performed a good deed – headed for home peacefully.

I flew over the defensive position the English had laid out around the Loosbogen. The great canal to Douai had come into view ahead of me, blinking brightly in the morning sun, when I saw a lonely, unmolested airplane circling around over the front a little below me. I looked at it and couldn't believe my eyes because I thought I saw a B.E., an ancient slow crate. I pushed towards him, and he took no notice of me at all, probably because I was coming so far back from 'England'. When I got within a few hundred meters, I could clearly see the type and insignia. It really was a B.E. The 'Franz' [i.e., observer] *was hanging half out of the crate, diligently observing the ground. The 'Emil'* [i.e., pilot] *was probably napping. Suddenly, they saw me. The observer, startled, stood straight up, and went for his machine gun. Then the good old B.E. made an incredibly bold turn and came right at me! What cheek! I let him slip away under me, made a short turn on my hindquarters and had him right in front of me. After barely 100 shots, his fuel tank burst, and he went down in a mighty trail of flame and smoke. . . The flak guns progressively shot at me and barked out their anger and frustration in countless shrapnel shells sent after me.*

When I landed, I was greeted by congratulatory comrades. . . the Flak Group had just reported that a yellow single-seat Albatros with a black tail had flamed an Englishman over the front."[24]

The aircraft was No. 2 Squadron's BE.2c 4541 that had been ordered to observe for the artillery. Both crewmen, 2Lt.

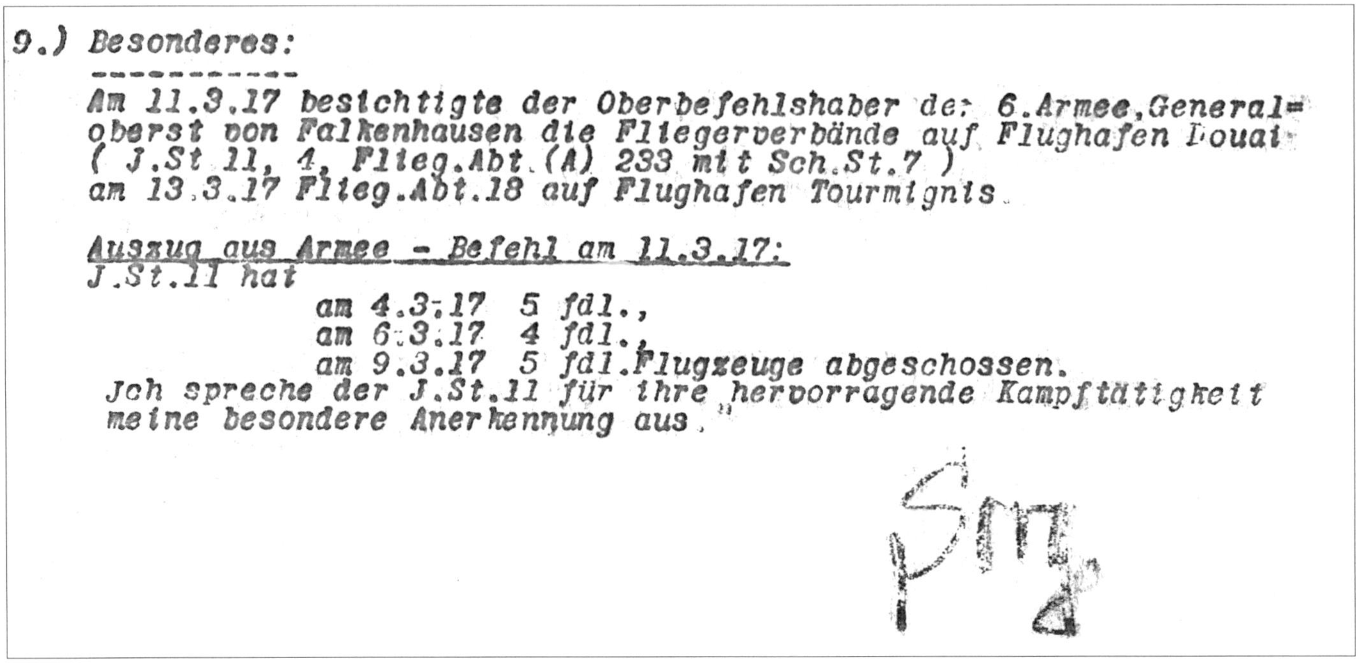

9.) Besonderes:

Am 11.3.17 besichtigte der Oberbefehlshaber der 6.Armee,Generaloberst von Falkenhausen die Fliegerverbände auf Flughafen Douai (J.St 11, 4, Flieg.Abt.(A) 233 mit Sch.St.7)
am 13.3.17 Flieg.Abt.18 auf Flughafen Tourmignis.

Auszug aus Armee - Befehl am 11.3.17:
J.St.11 hat
am 4.3.17 5 fdl.,
am 6.3.17 4 fdl.,
am 9.3.17 5 fdl.Flugzeuge abgeschossen.
Jch spreche der J.St.11 für ihre hervorragende Kampftätigkeit meine besondere Anerkennung aus."

Above: Weekly activity report by Kofl 6, covering the period 8 to 15 March 1917. It records the visit of Generaloberst von Falkenhausen to Jagdstaffel 11 and 4, Fliegerabteilung (A) 233 and Schutzstaffel 7 at the Douai airfield, on 11 March 1917. An excerpt of the daily Army Order of 11 March 1917 commending the successes of Jagdstaffel 11 on 4, 6 and 9 March is explicitly mentioned here. (Bayerisches Hauptstaatsarchiv, Abt. 4 München)

G.C. Hoskins and 2Lt. G.W.B. Hampton, were killed.

Flak Drocourt confirmed that a hostile aircraft crashed down in flames following air combat between Lens and Hulluch.

Flak Platoon 72 additionally reported: *"11:18, following a dogfight with an Albatros, a B.E. burning strongly between Hulluch – Vermelles, slowly descending behind enemy lines."*[25]

In the meantime, the Schwarm under Manfred von Richthofen's leadership had headed off toward Lens. Lt. Konstantin Krefft reportedly attacked and brought down an F.E. single-seater at the front around 11:45 a.m. for his first confirmed kill, but the precise location of its crash site is unclear from available documents. Kofl 6's weekly report gave it as "west of Souchez, across the lines" whereas the *Nachrichtenblatt der Luftstreitkräfte* said it was *"between Givenchy [-en-Gohelle] and Vimy."* In any event, British records do not list an F.E.8 or any other aircraft that had been downed around that time or at either of those locations, so it would seem that Krefft's alleged victim actually escaped.[26]

Richthofen, who had become separated from his Schwarm, flew alone toward Vimy where he spotted a British airplane around 11:50 a.m.:

"11 March 1917
B.E. two-seater.
12:00
South of La Folie Forest near Vimy.
Occupants: Lt. Byrne and Lt. Smyth, 40th [sic] *Squadron*
Plane: No. 6232

Details of motor not at hand, as motor dashed into earth; cannot be dug up as locality under heaviest artillery fire.

I had lost my squad and was flying alone, and had been observing for some time an enemy artillery flyer. In a favorable moment, I attacked the B.E. machine, and, after 200 shots, the body of the machine broke in half. The plane fell smoking into our lines. The plane is lying near the forest of La Folie, west of Vimy, only a few paces behind the trenches."[27]

Richthofen reported the wrong squadron number, but the airplane's serial number makes it clear that he had shot down B.E.2d 6232 from No. 2 Squadron, killing its crew 2/Lt. J. Smyth and 2/Lt. E.G. Byrne.

Less than 15 minutes later, Vzfw. Josef Baierlein claimed that he had brought down a B.E. near Villers-au-Bois; however, it went unconfirmed.

Jasta 11 was praised in the 6th Army's Tagesbefehl (Order of the Day) report of 11 March:

"Jagdstaffel 11 has:
on 4 March 1917, shot down five enemy airplanes.
on 6 March 1917, shot down four enemy airplanes.
on 9 March 1917, shot down five enemy airplanes.
I would like to express my special appreciation to Jagdstaffel 11 for their outstanding actions in battle.
Signed, von Falkenhausen."[28]

The pilots of Jasta 11 enjoyed another break from action over the next three days, 12 - 14 March. Rain, fog, and low-hanging clouds brought flight operations to a standstill on both sides of the front. Rain and poor weather allowed only a few flight activities on 15 and 16 March, and the only other occurrence worth mentioning is that Lt. Albert Oestereicher was transferred out to Bavarian FEA 1 at Schleißheim at

Above: Aerial view of the well-fortified German "Siegfried Line" near Bullecourt.

Above: The remains of 25 Sqn's F.E.2b A5439, shot down by Manfred von Richthofen southeast of Oppy on 17 March 1917. Lt. A. E Boultbee and 2/AM F. King were killed in the crash. (R. Absmeier)

this time. Oesterreicher began to train new pilots at Bavarian Military Aviation School 5 on 7 April 1917 and continued to do so until he was tragically killed in a crash on 8 November 1918 – only three days before the end of the war.[29]

By 16–20 March, the German Army had finished its retreat to the new, stronger "Siegfried-Linie" (or "Hindenburg Line") without the British and French headquarters being able to track much of its progress. The concealment of "Operation Alberich" was a complete success with the support of the German Fliegertruppe playing an important role in the operation.

The ground fog dissipated on 17 March, leaving a clear,

Left: The wreckage of the Sopwith Strutter A 1111, No. 43 Sqn. RFC., shot down on 17 March 1917 by Carl Allmenröder.

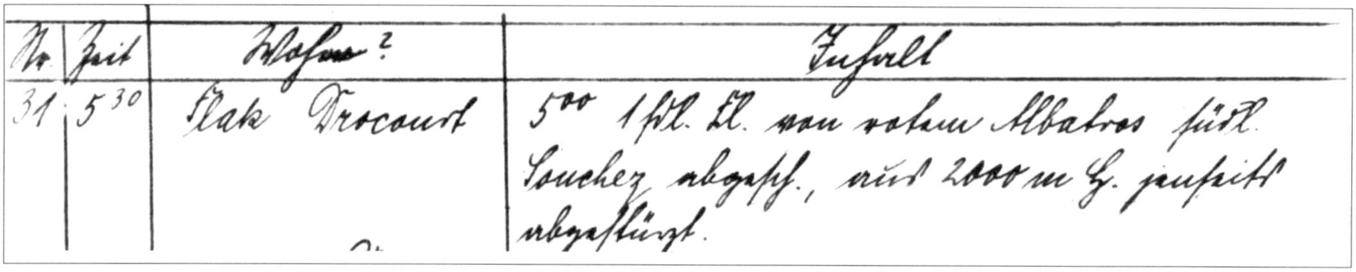

Nr.	Zeit	Wohin?	Zufall
31	5 30	Flak Drocourt	5 00 1 fdl. Fl. von rotem Albatros südl. Souchez abgeschoss., aus 2000 m H. jenseits abgestürzt.

Above: Report of the Flak Drocourt of 17 March 1917 in the Flugmeldebuch: "5:00 (p.m.): An enemy plane shot down by red Albatross south of Souchez, crashed from 2,000m altitude beyond." (Bayerisches Hauptstaatsarchiv, Abt. 4, München)

beautiful day for flight operations. Nine F.E.2b aircraft from 25 Squadron and nine Sopwith Strutters from 43 Squadron flew over the German lines at Givenchy and headed toward La Bassée at 11:20 a.m. Three of the F.E.s had been ordered to photograph the German positions from Annœullin to Vitry while the remaining six F.E.s and the nine Sopwith Strutters provided cover. The formation's approach was observed from La Brayelle airfield, whereupon Richthofen immediately took off with nine of his men and flew in the direction of Lens. The Staffel, constantly gaining altitude, intercepted the British group south of Pont-à-Vendin and attacked. Kurt Wolff described what happened in his diary:

"On the 17th, we saw a Geschwader of 16 F.E.s and Sopwiths from our field and took off as nine [sic] *machines. We flew right to the front, reached the Englishmen's altitude, and got in their way when they wanted to go back. I went after an F.E. first, but when Richthofen got there I left the two of them alone and Richthofen shot him down.*[30]

Richthofen's combat report stated:

"17 March 1917
Vickers Two-seater
11:30 a.m.
Oppy
Occupants: Both killed, no identity discs, names found on maps were Smith and Heanly.
Plane: A3439 [sic]

About 11:30 I attacked an enemy squadron of 16 units with nine of my machines. During the fight I managed to force a Vickers two-seater aside which I then, after 800 shots brought down. In my machine-gun fire, the machine lost its open work fuselage. The occupants were killed and taken for burial by the local commander at Oppy."[31]

The unfortunate airmen were Lt. A.E. Boultbee and 2AM F. King flying F.E.2b A5439 from No. 25 Squadron.

While Richthofen's fight was going on, Kurt Wolff attacked one of the protecting Sopwiths. He continued in his diary:

"Then I attacked a Sopwith. When I was an airplane's length away from the Englishman, a large flame leapt out after comparatively few shots. I caught all the smoke and stench in my face and escaped quickly by sharply turning. The Sopwith, burning brightly, went down vertically and its wings broke off during the fall. Al[l]*menröder also shot down a Sopwith, which burst apart in the air. Both apparatuses crashed near Le Point du Jour, east of Arras. A flak battery quickly claimed them both – shameless dog. When an Artillerie-Flieger-Abteilung claimed to have shot down a lattice-tail, also said to have crashed in flames, I set off and drove to the crash site on the evening of the 23rd. It of course was my Sopwith, and boy how it looked! Everything was completely*

Above: Manfred von Richthofen climbs into his red-painted Albatros D III.

incinerated and only one cylinder of the engine stuck out from the ground…

I was lucky on the 17th. A shrapnel shell exploded close to me over Arras, and I felt a powerful blow to the right shoulder. As it later turned out, a shrapnel ball had hit me but had bounced off the buckle of my seatbelt. After that, I only shot it up a little and without success."[32]

Kurt Wolff's unfortunate opponent had been Sopwith Strutter A1097 from No. 43 Squadron, manned by 2/Lt. A.L. Constable and 2/Lt. C.D. Knox.

As they were falling, Carl Allmenröder was dogfighting another Sopwith Strutter from No. 43 Squadron. After a brief exchange of fire, the machine, which was also being shot at by flak, crashed along the northern edge of Athies. It was Sopwith Strutter A1111, occupied by 2/Lt. J.C. Rimer and 2/Lt. R.H. Lownds. Both Sopwith crews were killed in action.

Carl Allmenröder wrote about this air fight in a postcard:

The 25 Squadron report of this encounter read:

"A formation of 6 F.E.2b's with 9 Sopwiths were escorting 3

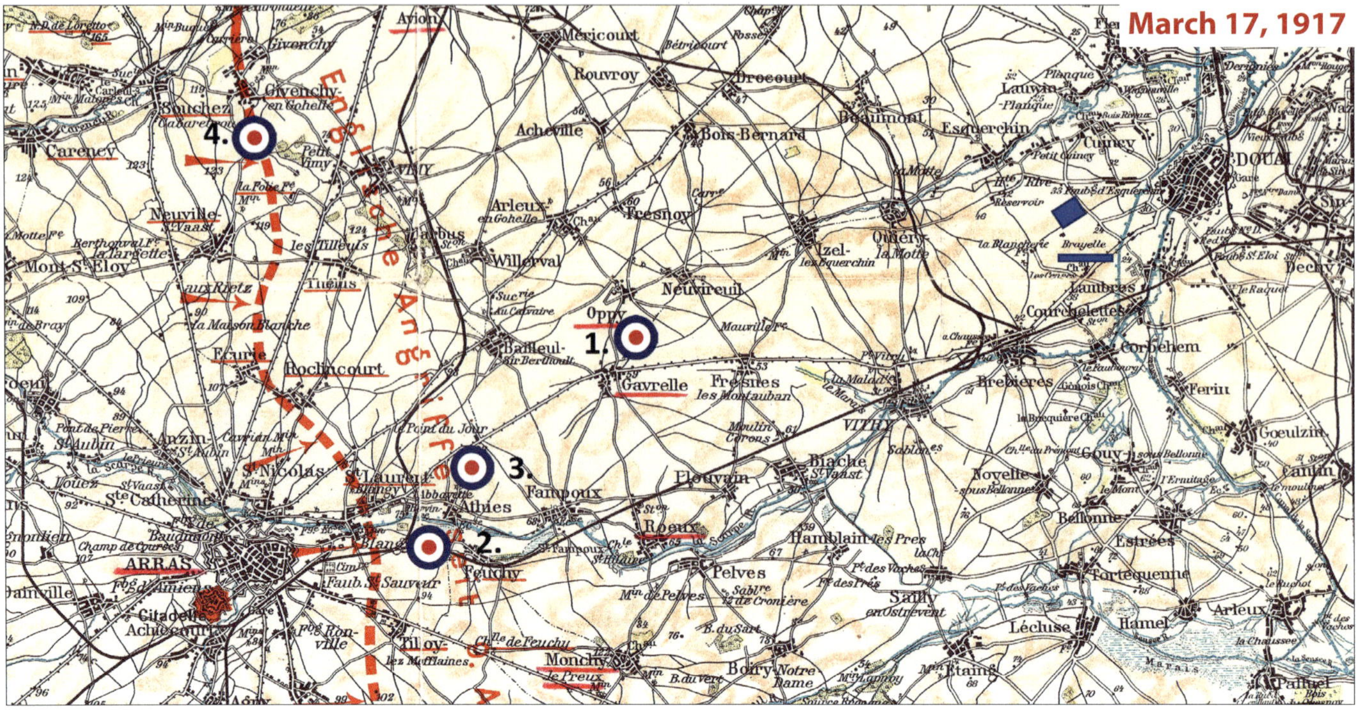

Above & Below: Map and details of the aerial victories of Jagdstaffel 11, March 17, 1917.

Nb.		Hour	Pilot	Location	Aircraft	Line
1.		11:40	Lt. Manfred von Richthofen	Oppy Forest	F.E. 2b A5439, 25. Sqn.	German
2.		11:43	Lt. Kurt Wolff	South-west of Athies	Sopwith, A1097, 43. Sqn.	German
3.		11:45	Lt. Carl Allmenröder	North of Athies	Sopwith, A1111, 43. Sqn.	German
4.		17:00	Lt. Manfred von Richthofen	South of Givenchy	B.E. 2e, A2814, 16. Sqn.	betw. lines

F.E.2b's to take photographs of Annoeullin-Vitry District. While over south of Pont a Verdin, the F.E.2b's were attacked by 9 hostile machines and these later, were reinforced by further 8–10 hostile machines coming from Douai. A running fight at first took place, before combat became general over Oppy. On closing in, Capt. Woollven's observer, Sergt. Boot fired a double drum into a scout (rotary engine) [sic] *which flew past his nose at 10 yards range. This scout slipped away and dived vertically and was last seen, still diving completely out of control at less than 1,000 feet.*

Capt. Richardson attacked a hostile machine and his observer, 2/Lt. Wollen, fired part of a drum into same, which immediately burst into flames and was last seen breaking up in pieces. A hostile machine attacked Capt. Richardson at this time but was driven off by a Sopwith. Several machines were fired on.

A hostile machine also attacked 2/Lt. Malcom and his observer, Lt. Wilson, opened fire with his front gun. The hostile machine was seen to half side-slip and nose dive. 2/Lt. Malcom followed the hostile machine at 50 yards, his observer firing continuously, down to 4,000 feet when the hostile machine was in a spinning nose-dive and completely out of control. Lt. Wilson saw machine crash.

A Halberstadt [sic] *attacked Lt. Davis from the rear, but swinging the F.E.2b round, his observer, 2/AM Emsden, fired 40 – 50 rounds into the hostile machine, which burst into flames and was last seen falling out of control and still in flames.*

Lt. Dunlop and 2/Lr. Weir (Camera Machine), 2/Lt. Parnell with 2/AM Taylor were both heavily engaged, but with no decisive result. 2/Lt. Parnell was wounded early in the fight but brought his machine safely back to aerodrome.

Sergt. Green with 2/Lt. Morrison (Camera Machine) had a hostile machine dive on their tail, but, turning round sharply, 2/Lt. Morrison fired half a drum into same nose on. The hostile machine was observed to go down in a steep nose dive, but landed, apparently under control south of Cite de Godault.

2/Lt. Bell with 2/Lt. Green (Camera Machine) was also in the fight and drove down a hostile machine, but were shot through the engine and had to return, landing at Estree Cauchie…

Lt. Boultbee with 2/AM King were seen during the combat but have not returned to the aerodrome. Further details are lacking. The hostile machines broke off the fight and the formation returned to our lines and continued the patrol, except for camera machines, which returned to the aerodrome.[34]

Major Douglas, Commander of 43 Squadron, wrote the following account based on his airmen's testimony:

"Capt. D C Rutter closely followed the F.E.2bs on a photographic patrol. At about 10:45 when over Beaumont the patrol was attacked by about 15 hostile aircraft, who manoeuvred into the sun and dived.

1. *Capt. D C Rutter closely followed the F.E.2b's whose leader started to circle. Seeing a hostile aircraft attacking an F.E. Capt. Rutter stalled his machine up and fired several bursts at the hostile aircraft; he saw bullets hitting the hostile aircraft. Lt. Venn his observer drove off several hostile aircraft who were diving on to his machine from rear.*
2. *Lt. Kitto engaged a hostile aircraft and fired a burst of about 30 rounds into him at close range. He saw him fall out of control but was unable to follow him down owing to the presence of other hostile aircraft.*
3. *The rest of the formation followed Capt. Rutter. Major Dore was dived on by two hostile aircraft, but Sergt. Cubberle his observer drove them off with several bursts. Lt. Thornton and Blackburn also engaged hostile aircraft.*
4. *The hostile aircraft followed the patrol back to the lines,*
5. *Two of our machines are missing. One was seen falling in flames."*[35]

The final report of the Royal Flying Corps said this about the battle:

"Eighteen machines of 1st Brigade taking photographs of Courrieres line encountered nineteen hostile machines. As a result of the fight, three hostile aircraft were destroyed, three driven down damaged, one driven down under control. Three of our machines did not return. The hostile formation was broken up and the photographic machines carried on with their photography."[36]

Contrary to the Royal Flying Corps report, which mentioned three destroyed and three damaged German machines, Kofl 6th Army's weekly report recorded no German losses that day.[37] Accordingly, the German machines reported by the British as destroyed or out of control were probably British aircraft mistaken for German ones and German fighters that had escaped by spinning down.

Such reports were produced after the participants had returned to their airfield, and the British crews stated what they had seen, or believed they had seen, during the battle. When the reports were viewed in combination, they could give the impression that considerably more enemy aircraft had been lost or damaged than was actually the case. But when we see today that such claims do not match what was reported in German records, it would be doing the British crews a grave injustice if we were to dismiss their reports as exaggerations or even braggadocio. The author has spoken with many former German, and some former British, pilots who could still vividly describe their air combat experiences. The nervous tension, excitement, and concentration that was involved was incredible. After all, it was a matter of life and death. This was especially true when, as in this case, a dogfight occurred between 20 or more aircraft, and any situation could change within seconds. A participant might suddenly see an enemy machine before his machine guns and fire. Impacts might even have been observed in the enemy's fuselage, but a moment later the attacker might have had to turn away to avoid another opponent. Dogfights were literally compared to roller coaster rides with loops and dives and other swirling maneuvers. Moments later, an airplane might be seen going down that was assumed to have been one you had shot at. During such aerial melées, the participants – under extreme stress – were usually only able to see and perceive fragments of the overall events that changed at lightning speeds.[38]

The air battle described above took place over German territory and it was therefore impossible for the British crews to verify their observations and claims. This was true for any fight that took place over enemy territory or over contested portions of the front. There is little question that German aircraft were often hit by British bullets during such fights. But this was not normally mentioned in the German weekly reports which usually related only when German airmen were wounded or killed. In this context, mechanics like Hermann Lohmeyer and Xaver Leinemüller said that machines returned with hits that had to be repaired after almost every dogfight. However, the damage was usually only minor and could be serviced at the airfield. This almost certainly had been the case for the Germans above. According to existing Armeeflugpark 6 records, no Jasta 11 fighters had to be returned to it for repair at that time.[39]

On the afternoon of 17 March, Manfred von Richthofen took off on a solo flight toward the front at Vimy and spotted a British B.E. directing artillery fire south of Givenchy-en-Gohelle:

"17 March 1917
B.E. Two-seater
5:00 p.m.
Above trenches west of Vimy
No details, as plane landed between lines.

I had spotted an enemy infantry flyer. Several of my attacks, directed upon him from above, produced no result, especially as my adversary did not accept fight and was protected from above by other machines. Therefore, I went down to 700 meters and attacked my adversary who was flying at 900 meters from below.

After a short fight, my opponent's plane lost both wings and fell. The machine crashed into no man's Land and was fired at by our infantry."[40]

2/Lt. G.M. Watt and Sgt. E.A. Howlett, flying No. 16 Squadron's B.E.2e A2814, were Richthofen's victims and both men were killed. Shortly afterwards, the downing was confirmed by Flak Drocourt:

"5:00: An enemy aircraft shot down by red Albatros south of Souchez, crashed from 2,000 meters on the other side.[41]

2/Lt. Watt's death prompted his father to write an angry letter to 16 Squadron's Commander, Major Maltby, that said:

"I had a letter from him on the 16th of March saying that unless the Government gave them new machines, the whole

Squadron would be wiped out. It is most distressing to think that our flying men with such machines have no chance with the Germans, and as my son said are thus 'really at their mercy.'"[42]

Interestingly, all of the German pilots with whom the author spoke expressed great respect for their British opponents but generally held the leadership of the British air forces in contempt because of the way they treated their own men. The British leadership was universally described as inhumane since they ruthlessly sent their airmen over German lines, even when they had inferior aircraft. Words like "murderer" and "criminal" were applied to the heads of the RFC/RAF.

18th March 1917 – 31st March 1917

On 18 March, there was moderate air activity in the 6th Army sector due to cloudy weather until the afternoon. And even then, only the southern section of the German 6th Army experienced increased enemy air activity toward the evening.

Eddy Lübbert returned from the hospital this day and was warmly welcomed by his comrades. Although he had only partially recovered from his wounds, he nevertheless – according to his flight log – immediately got back into an airplane:

"*Wanted to go to Abteilung 18* [Fliegerabteilung 18 (LB), his old unit] *in a Halberstadt. Piston rod break at 600 meters above Douai. Smooth emergency landing in a meadow.*"[1]

Following this emergency landing, he got back into his "half-blue, half-yellow" Albatros D III and flew to the front: "*6:10 to 6:40 hunting fight to Lille over clouds. Nothing seen.*"[2]

The next day, 19 March, brought good flying weather that resulted in a lot of flying activity along the front in the morning. Lübbert's log book shows that Jasta 11 was no exception:

"*8:00 to 8:45 a.m. hunting flight to Péronne. Air fight near Arras against 3 F.E.s who escaped, however. 10:57 to 11:45 hunting flight to Péronne and back. Nothing.*[3]

All that remains to be added for this day is that Jagdstaffeln 27 and 28 were transferred from the 4th Army to the 6th Army. Jasta 27 was based at an airfield at Tourmignies, south of Lille, whereas Jasta 28 was situated near Lille outside of Wasquehal.

"Fliegerwetter" returned again on 20 March when thick cloud cover, rain, and storms made it impossible to fly. Nevertheless, the pilots and groundcrews soon became startled by the ever-increasing thunder of cannons at the front. The bombardment ordered by British High Command in preparation for their major offensive had begun and was targeting the planned area of attack between Loos in the north and Fontaine-lès-Croiselles in the south.

21 March began with clouds and rain as well. Although no hostile flight activities were reported from the front, the German trenches near Vimy remained under heavy artillery fire that was being guided by captive balloons. As a result, Eddy Lübbert decided to exploit the cloud cover to try attacking one of the balloons. Hidden by the overcast conditions, he crossed over the front and penetrated about seven kilometers into British airspace. He briefly noted in his log book:

"*21 March 1917, first flight, tried to shoot down captive balloon at Villers-au-Bois. Went over and through clouds three times: no captive balloon, only machine gun fire and plenty of flak.*"[4]

After his return, the Staffel took off for the front several times, about which Lübbert wrote:

"*Several B.E.-hunting flights. 2nd flight nothing seen.*"

Lübbert also participated in a patrol with the Jasta at noon, but it was uneventful since again no enemy machine was spotted.[5] These entries demonstrate that not every frontline flight resulted in an encounter with enemy aircraft. "Working aircraft", i.e. artillery spotters, air photographers, infantry supporters, in particular avoided combat with single-seaters and retreated behind their own lines whenever possible.

At around 3:40 p.m., Jagdstaffel 11 embarked on another flight to the front from Lens to La Bassée. Carl Allmenröder attacked a Sopwith two-seater over the lines and after a brief exchange of fire, his opponent went down in flames at 3:45 south of Vermelles. Lübbert mentioned in his flight log:

"*Third flight, Allmenröder brought down a B.E.*"[6]

The downing was observed from the ground by Flak Faschode which reported:

"*15:46: B.E. shot down in flames by 1 Albatros; lies near Vermelles toward Grenay.*"[7]

Just a few minutes later, Flak Drocourt reported:

"*15:50: A B.E. shot down by Albatros at 800 meters' height north of Loos.*" After that, Jasta 11 recorded: "*B.E. shot down by Lt. Allmenröder.*"[8]

Allmenröder's victim was Sopwith Strutter A2390 from 43 Squadron. According to British documents, its pilot, Lt. H.W.L Poole, managed to land the burning plane and escaped with only minor burns, whereas his observer, 2/AM A.J. Ball, died from severe burns the next day.

The casualty report logged by 43 Squadron read:

"*Left aerodrome at 1,45 p.m., landed at 2,45 p.m.... Engaged hostile aircraft. Petrol tank set alight. Machine entirely destroyed by fire, except engine and Vickers gun... Pilot slightly burnt. Observer died of burns.*"[9]

"*Dear Lies,*

I shot down my 4th Englishman today. The airplane began to burn and went straight down. What do you think about Willi's incredible success? I long for a letter from you."[10]

Later that afternoon, Jasta 11 received reports from the forwardmost lines that numerous enemy aircraft were active there at about 1,000 meters despite the bad weather and

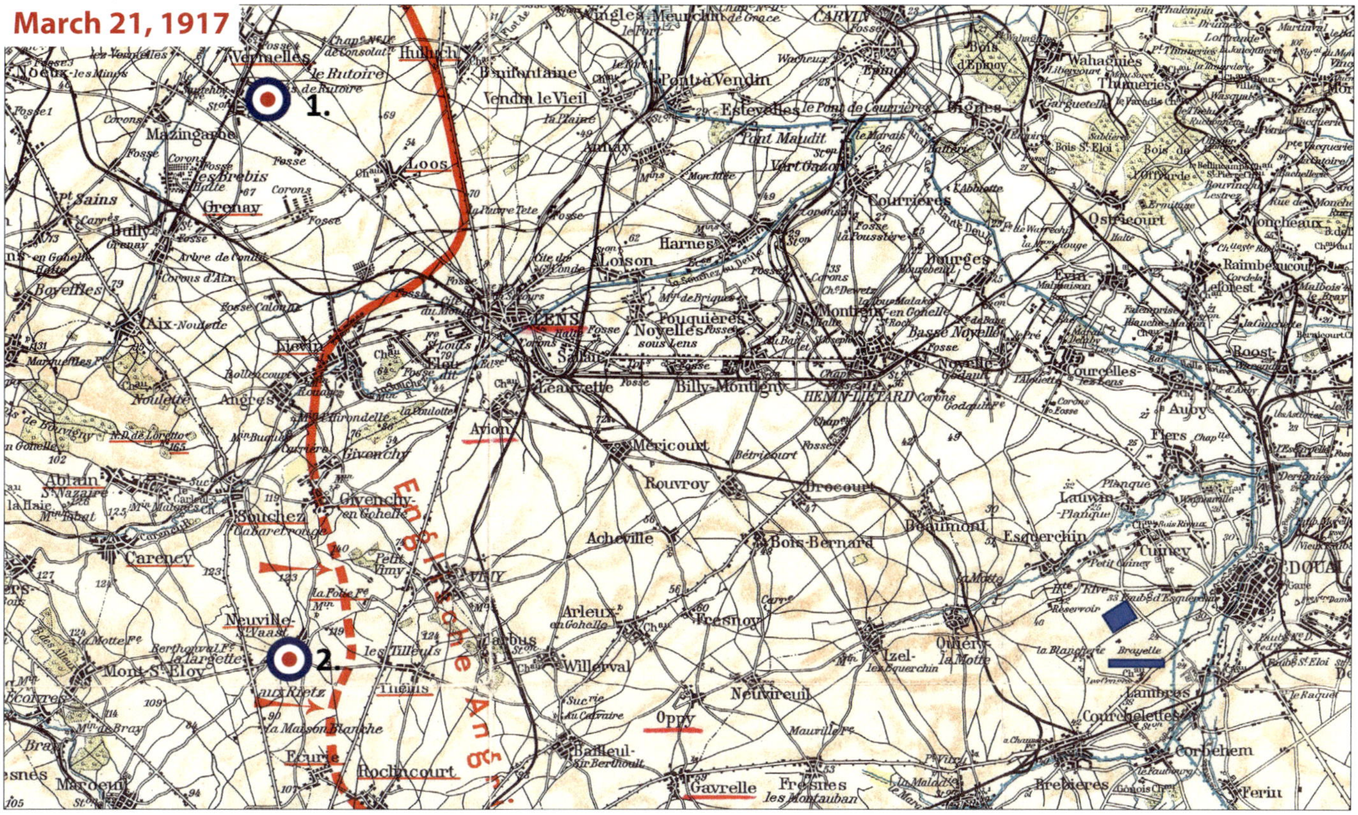

Above & Below: Map and details of the aerial victories of Jagdstaffel 11, March 21, 1917.

Nb.		Hour	Pilot	Location	Aircraft	Line
1.		14:50	Lt. Carl Allmenröder	South of Vermelles	Sopwith A2390, 43. Sqn.	Allied
2.		17:40	Lt. Manfred von Richthofen	Neuville St. Vaast	Sopwith A1097, 43. Sqn.	Allied

strong easterly winds. As a result, Richthofen took off alone since the appearance of an entire Staffel had often led to the British artillery and observation planes withdrawing behind their own lines only to return and continue their work once the Staffel had departed. After about an hour's uneventful flight at the front, he saw several enemy artillery airmen across the lines at about 800 meters' altitude:

"21 March 1917
B.E. Two-seater
5:30 p.m.
Hill 123 north of Neuville
Plane details unknown, as plane came down on enemy's territory.

Messages came through that enemy planes had been seen at 1,000 meters' altitude in spite of bad weather and strong east wind.

I went up myself intending to bring down an infantry or artillery flyer.

After one hour I spotted at 800 meters a large number of enemy artillery flyers beyond the line. They sometimes approached our front, but never passed it. After several vain attempts, I managed, half hidden by clouds, to take one of these B.E.'s by surprise and to attack him at 600 meters 1 kilometer beyond our lines. The adversary made the mistake of flying in a straight line when he tried to evade me, and thus he was just a wink too long in my fire. (500 shots). Suddenly he made two uncontrolled curves and dashed, smoking, to the ground. The plane was completely ruined."(11)

No. 16 Squadron's B.E.2f A3154 and its crew, FSgt. S.H. Quicke and 2/Lt. W.J. Lidsey, had become Richthofen's 29th victory.

16 Squadron's casualty report read:

"The machine left the aerodrome at 3,00 pm and at 4,30 pm a message come through from the batteries stating machine had been shot down. The pilot (FSgt. S.H. Quicke) was killed in the air and the observer (2/Lt. W.J. Lidsey) died the following morning. The machine was subsequently shelled and utterly wrecked. Two guns, petrol-gauge, rev-counter, 1 watch was salved."(12)

Due to the constantly increasing flight activity of the allied air formations, Jasta 3 was redeployed from the German 2nd Army's sector to the 6th Army's before the day was over. The unit occupied Guesnain airfield, just under four kilometers southeast of Douai. As a result, six Jagdstaffeln

Above: Oblt. Paul-Henning von Osterroth and Manfred von Richthofen in front of the pilot mess of Jagdstaffel 12 at Epinoy on 23 March 1917.

were now stationed in the 6th Army sector:

Jagdstaffel 3 at Guesnain
Jagdstaffel 4 at La Brayelle
Jagdstaffel 11 at La Brayelle
Jagdstaffel 27 at Tourmignies
Jagdstaffel 28 at Wasquehal
Jagdstaffel 30 at Phalempin

22 March brought cloudy weather with overcast skies that only cleared up in certain spots that afternoon. Accordingly, flight activities were at a standstill in the morning and only active in isolated places later on.

Eddy Lübbert took off for his first flight of the day at 2:10 p.m., but after just a few minutes had to go back to base at 2:29 p.m. due to poor visibility. He was able to try again in the early evening when his Kette left for the front at 6:22 p.m. This time, he made it to the front but did not encounter any enemy aircraft and returned at 7:17 p.m.(13)

The next day, 23 March, was similar in that it began with clouds and overcast conditions that only partly improved in the afternoon. Flight activity was therefore low. Nevertheless, the Staffel's Ketten took off for the front on several occasions to try to interfere with the ongoing artillery bombardment that was being directed by enemy aircraft.

Eddy Lübbert reported:

"23 March 1917: 8:36 to 9:32. Hunting flight with Schäfer, Allmenröder, Wolff, Baierlein. 13 F.E.s and DD 2s [two-seater biplanes] *sped off to the west. Went home because thick cloud cover came up."*(14)

This was Vzfw. Baierlein's last mission with Jasta 11. His transfer to Bavarian Flieger-Ersatz-Abteilung 1 had come through the day before.(15)

After lunch, Richthofen got into Albatros D III D.2006/17 and flew over to Jasta 12's Epinoy airfield, some 18 kilometers south of La Brayelle, to visit his friend Oblt. Paul-Henning von Osterroht, the unit's CO.(16) The reason might have been because it was at this point that Jasta 12 was placed under the tactical control of the 6th Army.

Following their leader's return from Jasta 12, a Jasta 11 Kette went to the front as reported by Eddy Lübbert:

"17:30 to 18:39. Hunting flight with Richthofen and Wolff. Saw 3 F.E.s but couldn't get to them due to strong easterly wind."(17)

News arrived at the Staffel the same day that Richthofen had been promoted to Oberleutnant ahead of the normal protocol in recognition of his achievements. His promotion was celebrated extensively in the Officers' Mess that evening.

Kurt Wolff had a reason to celebrate, as he wrote in his letter from the 23 March 1917:

Right: Members of Jagdstaffel 12 with Manfred von Richthofen, in front of the aircraft tents at Epinoy. From left: Lt. Müller, Offz. z.b.V., Vzfw. Schorisch, Oblt. von Osterroth, Vzfw. Grigo, Oblt. von Richthofen.

"I also experienced something very good. I have been awarded the cup of honor with the inscription "To the Victor in Aerial Combat" by the Commanding General of the Air Forces. This is a large, very nicely crafted silver cup, but so far, I only have the diploma. The cup is probably still being manufactured."[18]

As a final note, Jasta 11 was also informed that Lt. Hans Hinsch had been transferred to Jasta 11 that day.[19] But since he was on home leave with his family to celebrate his 23rd birthday, he did not actually arrive at the Staffel until 1 April.[20]

The weather on 24 March was cold but clear. Accordingly, flight activities on both sides ramped up and were quite lively by the afternoon. Richthofen and a Schwarm consisting of Schaefer, Allmenröder, Wolff, and Lübbert took off toward Arras at 10:36. Once again, their targets were the enemy's artillery fliers who had to be shot down or at least prevented from directing artillery fire onto the German trenches. Along the way, Lübbert's half-blue and half-yellow Albatros D III developed engine problems and he had to return to La Brayelle, where he landed at 11:11.[21]

The remaining four pilots continued their flight, and upon their arrival at the front Schaefer surprised a "Sopwith" over the eastern edge of Arras that he said he shot down in flames at 11:00. No downed Sopwiths suit his claim, but it is possible that his victim was No.8 Squadron's B.E.2e 7248, which was reported as seen going down in flames at 11:00 near Achicourt, a place located only a few kilometers south of Arras. The crew, consisting of 2/Lt. H.W. Tagent und 2/Lt. G.T. Gray, were killed.

After the Staffel's Albatros D IIIs had returned to La Brayelle, mechanics refueled and rearmed them after checking for any damage while the pilots enjoyed a quick lunch. About one hour later, the aircraft propellers started running again, the pilots turned into the wind, rumbled over the grass runway, and took off in the direction of Vimy. They encountered a group of British aircraft that were about to enter German airspace west of Acheville. Among them were three SPAD VIIs from 19 Squadron – a type that had only recently been introduced at the front. Richthofen attacked one of them and later reported:

"24 March 1917
11:55 a.m.
Givenchy
Plane: A Spad with Hispano Motor. The first here encountered.
Plane: No: 6607 [sic]
Motor: Hispano-Suiza 140 H.P.
Machine-gun No. Maxim 4810

I was flying with several of my gentlemen when I observed an enemy squad passing our front. Flying along with this squad were two new one-seaters with which I was unfamiliar.

I attacked one of them and ascertained that my machine was the better one. After a long fight, I managed to hit the adversary's tank. The propeller stopped running. The plane had to go down. As the fight had taken place above the trenches, adversary tried to escape, but I managed to force him to land behind our lines near Givenchy. The plane turned upsidedown in a shell-hole. The plane was taken by our troops"[22]

Above: Oblt. Manfred von Richthofen and his old friend Oblt. von Paul-Henning von Osterroth at Epinoy airfield. The nose of von Richthofen´s mount, D.2006/16 can be seen in the photo, clearly showing the red-painted spinner.

Richthofen had shot down SPAD VII A6706. The pilot, Lt. R.P. Baker, had been wounded and was taken prisoner by German troops.

Lt. Baker was questioned after his capture and a portion of the interrogation report read:

"27 March 1917

1. Personnel:

1Lt. Backer, from Vancouver, Canada, worked in a shipping business, joined the 11th Battalion Canadian Mounted Infantry soon after the outbreak of the war. The battalion was converted into an infantry battalion in 1916 and came to France. The prisoner left it in September 1916 before it was at the front and joined the R.F.C.; educated in England; in the field as a pilot since March 1917...

3. Orders for Last Flight

To fly a blockade flight with a second airplane. He was surprise-attacked and immediately wounded in the knee. Soon after that, his engine stopped too, probably due to gunshots. There was nothing left to him but to evade the enemy shots as best as he could and land. At the end, he no longer had any control over his machine. Since he was preoccupied with his own situation and didn't get around to firing a single shot, he could neither give an opinion about the victorious machine nor compare it to his own because he was unable to pit himself against the opponent. If he had gone down in a glide to land, his opponent could easily have pursued and shot him down. Therefore, he performed a kind of loop to avoid being fired at. He had little frontline experience due to the many foggy days this month. The fight that ended with his being shot down was actually his first air combat.

Signed, Freiherr von Karg-Bebenburg

Commander of Flak 6"[23]

2/Lt. F.L. Harding, the pilot of 19 Squadron's SPAD VII A263, stated in his combat report:

"When on patrol at 12,000 feet I saw two enemy aircraft 1,000 feet below me and dived down to get under their tails. I was at once attacked by about 10 hostile scouts but owing to gun stopping continually, I could not put up a fight so returned home. Time 10:50."[24]

Finally, there is the report from Lt. H.A. Orlebar who in SPAD VII 6663 had a fight with a *"Halberstadt...dark colored with very white background for its cross"*:

"When first seen the machine was a few miles east of Lens flying South. I was considerably above him and tried to dive down but could not reach him owing to falling temperature. He turned east

Above: Conversation between Oblt. von Osterroth and Oblt. von Richthofen before the latter's return flight next to the wing of von Richthofen´s Albatros D III D.2006/16, most of the military number of the aircraft being just visible behind the "V" interplane strut.

and I dived slowly with the sun as far as possible behind me.

I opened fire at about 100 yards and had a No.1 which I rectified, after about a couple of shots I had a No. 3 which I also corrected and then got a burst of about 12 shots, after which I have to break off owing to a No. 3 which I could not correct. The machine was in my sights and the tracers seemed to be entering the fuselage, but they had no visible effect.[25]

After the first encounter with the SPAD VII the Albatros flight returned to La Brayelle. Richthofen led a Schwarm consisting of Schaefer, Wolff, Allmenröder, Lübbert, and brother Lothar to the front again at 3:30 p.m., but the patrol went without incident.

There were variable clouds on 25 March and airspace observers initially reported moderate air activity that became somewhat more lively in the evening. Despite the morning's rather limited enemy flight activity, a Staffel Schwarm that did not include Richthofen set out for the front at 8:06 a.m. Upon reaching the front lines, they saw a formation of Sopwith Strutters, F.E.8s, and Nieuport fighters that Eddy Lübbert described as:

"25 March 1917: 8:06 to 8:55 (a.m.), hunting flight with Allmenröder, Wolff, Richthofen II [Lothar]*. Formation of Sopwith two-seater biplanes, F.E. single-seat biplanes, and Nieuports. Attacked about eight machines, had a large F.E. in my sights, but then a gun jam."*[26]

Lothar von Richthofen wrote a more detailed description of the encounter:

"[Our] *group had an early start that day, which means we had to be ready to take off at a moment's notice at dawn; that was very early, between 4:00 and 5:00. We had just gotten up and were sitting in the ready house when the telephone rang: 'Six Bristols* [sic] *from Arras to Douai.' Well, we were already up, so off we go. There was broken cloud cover at 3,000 meters' altitude.*

As we were about to take off, we saw the Englishmen already near our airfield just under the cloud cover. My brother's red bird stood ready for takeoff on the field, his mechanics next to it. Nothing was seen of my brother. We still got hold of the Englishmen, but the fellows flew so skillfully in the cloud layers

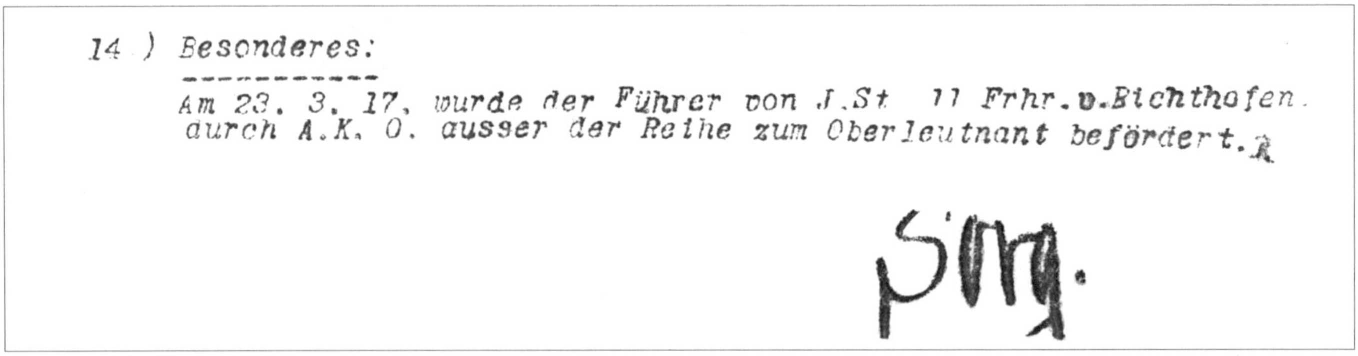

14) Besonderes:

Am 23. 3. 17. wurde der Führer von J.St. 11 Frhr.v.Richthofen. durch A.K.O. ausser der Reihe zum Oberleutnant befördert. R

Sorg.

Above: Excerpt from the report of the commander of the airforece (Kofl) of the 6th Army concerning the early promotion of the commander of Jagdstaffel 11: "On 23 March (19)17, the leader of Jagdstaffel 11, Frhr. von Richthofen, was promoted by A.O.K. to Oberleutnant (first lieutenant) out of order, signed: Sorg". (Bayerisches Hauptstaatsarchiv, Abt. 4, München)

Above: Lt. Hans Hinsch (T. Pietsch, Frontflieger.de)

Above: The emergency-landing of SPAD VII A6706 tore off its landing gear, leaving it in less than perfect condition. The pilot, Lt. R.P. Baker, was wounded and became a prisoner of war. (C. Huston)

that none of us could shoot one down. Whenever you got in range behind one, it disappeared up or down into the clouds. This was my first air combat, and I was quite proud when I got one before me and he started to 'stink' [i.e., emit a smoke trail] *under my machine gun fire. I had pierced a fuel tank; but the next moment he disappeared again in the clouds. Since almost all machines have a reserve tank, he had probably switched over to the other one. The fellow at least kept on flying.*"(27)

Shortly after this Kette had taken off, Jasta 11 received a report that an enemy formation was about to cross over the lines at Arras. This prompted Manfred von Richthofen to set out alone to take up the pursuit. He caught up to the last machine, a Nieuport, and opened fire:

"25 March 1917
Nieuport One-seater
8:20 a.m.
Tilley
Inmate: Lt. Grivert [sic] *– English*
Plane: burnt

An enemy squadron had passed our lines. I went up, overtaking their last machine. After only a very few shots, the enemy's propeller stopped running. The adversary landed near Thilloy, thereby upsetting his plane. I observed that some moments later the plane began to burn."(28)

The pilot of 29 Squadron's Nieuport XVII A6689, 2/Lt. C.G. Gilbert, managed to land the burning aircraft and escape the cockpit in time. Unharmed, he was captured by German soldiers. He was eventually questioned by German intelligence officers, whose report read:

"1. Personnel

1Lt. Guy Gilbert, in the field with the Imperial Yeomanry at the outbreak of the war, took part in the retreat from Mons, then transferred to the Dorsetshire Infantry Regiment; about 4 or 5 years as a pilot with the R.F.C. (six brothers also in the field, one of whom was killed)

2. Aviation Associations

He refused to make any statements about his aviation association

March 24, 1917

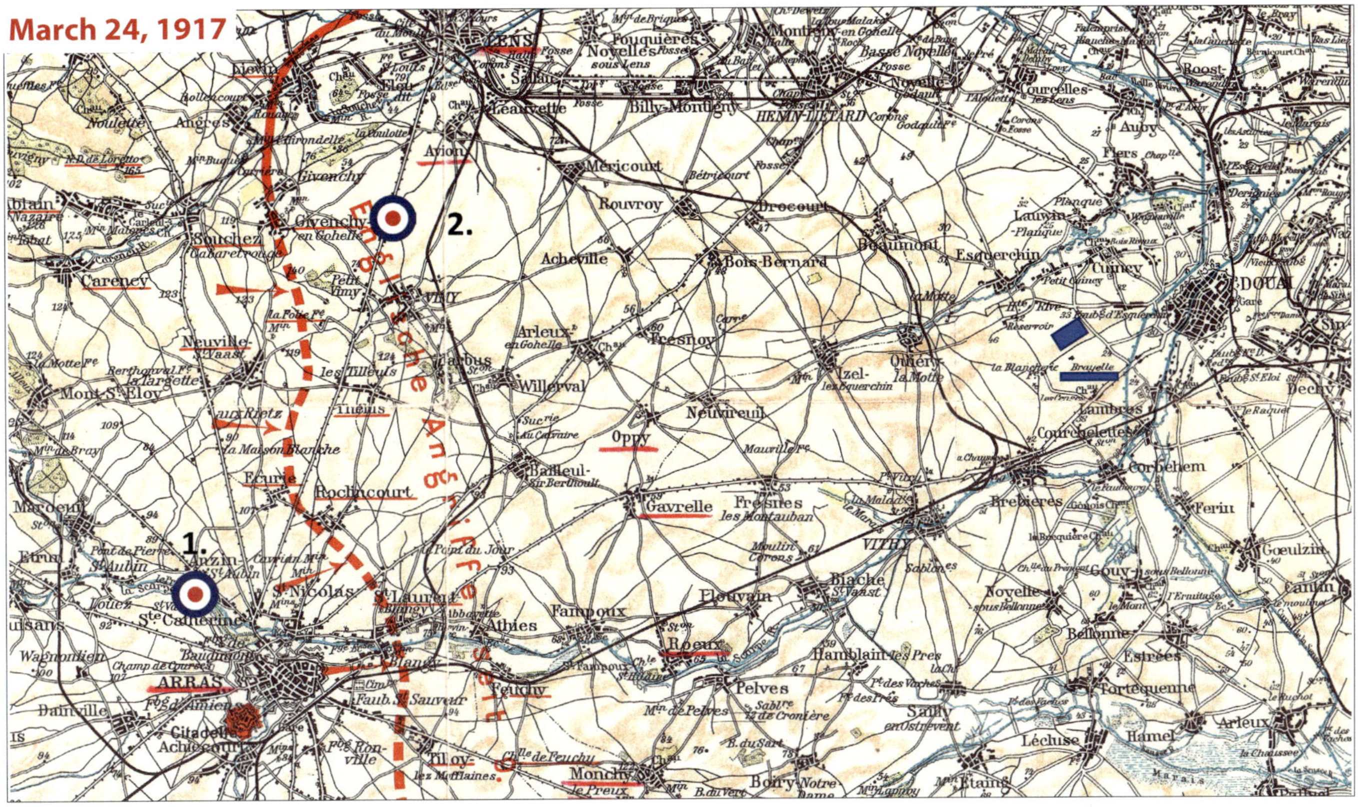

Above & Below: Map and details of the aerial victories of Jagdstaffel 11, March 24, 1917.

Nb.		Hour	Pilot	Location	Aircraft	Line
1.		11:00	Lt. Emil Schäfer	Anzin, north of Arras	B.E. 2e 7248, 8. Sqn.	Allied
2.		11:55	Lt. Manfred von Richthofen	Givenchy	Spad VII A6706, 19. Sqn.	German

or airfield.

3. Airplane:

Nieuport single-seat biplane, produced in France, prisoner considers the German Albatros single-seater to be more capable in climbing. However, the Nieuport is by no means the R.F.C.'s best and most modern type.

4. Orders for Last Flight:

To accompany and cover, along with two other Nieuports, four F.E. two-seaters on a reconnaissance flight to Douai airport. During the return flight, the prisoner was attacked by an Albatros and forced to land after being shot in the fuel tank; his machine gun had a loading jam and was out of action. Prisoner burned airplane after landing.

Signed, Frhr. von Karg-Bebenburg

Commander of Flak 6"[29]

After the war and his return to England, Gilbert testified:

"Escort to No. 11 Sqn. F.E. (Photography). 5 Nieuports should have accompanied the photographic squadron. 4 however failed to leave the ground. I therefore proceeded alone… Surrounded and cut off by Richthofen Squadron [sic] *over Douai. Shot down – petrol tank hit – controls shot away over Douai. Not wounded."*[30]

For those critics who have speculated that Richthofen was a bloodthirsty killer, it should be noted here that he easily could have continued firing at both Richard Baker and Christopher Gilbert after their engines had stopped. Instead, he allowed them to land without further attacks. Therefore, we might conclude from these two cases and similar ones that will be encountered later, that Richthofen found it incompatible with his self-image as a soldier to shoot down defenseless opponents. Moreover, such behavior was imitated by most of his pilots.

Manfred von Richthofen returned to La Brayelle airfield before his brother Lothar's Kette. Lothar recalled:

"None of us managed to shoot anyone down, and after about an hour we landed back at our airfield. My brother's red bird was on the ground again, but one could tell from a distance from the machine's position and the mechanics working on it that he had been out and about. We were also immediately informed: Yes, Herr Rittmeister [sic – he was still an Oberleutnant at the time] *had taken off about five minutes after we had. He had been in bed when the report had come in. Then it was flying clothes quickly over his pajamas and off and away. He returned after about 20 minutes and during that time had shot down an*

Left: Lothar von Richthofen talking to his brother on the snow-covered airfield at La Brayelle. (Taktisches Luftwaffengeschwader 71 Richthofen)

Englishman this side of the lines. Now, after our return, he was back in bed and sleeping as if nothing had happened. Only a few hits to his machine and incoming reports concerning the airplane he had brought down testified to his flight. We were all somewhat ashamed. Five of us had taken off before and landed after him, yet we had brought nothing down. When we gathered again for a second takeoff around 8:00 [sic], *my brother appeared. He grumbled about the English, those nocturnal disruptors who drum peaceful people out of bed in the middle of the night. We heartily congratulated him…"*[31]

Eddy Lübbert's flight log tells us that he and Carl Allmenröder took off again at 10:34 on a flight to the northern part of the front near Armentières but did not encounter any enemy aircraft. In response to another report, both men took off again at 15:05 to pursue British aircraft; however, they returned at 15:32 after another futile hunt. Lübbert was in the air yet again at 16:32 and headed to the front with Carl Allmenröder, Konstantin Krefft, Kurt Wolff, and Lothar von Richthofen. Once again, the Kette returned without success at 17:10.[32]

Flight operations in the 6th Army sector came to a standstill on 26 and 27 March due to rain, low-hanging clouds, and snow. This gave the pilots some time to recover from the strenuous missions of the past few days. Manfred von Richthofen used the break to write to his mother on the 26th:

My Staffel is coming along, and I'm happy with it. Lothar had his first air combat yesterday. He was quite pleased because his opponent was shot and wounded. We say 'he stank' because it left a trail of black smoke behind it. Of course, it still didn't go down, because that would have been too big a stroke of luck on the first time. Lothar is really well and will acquit himself well."[33]

Despite the falling snow, Eddy Lübbert took advantage of the "Fliegerwetter" to visit his old unit, Fliegerabteilung 18 (Lb) at 14:30 on 27 March. He set course for Jasta 30's airfield at Phalempin, and from there flew to Fliegerabteilung 18 Lb which was stationed at Tourmignies airfield together with Jasta 27. After a nice afternoon with his old comrades and a good coffee break, he returned to La Brayelle around 18:30 that evening.[34]

The morning of 28 March began with overcast conditions that gradually cleared up, which in turn led at times to quite lively flight activities in the 6th Army's southern region. When the weather had sufficiently improved, five Albatros D IIIs of Jasta 11 took off around 16:30 and headed toward the south of Arras. Then they followed the frontline towards the north. As the German pilots were approaching Vimy, they recognized a British formation. It consisted of three 25 Squadron F.E.2b aircraft flying cover for two 18 Squadron B.E.2e airplanes whose mission was to photograph German positions.

25 Squadron's combat report noted what then occurred:

"25. Sqn. F.E.2b's No. 4925, 7025 and 7715. Pilots: Lt. J. Whittaker, 2/Lt. B. King, 2/Lt. N.L. Knight, observers: Lt. C.W. Wilson, Lt. D.E. Holmes. 2/Lt. A.G. Severs. Hostile machines: 5 Halberstadt [sic] *and Albatros scouts.*

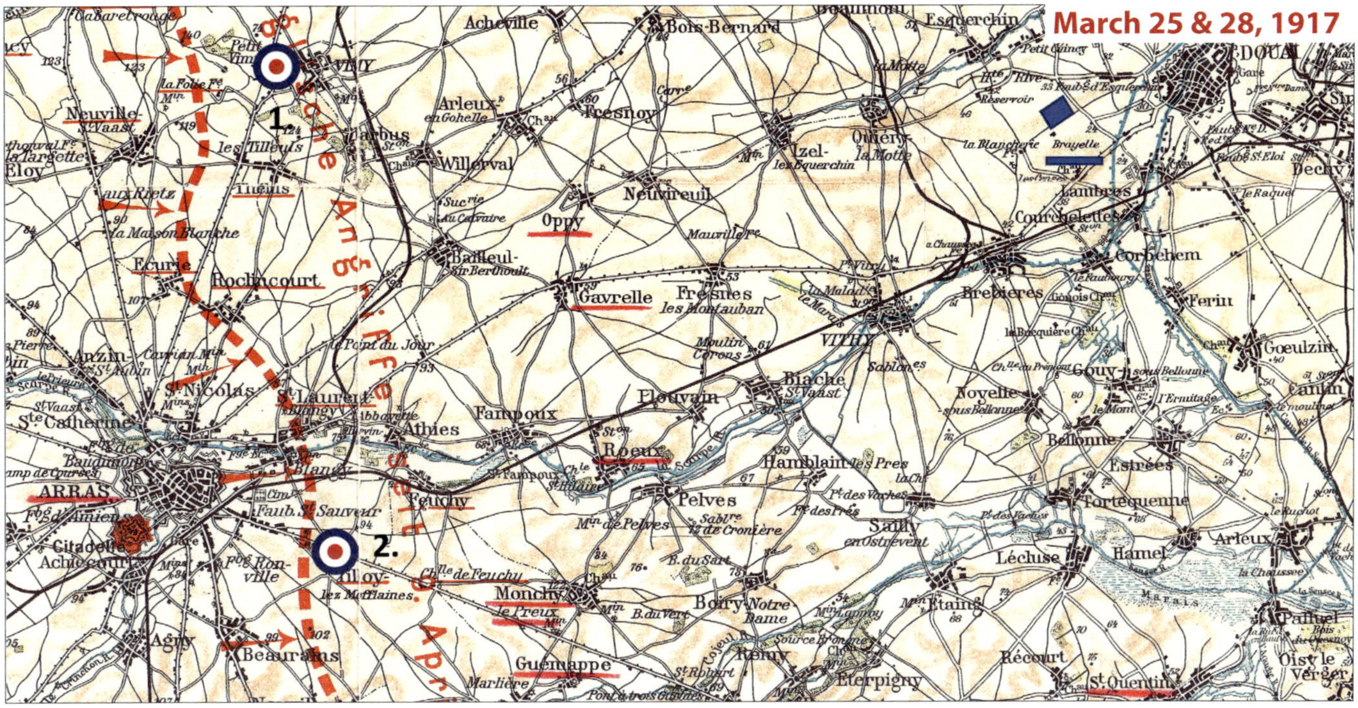

Above & Below: Map and details of the aerial victories of Jagdstaffel 11 on March 25 & 28, 1917.

25 March 1917

Nb.		Hour	Pilot	Location	Aircraft	Line
1.		8:20	Lt. Manfred von Richthofen	Tilloy-les-Mofflaines	Nieuport XVII A6689, 29.Sqn.	German

28 March 1917

Nb.		Hour	Pilot	Location	Aircraft	Line
2.		17:20	Lt. Lothar von Richthofen	Vimy Trainstation	F.E. 2b 7715, 25. Sqn.	German

The B.E`s of No. 18 Sqn. were doing Artillery work protected by the Line Patrol. Two hostile scouts made two attempt to approach and attack the B.E's but were driven off by the escort.

Shortly before 5 p.m. two scouts dived at one of the B.E's which puts itself in a spin. 2/Lt. King at once dived at and engaged the scouts. He got a bullet in his main tank but drove them off: He had to return home. The other two F.E's continued the patrol. A few minutes later two scouts with two or three more close by attacked the B.E's again. Lt. Whittaker, followed by 2/Lt. Knight, attacked the scouts and drove them away: While doing so, Lt. Whittaker had one on his tail. This was driven off by 2/Lt. Knight, but Lt. Whittaker's radiator was hit and the water poured out. He therefore come back and landed near Anzin. 2/Lt. Knight continued patrol above watching the B.E's. He was attacked by two scouts and while engaging them drifted over the lines and was set upon by three more and driven down. The machine caught fire on the way down, but the fire was out before it reached the ground. Ground observers think that the machine flattened out and landed near La Folie Farm, in hostile territory. All the B.E's returned to their aerodrome.[35]

One of the B.E. crews issued this report:

"No 16. Sqn. BE2e 2876, Pilot: 2/Lt. Crow, observer: 2/Lt. Durham.

Hostile machine: 4 Halberstadts [sic]

At 5 pm 2 Halberstadts attacked 3 F.E.2b's one of which remained behind. Went to his assistance and from about 100 yards opened fire from mounting, getting off about 30 rounds. Halberstadts retired, returning again at 5:10 with two more. All four dived on F.E.2b's. We again went to his assistance and at about 150 yards opened fire, getting off 8 rounds and observed about 20. We were a considerable distance over the lines and at a very low altitude. F.E. was in great difficulty and seeing we would be of no future use we returned. F.E. descended and engine caught fire about 150 feet from the ground. He was seen to land successfully at approx. S.17.Central."[36]

It was Lothar von Richthofen who had forced F.E.2b 7715 to land east of the Vimy train station. Its wounded pilot, 2/Lt. N.L. Knight, was taken prisoner, but observer 2/Lt.

Above: Snapshot of Eddy Lübbert possibly taken at La Brayelle airfield. (L. Bronnenkant)

Above: The sad wreckage of Nieuport XXIII A6780, 40 Sqn. RFC, shot down by Lt. Allmenröder, also on 30 March. The impact was so severe that the Le-Rhône engine was half-buried in the ground. Lt. D. M. F. Sinclair was killed as a result of the same engagement that cost Eddy Lübbert his life.

Above: Lt. Kurt Wolff, with his hands in gloves, resting on the trailing edge of the lower wing, inspects the wreckage of Nieuport XVII A 273. This was an aircraft from 60 Sqn. RFC, which he shot down on 30 March 1917. It was from this point that Kurt Wolff observed the dogfight leading to the fatal crash of Eddy Lübbert. He was the first member of Jagdstaffel 11 to be killed in combat. (R. Absmeier)

Above: The tombstone marking the grave of Eduard Lübbert in Hamburg. Curiously, it carries his nickname, rather than his official first name.

A.G. Severs had been killed in action. It was Lothar's first confirmed victory.

2/Lt. Knight's interrogation report stated:

"1. Personnel:

Lt. Leslie Knight, 18½ years old joined the RFC one year ago and soon afterwards was assigned to a squadron in Flanders... Transferred to his current squadron 5 days ago, shot down on his second flight; pilot. Two brothers in the field, one of whom died as an English anti-aircraft officer in Flanders.

2. Aviation Associations:

Avoided naming his squadron and airfield; F.E. two-seater.

3. Orders for Last Flight:

Along with two other F.E.s, fly a blockade flight north of Arras and cover airplanes directing artillery.

According to his orders, the blockade flight was supposed to last up to 5:00 a.m. The other two F.E.s got lost five minutes before 5:00 a.m., but as the youngest aviator he in no way wanted to return home before the others. Just as he was about to break off the flight at 5:05 a.m., he saw two Albatroses coming towards him. Since they were superior to him in speed and escape was unthinkable, he decided to attack; an Albatros flew over him at a distance of less than 100 meters and his observer could easily have shot him down if it had not been for jams in both machine guns. Was then suddenly attacked from behind by a third Albatros that had not

11.) Eigene Verluste:

Am 30. 3. 17 2,15 nachm. Lt.Lübbert von J.St. 11 im Luft= kampf abgeschossen und bei Gavrelle töd= lich abgestürzt.

Above: Excerpt from the Kofl-Report dated 30 March 1917:
11) Own Losses. On 30 March [19]17, 2:15 p.m., Lt. Lübbert, Jasta 11, was shot down in air combat and fatally crashed near Gavrelle." (Bayerisches Hauptstaatsarchiv, Abt.4, München) The Kofl reports listed all German losses where a pilot was either killed or wounded enough to be put out of action.

Lt. A. Strange stands in front of his crash-landed aircraft. (G. VanWyngarden)

Above: German soldiers seize the opportunity to have their picture taken next to the downed F.E.2b 7691. This view vividly illustrates the size of this type of aircraft. (L. Bronnenkant)

Right: The cockpit of F.E.2b 7691 was marked with the inscription "New South Wales No 4. - The F J White Sumarez and Baldblair" below the cockpit. (G. VanWyngarden)

Above: As was usually the case, F.E.2b 7691 was guarded by German soldiers until it was transported to the Armeeflugpark. Apparently, strong winds had flipped the FE2b over onto it back by the time this photo was taken. (G. VanWyngarden)

been seen before; observer, Lt. Severs, was killed immediately and the airplane was set on fire; The machine overturned upon landing and the prisoner was unable to extricate the observer, who was incinerated. Prisoner is slightly wounded in the arm and foot and has burns to face.

4. Airplane:

Prisoner described the F.E. as a terribly slow airplane; however, it flies itself.

Tattenburg [?]

Hauptmann and Intelligence Officer."[37]

One of the attacking machines was described as a *"red Halberstadt"* in another report submitted by one of the 16 Squadron B.E. crews. Since Manfred von Richthofen was flying the Jasta's only red-painted airplane at the time, it appears that he was involved in this dogfight.

It was "Fliegerwetter" again on 29 March with low clouds and rain, and flight operations were therefore suspended on both sides of the front. Eddy Lübbert used the opportunity to write to a friend:

"Dear Gustav,

Thank you very much for your letter, which I can only answer today since I was wounded again in the interim. During an air battle against superior forces, they shot me through my left chest and right arm. But I'm ready for action again now. By the way, I'm in Douai. It would be great if we could see each other again sometime. Telephone me in the evening between 7:00 and 8:30. My Staffel leader is the brilliant Oblt. von Richthofen.[38]

When Jasta 11's pilots looked out their windows on the morning of 30 March, they saw low-hanging clouds and

Above: The pilot of F.E.2b 7691, Lt. Leslie A. Stange, photographed in September 1916. (G. VanWyngarden)

intermittent rain. Accordingly, the front initially reported relatively light flight activity. Nevertheless, Jasta 11 sent up a Kette of five aircraft at 11:30 a.m. to patrol along the Douai–Arras railway line. Lt. Kurt Wolff eventually became separated from his comrades but then spotted a single

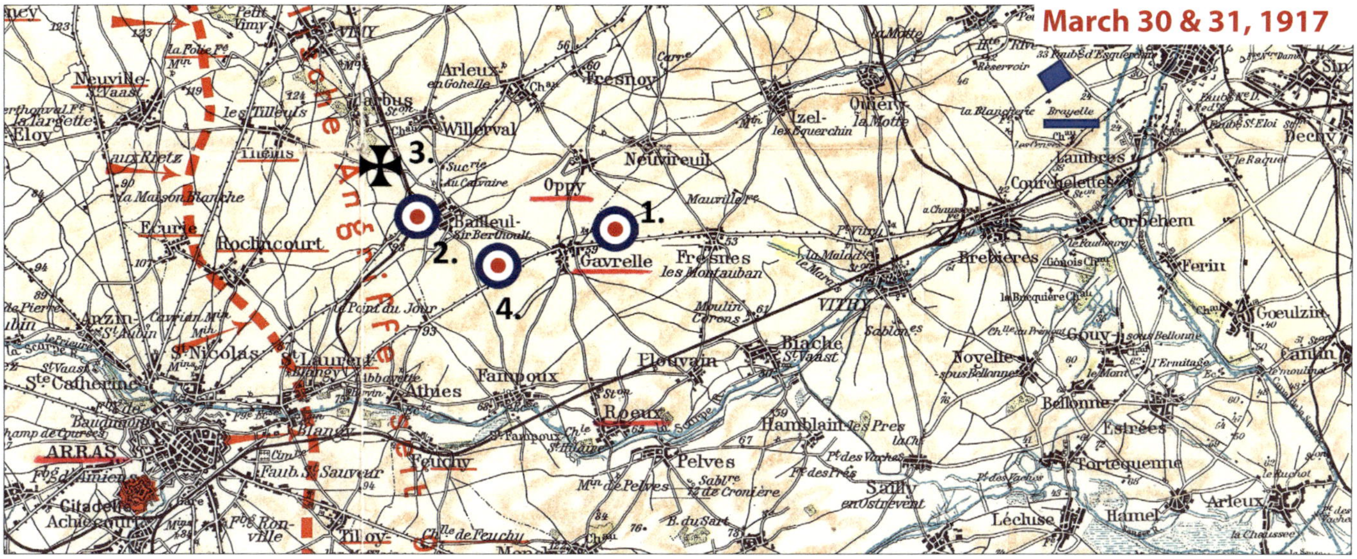

Above & Below: Map and details of the aerial victories and loss of Jagdstaffel 11 of the March 30 and March 31, 1917.

March 30, 1917

Nb.		Hour	Pilot	Location	Aircraft	Line
1.		11:45	Lt. Kurt Wolff	East of Gavrelle	Nieuport XVII A273, 60. Sqn.	German
2.		14:15	Lt. Carl Allmenröder	Bailleul-Sir-Berthould	Nieuport XXIII A6780, 60. Sqn.	German
3.		14:15	Lt. Hans-Georg Lübbert	Between Thélus and Bailleul-Sir-Berthould	Albatros D III	German

March 31, 1917

Nb.		Hour	Pilot	Location	Aircraft	Line
4.		7:50	Lt. Kurt Wolff	Sugar-Factory south-east of Bailleul-Sir-Berthould	F.E. 2b 7715, 25. Sqn.	German

Nieuport fighter, also on his own, near Gavrelle. The pilot of Nieuport XVII A273, Lt. W.P. Garnett, had been part of a flight of six Nieuports belonging to 60 Squadron, but like Kurt Wolff, he had also lost contact with his comrades. A dogfight ensued and Wolff succeeded in downing the Nieuport at 11:45. He later wrote in his diary:

"On the 30th, I encountered a Nieuport east of Arras; we were both alone. After a long dogfight, the Englishman dove, and a wing broke away in the dive. The wreckage lies at Gavrelle. The pilot had been shot in the thigh, the fuel tank had not been hit, and nothing else could be determined from the wreckage. Machine No. A273, pilot Lt. Garnett (60 Squadron), M.G. Lewis 11711. It was my fourth shoot down."[39]

While this was occurring, Wolff's four Jasta 11 comrades caught up with Garnett's squadron mates: Lts. Binnie, Townesend, Bower, Bishop, and Maj. J.A.A. Milot. Bishop reported:

"While on offensive Patrol, two machines engaged 5 Albatros Scouts and chased another. Cloud and wind prevented decisive results. Lt. A. Binnie fired on two, Lt. W.A. Bishop on three Albatros Scouts, Major J.A.A. Milot dived twice on hostile aircraft, but his gun jammed. Lt. E.J.D. Townesend chased one hostile aircraft but was unable to get within firing distance.

Lt. W.P. Garnet and 2/Lt. Bower have not returned."[40]

German flak fired at the 60 Squadron Nieuports as they were returning home, and a British report stated:

"… Lt. F. Bower… was hit from behind by an explosive bullet and his aircraft went into a spin. He recovered and, in a state of shock, with blood pouring from his gaping wounds and holding in his intestines with one hand, he flew west until he no longer could hear anti-aircraft fire. He then landed at Chipilly, climbed out of his Nieuport and walked forty yards before collapsing. He died in hospital the next day…"[41]

That same afternoon, Eddy Lübbert and Carl Allmenröder set off together on a flight to the front at around 2:00 p.m.. Lübbert's usual half-blue, half-yellow machine was still not ready for use as an engine change had to be carried out; so just as he had in recent days, he borrowed another Albatros from a Staffel comrade.[42] The two men turned west toward the front and passed Fresnoy. When they had reached the area south of Vimy and were still gaining height, they were surprised by the diving attack of a formation of No. 40 Squadron Nieuports. Allmenröder was able to evade their first pass with a skillful maneuver and, for his part, got behind Lt. D.M.F. Sinclair in Nieuport XXIII A6780. Allmenröder opened fire and saw his bullets strike the fuselage. The Nieuport turned over, nosedived, and hit the ground west of Fresnoy at 14:15. Lt. Sinclair, his opponent, lost his life as a result.

In turn, Eddy Lübbert was attacked by Capt. W.R. Gregory in Nieuport XXIII A6680. Gregory reported after his return:

"Flying at 15,000 feet saw two Halberstadts [sic] *flying among clouds at 12,000 feet south of Vimy at 14:15.*

I dived on hostile aircraft who was heading East. I got on tail of hostile aircraft and fired one large drum at very close range. I started firing at 40 yards, finishing at about 15 yards.

Hostile aircraft turned over and was last seen side slipping into a cloud near Bailleul. Fight finished at 5,000 feet."[43]

Capt. Gregory likely hit his opponent's fuel tank, as the Albatros D III began to burn in the air before it smashed into the ground between Thélus and Bailleul-Sir-Berthoult.

After Allmenröder had shaken off his opponents, he began to search for Lübbert whom he had befriended during the past weeks. With his visibility obstructed by low-hanging clouds, he circled for a long time in the dogfight's area but could not find Lübbert. When his fuel began to run low, he flew back to La Brayelle, hoping that his friend had already landed there. It was only then that he discovered that Lübbert was missing. Shortly afterwards, at around 14:40, the Staffel received the following message from Flak Vitry:

"1 Albatros crashed in flames, Bailleul – Thelus; occupant dead, apparatus completely destroyed."[44]

While visiting the crash site of the victory he had scored that morning near Gavrelle, about three kilometers east of Bailleul-Sir-Berthoult, Kurt Wolff witnessed Lübbert's final air battle from the ground, and later wrote in his diary:

"When I was at the crash site that afternoon, I saw a fight between two Albatros single-seaters and three Nieuports overhead. After a short fight, a smoking Albatros plunged down vertically close to the village of Gavrelle. It was Lübbert. The crash looked awful, the engine completely buried, the corpse completely mangled deep in the earth, the machine incinerated up to the rudder. At least Al[l]*menröder shot one of the Nieuports down at Fresnoy* [-en-Gohelle], *also a fatality, machine totaled."* [45]

It was a strange twist of fate that Hans-Georg Lübbert had been killed by the very same man – Capt. W.F. Gregory – who had wounded him before on 6 March.

Hans-Julius Lübbert, Eddy's father, who at the time was the Director of the German Military Administration in Romania, was sent the following telegram:

"Lt. Lübbert killed in action in air combat today, Jagdstaffel 11."[46]

Jagdstaffel 11 had suffered its first fatality more than five and a half months after its formation. Pilots had been wounded in combat before, but for the first time, the pilots of Jasta 11 were confronted with the loss of a beloved comrade and friend, and there was much grief. The sensitive Carl Allmenröder was deeply affected by the death of his friend, even though he was not to blame for it. Without saying a word, he retired to his room and wrote a letter to his fiancée. Manfred von Richthofen went to see him that evening and privately spoke with him far into the night to raise his spirits again.[47]

Richthofen, who greatly valued Lübbert, was also touched by his death. But his nature and upbringing discouraged him from displaying his feelings to the outside world. Besides, his job was to support his men and help them to overcome their grief. In such situations, he believed in the rider's motto: *"After a fall, you get right back up on the horse."* So, he kept his men busy to distract them and the next morning, the Staffel went back into action.[48]

If Hans-Georg "Eddy" Lübbert had not fallen for his fatherland on 30 March 1917, he would have faced persecution as a Jew some 20 years later. This is because his family, despite having converted to Christianity, were nevertheless viewed by Nazi authorities as having "Jewish blood" and he would have been exposed to the persecution of Jews in the 3rd Reich, just like his younger brother and the whole family.[49]

The weather on 31 March was no better than the previous day's – low clouds moved across the sky, and it was raining. As a result, flight activities were low to moderate, with elevated enemy activity being reported only south of Lens. Nevertheless, Jasta 11 received an early request from the front to which it responded by sending off one of its Ketten toward Arras shortly after 7:00 a.m. Manfred von Richthofen, who assigned Carl Allmenröder to this early mission, urged him to stay close by his side since Carl was still shaken by Lübbert's death. As they approached Arras to turn north along the front, the Kette encountered a British formation that consisted of No.11 Squadron FE.2s conducting reconnaissance and their escort, Nieuport fighters from No. 60 Squadron. Richthofen and his men went on the attack northeast of Arras. Allmenröder was the first to reach the enemy formation and went after a Nieuport at 7:35 a.m. only to have to let him go when another Nieuport attacked him in turn.

Lt. Bishop, one of the 60 Squadron pilots, later reported:

"While on escort duty I went to the assistance of another Nieuport which was being attacked by an Albatros scout. I opened fire twice, the last time at 50 yards range. My tracers were seen to hit his machine in the center section. The Albatros seemed to fall out of control, as he was in a spinning nose dive with his engine on. Hostile aircraft crashed at 7:30 south of Gavrelle."[50]

2/Lt. Leckie, an eyewitness in the air to Bishop's fight, added:

"I was behind Bishop and saw the Albatros scout go down in a spinning nosedive, seemingly out of control."[51]

It seems that Bishop had attacked Allmenröder's Albatros D III, which spun down to escape. But it was not shot down. There were no German losses that day anywhere in the 6th Army area.[52]

In the meantime, Kurt Wolff approached 11 Squadron's FE.2b 7691 at 7:40 a.m. and opened fire. After several volleys, the observer's machine-gun fire fell silent. Wolff stayed behind his opponent, but instead of firing again, forced it to land near Gavrelle's sugar factory. German soldiers rushed to the spot and captured the unwounded pilot, Lt. L.A.F. Strange. His observer, 2Lt. W.G.T. Clifton, had suffered a fatal shot to the stomach.

Wolff's diary entry read:

"I got hold of an F.E. two-seater the next day at the same location. I severely wounded the observer with the first shot – he died that evening – and then I 'drilled' [i.e. forced] *the machine down. It landed near Gravelle, close to Lübbert's crash site. Machine no.7691, Engine Bear*[d]*more 918, M.G. Lewis 17989 and 12889, pilot. Lt. Strange and observer Lt. Clifton (dead). That was number 5."*[53]

Here is an excerpt from Lt. A. Strange's interrogation report:

"1. Personnel:

Lt. A. Strange, an engineer with the London Underground before the outbreak of the war, in an infantry regiment in 1915, went to England in August 1916 to train as a pilot at a military aviation school, and in mid-January 1917, after just two days with the St. Omer Aircraft Depot, was with his squadron as a pilot...

3. Orders for Last Flight:

To fly cover for one of the squadron's photography aircraft along with a third F.E.; Nieuports should have been there for cover too, but were not seen once again. The three aircraft obviously were flying low because of their orders. The Albatros' first shot seriously wounded the observer, Lt. W.G.T. Clifton. Since both of the aircraft's machine guns were mounted with the observer, the prisoner was therefore utterly defenseless and had to land, particularly since his fuel tank was also shot through; his clothes were riddled with bullet holes.

4. Airplane:

The large F.E. is considered quite inferior. It is too slow and too cumbersome, flies 70 miles per hour. The Spad is considered the best and newest machine. Prisoner persistently inquired whether Lt. Baker's Spad machine had fallen into German hands undamaged.

A red Albatros is known among English aviators as particularly dangerous.

Tattenburg [?]

Hauptmann and Intelligence Officer"[54]

Leslie Stange, following his return to England on 31 December 1918, reported:

"Yes, attacked by enemy aircraft when in rear position in formation. Observer mortally wounded with first burst, engine and petrol tanks hit... Unable to destroy machine as observer could not be removed before the arrival of German soldiers who came up at once.[55]

Lt. E.T.C. Brand and Lt. H.W. Guy, who had been flying F.E.2b A808, stated:

"Over Boiry an Albatross scout, one of a formation of 4 hostile aircraft attacked an F.E. of the formation.

I fired one double drum at 300 yards range, several tracers appeared to hit, the hostile machine went down under control."[56]

F.E.2b 7691, forced down by Wolff, was a so-called "presentation aircraft" with the label: *"New South Wales No 4. The F J White Sumarez & Baldblair"*: These were airplanes purchased for the British Air Force through donations from cities, municipalities, companies, or even individuals from across the Commonwealth that bore the donor's name on their fuselages. The British Air Force received about 900 of these sponsored aircraft – a resource that the Germans did not have.[57]

This was Jagdstaffel 11's final victory for the month of March 1917. Just under 10 weeks from the time that Manfred von Richthofen had assumed command, the Staffel had achieved 36 confirmed air victories and lost only one pilot in air combat. This surpassed Richthofen's former unit, Jagdstaffel 2 "Boelcke", which compiled 30 victories during the same timeframe.[58]

During the month of March 1917, a total of 38 enemy aircraft were confirmed as victories within the German 6th Army's sector. Of these, 28 were attributed to Jasta 11, or about 74% of the total.[59] This can be considered an extraordinary achievement!

Manfred von Richthofen's leadership and training program had obviously made the difference. Starting with a group of unsuccessful beginners, he had formed the most successful Jagdstaffel for its time. The foundation for the Richthofen legend and that of Jagdstaffel 11 had been laid.

Hermann Lohmeyer recalled that as Jasta 11's successes increased, and it was mentioned and praised more often in the press, there was a commensurate rise in the public's awareness of the unit. This in turn led to an increase in the number of the so-called "Liebesgaben" – gifts sent to men at the front from members of the public – delivered to the Staffel, which of course was very much welcomed by them. Edible items like ham and quality canned food, or beverages such as beer or wine, were particularly popular. One large winery in the Rheingau area sent an entire truckload of excellent Rhine wine to the "heroes of the skies" without being asked. Some breweries insisted on honoring the Staffel with whole barrels of beer. When the Jagdstaffel received such allotments, they were shared with the ground crewmen and the pilots took special care that their mechanics did not come up short. A lack of food and drink – especially champagne, which the pilots loved – did not exist at Jagdstaffel 11. This was one of the more pleasant side benefits of winning fame from hard-fought air combat.[60]

The tutelage by their Staffelführer had paid off handsomely in March 1917, and the Albatros D III had proven to be a highly capable fighter aircraft in the hands of skillful pilots. Manfred von Richthofen's Jagdstaffel 11 was more than ready to meet the upcoming challenges of the major British offensive that was expected to commence in the near future.

	Stand am 3. 19		Abgang vom 1. Januar bis 3 28. Februar 17. 19			
	Fabrik	Nummer	Fabrik	Nummer	Grund	Verbleib
R G.			Ru.G.I.	116/16	28.1.Kriegsmaschine	A.Fl.P.1.
G D.	a)Schulflugzeuge		Halb.D.V.	1118/16	4.1. Kriegsmasch.	J.St.11.
	Halb.D.V.	414/16	Alb.D.III.	1993/16	21.1. "	" 11.
	"	801/16	" III.	1996/16	" "	" 11.
	"	805/16	" III.	1980/16	22.1. "	" 11.
			" III.	1991/16	22.1. "	" 11.
	b) Flugzeuge, die n/Deutschland verladen werden:		" III.	2006/16	24.1. "	" 11.
			Halb.D.V.	412/16	8.2. "	" 30.
			" V.	420/16	" "	" 30.
D	Halb.D.II.	116/16	Alb.D.III.	1985/16	9.2. "	" 11.
	"D.V.	412/16	Halb.D.V.	418/16	17.2.Kriegsunbr.	Märk.Flugzgw.
	"D.V.	415/16	" II.	107/16	" "	" "
	"D.V.	421/16	Alb.D.III.	2038/16	25.2.Kriegsmasch.	J.St.30.
	"D.V.	1100/16	" III.	2051/16	" "	" 30.
	"D.V.	1118/16	" III.	2054/16	" "	" 30.
			" III.	2062/16	" "	" 11.
			" III.	2015/16	26.2. "	" 11.
E D.			" III.	2016/16	" "	" 11.
			" III.	2036/16	" "	" 11.
			" III.	2014/16	27.2 "	" 11.
			" III.	2026/16	" "	" 11.
			Halb.D.V.	413/16	" Kriegsunbr.	Märk.Flugzg-werft.

Report filed by the Armeeflugparks La Briquette near Valenciennes dated 31 March 1917 listinging the delivery of aircraft to Jagdstaffeln 11 and 30, amongst others. (Bayerisches Hauptstaatsarchiv, Abt. 4, München)

Leutnant Frhr. v. Richthofen (im Pelz links),
der ausgezeichnete Fliegeroffizier, der bisher 26 feindliche Flugzeuge im Luftkampf besiegte.

Above: Newspapers and journals very quickly recognized the public's interest in Manfred von Richthofen. This is probably one of the first mentioning in a newspaper about his aerial victories. As Manfred von Richthofen achieved his 26th aerial victory on March 11, 1917, and was promoted to Oberleutnant on March 23, 1917, the publication must have been published within this time frame. (G. VanWyngarden)

Nummer 14. | DIE·WOCHE Bilder vom Tage | Seite 461.

Oberleutnant Frhr. von Richthofen.
Zu seinem 30. Luftsieg.

Above: Manfred von Richthofen's successes were already being celebrated in the German press in April 1917. Here is the edition of the magazine "Die Woche" No. 14 on the occasion of his 30th aerial victory, which he achieved on 24 March 1917. (Taktisches Luftwaffengeschwader 71 "Richthofen")

Aircraft Markings and Colors

When depicting the personal and unit markings applied to the aircraft of a single-seater unit, we must always keep in mind that most available photographs usually only document a part of the aircraft. The vast majority of these photos were taken by members of the respective units as a personal memory and migrated to private photo collections and photo albums after the war. Almost none of the German aviation enthusiasts who were in contact with numerous former fighter pilots in order to collect information in the 1920s and 1930s had photographic equipment at their disposal. As a result, their photo collections were usually rather small. In most cases, they relied on receiving a photo as a gift. Some of the former fighter pilots were very generous, such as Paul Bäumer or Josef Jacobs, who had the photos reproduced at their own expense and gladly provided them to the aviation enthusiasts who contacted them. A lot of private, but also official, photo material was irretrievably lost during the 2nd World War due to bombing or outsourcing, or simply left behind when the eastern part of the German Reich came under Russian occupation.

For these and other reasons, the available photos of the aircraft of a Jagdstaffel always depict only a small part of the aircraft that were actually used. Therefore, it is necessary and legitimate to also resort to written or oral information of the aircraft paint schemes provided by contemporary witnesses when they provide information not documented by available photos. This data must then be interpreted in the context of other available information, and presenting this information via reconstructions assures that it is not lost. This is done with the hope that photos providing further information might appear at a later date. Matching these "new" photos to contemporary descriptions will then enable enthusiasts to come up with more detailed depictions at some time in the future. Color profiles created on the basis of such information are therefore accordingly marked as "reconstruction".

The scientifically untenable "grey-scale interpretation" of the colors depicted by orthochromatic black-and-white photos as a sole method of photo interpretation was not resorted to.

The Halberstadt Era

Available photographic coverage of the Halberstadt fighters operated by Jagdstaffel 11 indicates that they all retained their respective factory camouflage schemes, with only small personal markings appearing in a few cases. And this is exactly where a major complication appears.

Jasta 11 was equipped with a single example of the Halberstadt-built D II (D.107/16) and and one D III (D.116/16). However, no photograph confirmed as showing either of them is known to the author. Furthermore, several examples of the Halberstadt D V in the range D.410/16 to 421/16, Hannover-built D IIs in the D.801/16 to 805/16 range, as well as three late-production D Vs in the D.1099/16 to 1118/16 range were used by Jasta 11.

To date, no official documents describing the color of the factory finishes applied to the various production batches of Halberstadt fighters have been discovered. By 1916, the Halberstadt company was very small in comparison to other aircraft companies such as Albatros, Aviatik or L.V.G. For this reason, the company was not in the position to order large amounts of a specific fabric for their aircraft. Rather, it had to procure what was available on the market. For this reason, the aircraft of the various D-Type batches produced by Halberstadt were covered in fabric colored in different ways.

The British and French reports on the few captured Halberstadt D-Types are equally inconclusive in this regard. Taking this lack of first-hand documentation into account, we must rely on the testimonies of contemporary eyewitnesses and attempt to painstakingly match them to available photographs.

According to statements made by former fighter pilots to German air historian Heinz Nowarra in the 1960s, the various types of Halberstadt fighters received different color schemes at the factories. Nowarra reported that he found out that some had a "light blue" overall finish, while others had a bluish-gray color on their upper surfaces. Accordingly, these two versions were presented in his book, "*Eisernes Kreuz und Balkenkreuz*".[1]

Rudolf Nebel, who was a member of Jasta 5 in the late autumn of 1916, confirmed Nowarra's research in a statement made to the author. He stated that some of the Halberstadt fighters used by Jasta 5 during his time there had a light blue fabric whereas others were "bluish field gray."[2]

Matching Nebel´s statements to available photographs, it becomes apparent that the light blue scheme had been applied to Halberstadt D IIs, while the "bluish field-gray" matches the darker color of the D Vs operated by Jasta 5 in late 1916.

In addition to these two descriptions, another "overall" color applied to Halberstadt fighters has been documented. In his book "*El Schahin* (*Der Jagdfalke*)", Hans-Joachim Budddecke referred to the Jasta 4 Halberstadt single-seaters as "our brown rats."[3] This indicated that these fighters were covered with a fabric colored in a greyish-brown color. Photographs document that during this period of time Jasta 4 operated examples of both the Halberstadt D III and D V variants.

The Halberstadt D II fighters built under license by Hannoversche Waggonfabrik AG were finished in a two-color camouflage scheme on their upper surfaces that may be interpreted as having been rust red-brown and light green, whereas the undersurfaces were light blue.[4] The

Hannoversche Waggonfabrik had only started its production of aircraft in 1916 by manufacturing aircraft under license such as the Rumpler C I. These aircraft appear to have been finished in a rust red-brown and light green camouflage paint scheme. It seems logical to assume that the Halberstadt D II (Han.) aircraft, built around the same time, were also finished this way.

In summary then, it appears that the fabric issued by the Halberstadt factory for its D II, D III, and D V fighters: light blue (D IIs and some early D IIIs), greyish-brown, likely consisting of two shades of brown (late D IIs, late D IIIs, and possibly some early D Vs), and bluish-gray (likely several D.Vs in the first batch and likely the final batch, too). The license-built D.IIs manufactured by the Hannoversche Waggonfabrik AG are thought to have been finished in a rust red and light green camouflage scheme on the upper surfaces, light blue on the lower surfaces..

An evaluation of the available photos depicting the Halberstadt fighters operated by Jasta 11 indicates that its Halberstadt D II fighters might have been greyish-brown (possibly consisting of two shades of brown), its D II (Han) fighters were rust red and light green on their upper surfaces and light blue on their lower surfaces, and its three D Vs in the 1099/16–1118/16 range were bluish-gray.

Oblt. Rudolf Lang introduced an aircraft identification system that he knew from his previous service in the Kampfstaffeln of the Kampfgeschwader -- that is, a white letter that normally represented the initial of a pilot's first or last name was applied to both sides of the fuselage.

1. Lt. Carl Allmenröder, Halberstadt D V

The aircraft had the fuselage and wings most likely in a factory-applied bluish-field-gray camouflage, consisting of two shades of blue. His personal marking was a white 'C' for his first name Carl, which was applied on both fuselage sides.

2. Sgt. Hans Howe, Halberstadt D II (Han.) D.805/16

The aircraft had a factory-applied rust red-brown/light green camouflage pattern on the fuselage and upper part of the wings, whose undersurfaces were light blue. The personal marking was a white 'H' on a black background on both fuselage sides. The aircraft was included in the 31 December 1916 aircraft inventory list of Jagdstaffel 11, which meant that it was serving with the unit on that date. On 28 February 1917, the aircraft was returned to Armeeflugpark 6 in exchange for an Albatros D III.(5)

3. Lt. Konstantin Krefft, Halberstadt D V

The aircraft had the fuselage and wings most likely in a factory-applied bluish-field-gray camouflage, consisting of two shades of brown. His personal marking was a white "K" for "Krefft" on both fuselage sides.

4. Lt. Carl Allmenröder?, Halberstadt D V

In one line-up photo of Jasta 11 Halberstadt fighters, a D V marked with the letter "C" on a black background can be seen. The aircraft was most likely covered in a bluish field-gray, perhaps in two different tones.

It would have been quite unusual for the aircraft of different pilots to have carried the same personal marking. We know that a Staffelführer, who commonly had two aircraft at his disposal, sometimes had them marked identically; however, no connection between a 'C' and Oblt. Rudolf Lang is apparent here. Therefore, it seems more likely that one of the two "C"-marked Halberstadt D Vs was an older aircraft flown by Lt. Carl Allmenröder that was due to be returned to the nearby Armeeflugpark.

5. Lt. Georg Simon, Halberstadt D II (Han.) D.801/16

The aircraft had a factory-applied rust red/light green camouflage pattern on the fuselage and the upper part of the wings and light blue undersurfaces. The personal marking was a white "S" on a black background on both fuselage sides. As with D.805/16 above, this aircraft was included in Jasta 11's 31 December 1916 inventory list and was later returned to Armeeflugpark 6 in exchange for an Albatros D III on 28 February 1917.(6)

6. Lt. Kurt Wolff, Halberstadt D II (Han.) D.802/16

The aircraft had a factory-applied rust red/light green camouflage pattern on the fuselage and upper part of the wings and light blue undersurfaces. The personal marking was a white "W" on a black background on both fuselage sides. The aircraft is listed in the aircraft inventory chart of Jagdstaffel 11 dated 31 December 1916, documenting that it was operated by the Staffel at that time. Like several other Halberstadt D-types, this particular plane was also returned to Armeeflugpark 6 on 28 February 1917.(7)

7. Lt. Carl Allmenröder, Halberstadt D V, February 1917

The aircraft is an example of the late-production D Vs (1099/16–1118/16) that were finished in bluish grey on the upper surfaces and light blue on the lower surfaces of the fuselage and wings. At this time, his personal marking was a black cannonball on both sides of the fuselage. This was a reference to his previous service with the artillery. As a former member, he regularly flew over the German artillery positions and waggled his wings to greet his brothers in arms. He flew this aircraft when he scored his first confirmed victory on 16 February 1917, and kept doing so until the grounding of the Albatros D IIIs was lifted.(8)

8. Albatros D I D.425/16

Albatros D I D.425/16 was the first and only Albatros D I delivered to Jagdstaffel 11 in the "Halberstadt" era and arrived there on 11 October 1916. The aircraft retained its "factory-finish" and did not receive any kind of individual markings. After having served with the Jasta for roughly two months, it left Jagdstaffel 11 on 14 December and was taken on strength by Jagdstaffel 10, which was stationed at the time near Aincreville, in the sector of the 5th German Army.

1. Lt. Carl Allmenröder,
Halberstadt D V

2. Sgt. Hans Howe,
Halberstadt D II (Han.)
D.805/16

3. Lt. Konstantin Krefft,
Halberstadt D V

4. Lt. Carl Allmenröder?,
Halberstadt D V

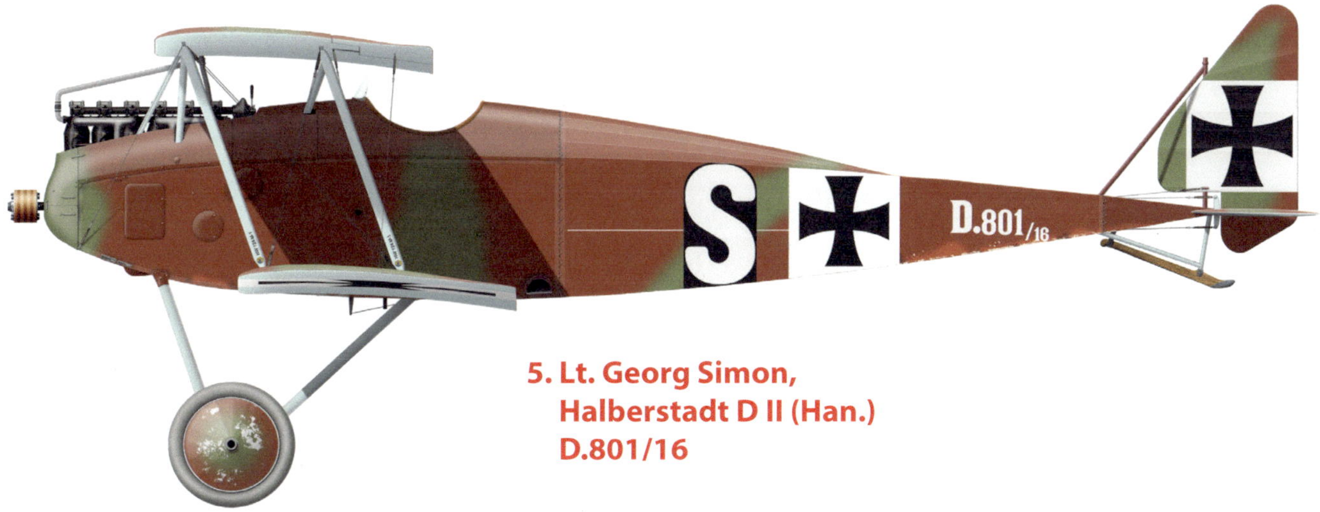

5. Lt. Georg Simon,
Halberstadt D II (Han.)
D.801/16

6. Lt. Kurt Wolff,
Halberstadt D II (Han.)
D.802/16

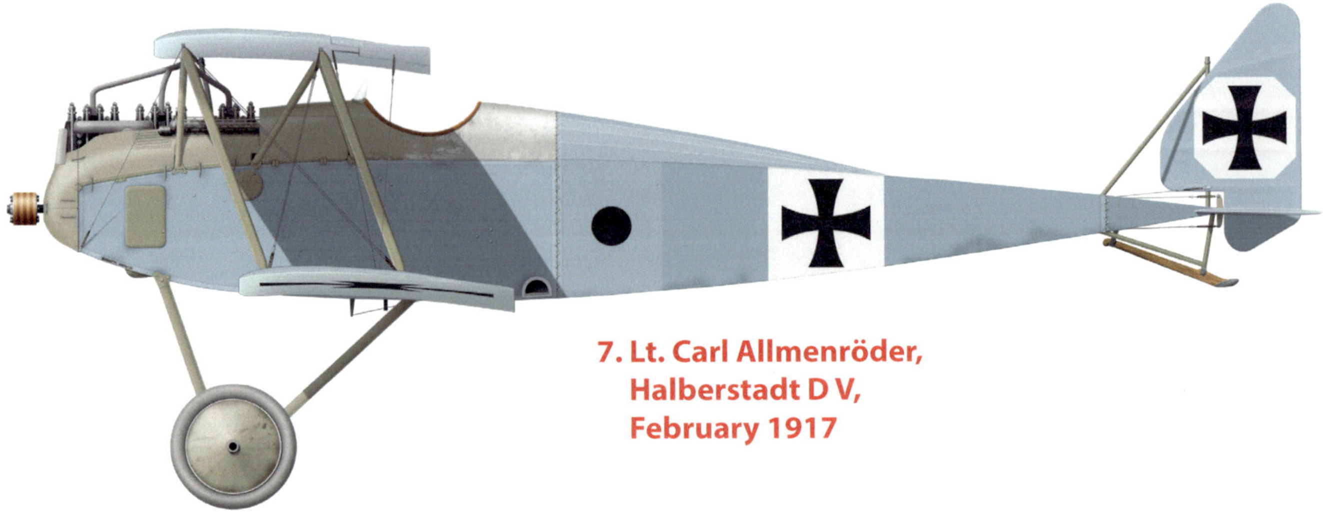

7. Lt. Carl Allmenröder,
Halberstadt D V,
February 1917

8. Albatros D I
D.425/16

9. Lt. Manfred Freiherr von Richthofen,
Albatros D III
Le Petit Rouge

9. Lt. Manfred Freiherr von Richthofen,
Albatros D III
Le Petit Rouge

9. Lt. Manfred Freiherr von Richthofen,
Albatros D III
Le Petit Rouge
Alternate Interpretation

The Albatros D III Era

The number of available photos of Jagdstaffel 11's aircraft during the period from October 1916 to March 1917 is surprisingly low. Those used herein came from Carl Allmenröder's photo album, Emil Schäfer's album as well as the Krefeld City Archive's collection of Schäfer photos, Hermann Lohmeyer's photo collection, Ed Wischnowski's negatives from Manfred von Richthofen's album, and other generous sources such as Lance Bronnenkant, Greg VanWyngarden, Reinhard Kastner, Reinhard Zankl, and Rainer Absmeier.

The most important written or oral information relating to Jasta 11's Albatros markings from January to March 1917 came from one of the unit's mechanics, Hermann Lohmeyer, and from Georg Simon, who related his memories to Alex Imrie.

The records in the Flugmeldebuch (flight reports book) of A.O.K. 6, as well as British air combat reports, also provided information in this respect.

Hermann Lohmeyer reported on the painting schemes applied to the aircraft in February and March 1917:

"After a machine arrived with us and was assigned to a pilot, it was painted according to his liking. Each machine had its own color. I think it was Richthofen himself who ordered different paint schemes to be applied to the aircraft, so that he could immediately recognize every machine in the air. So very soon [he later related that it was before the Battle of Arras in April/May] *we had colorfully painted planes, an impressive sight! It looked like a rainbow. Of course, I can't remember who flew which plane, except for the machines to which I personally applied the brush to."*[1]

Georg Wegener, a journalist for the *Kölnische Zeitung* magazine who wrote about Jasta 11's Albatros D III markings in mid-April 1917, confirmed Lohmeyer's statement:

"At the edge of the free airfield, five biplanes in a row were ready for departure. A sixth, that of Freiherr von Richthofen, a little forward to the side... ***and all painted differently. From a distance, they looked like iridescent giant insects, like a swarm of glowing butterflies that sunbathe on the ground with outspread wings.*** *The principle of making yourself as air-colored as possible was completely abandoned here. "Invisibility," they explained to me, "you can't achieve it, but you run the risk of confusion of enemy and friendly aircraft". These different markings on the fuselages are clearly visible in the air, you recognize yourself during the fight and can support yourself. That's why every flyer gave his personal machine ... a special marking that allowed his comrades to keep an eye on him in air combat and always know who is at the controls of the machine."*[2]

According to available documents and the evaluation of available photos, the upper wings of the first Albatros production batch D.1910/16 to 2309/16 had a light-green, rust-red, and dark green camouflage, while the lower part of the wings was light blue.[3]

9. Lt. Manfred Freiherr von Richthofen, Albatros D III *Le Petit Rouge*

Following his fight with FE.2b 6997 on 24 January 1917, Manfred von Richthofen made an emergency landing with *le petit rouge* due to lower wing failure. The aircraft, which he is thought to have brought over from Jagdstaffel 2 about one week before, ended up on its back on snowy ground.

The photo documenting the crash proves that the fuselage, the struts, the engine covers, the spinner, the landing gear, the wheel hubs, the tail skid fairing and the upper surfaces of both wings – and likely the lower surfaces too - were painted red. The fin was obviously also painted red, since no trace of either the military number or the tail cross can be made out. Interestingly, the white borders of the fuselage and upper wing crosses were carefully painted red, too. And not a trace of the white border remains visible in the photo, so a very good-quality paint was used. The rudder appears unusually light in the photo, and this may be interpreted in two ways: the first being that it was also painted red, matching the rest of the airframe and the light grey tone is a result of the low standing sun, the position of the rudder to the camera, and the reflection of sunlight from the snowy ground.

The other take is that the rudder was finished in clear doped fabric, maybe due to a repair, and only a narrow red stripe was applied at the bottom of the fabric.[4] Presented are two profiles illustrating both these theories. He flew this aircraft during his first and second aerial victory with Jagdstaffel 11.

In any case, it is noteworthy that the 23 January 1917 combat report from 40th Sqn. RFC simply described his aircraft as a red machine.[5]

He flew the same Albatros D III on the next day when he forced 25 Sqn RFC's F.E.2b 6937 to land on German held territory. According to his combat report, the captured British crew called his aircraft "*le petit rouge*".[6]

Close examination of the photo taken following the emergency landing of this Albatros D III on 24 January proves that this aircraft was not the aircraft that has up until now been called "*le petit rouge*", which still carried the Albatros company logo on the rudder after the rest of the fin and rudder were painted red. On both aircraft the fuselage, the struts, the engine covers, the spinner, the landing gear, the wheel hubs, the tail skid fairing and the control surfaces of the aircraft were painted red. The national markings on the fuselages of both aircraft were also overpainted. However, on the aircraft flown and crashed on 24 January 1917, the rudder shows no national marking at all. In any case, the amount of damages seen on the airframe, including wing breakage, broken landing gear and fin section, and, most importantly, the broken engine mounts, must have resulted in the aircraft being considered a complete write-off.[7]

Alex Imrie reported to the author that according to the information from Georg Simon, Manfred von

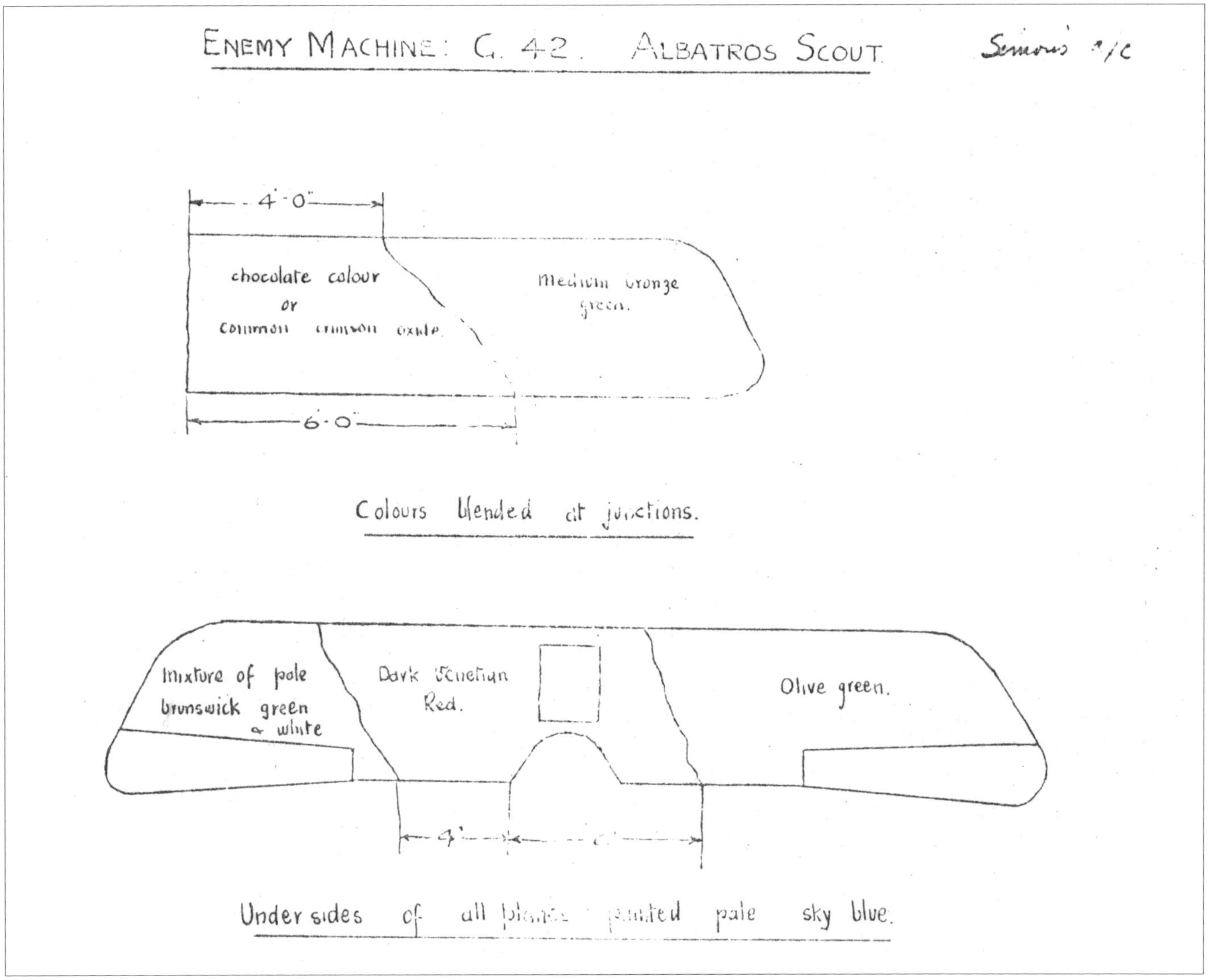

Above: Excerpt of G-Report G-42 describes the colors of the upper wing as: "mixture of pale Brunswick green and white, dark venetian red, olive green." The lower wings as: "chocolate color or common crimson oxide and medium bronze green". (A.Imrie)

Richthofenbrought along his new Albatros D III from Jagdstaffel Boelcke to Jagdstaffel 11. A look at the documents of Armeeflugpark La Briquette confirms the result of Alex's research:

The first two Albatros D III with the military numbers D.1993/16 and 1996/16 were at the Armeeflugpark La Briquette on 21 January 1917 and should have arrived at Jagdstaffel 11 in the (late) afternoon.[8] It is not known whether the aircraft have already been assembled at the Armeeflugpark, or whether this happened at Jagdstaffel 11.

According to aircraft fitters, new aircraft were first "put through their paces" following their arrival at the Jagdstaffel and, if necessary, adjustments and modifications were made. This included a thorough examination of the entire fuselage's construction, all bowden cables, all screw fastenings, and the fabric covering. Particular attention was paid to the workmanship of the wings and control surfaces. Then the machine guns were fitted and properly adjusted. It was only after this that the airplane's intended pilot began to fly the new machine. He would begin with cautious laps around the airfield and would gradually lead up to establishing the machine's limits. Each airplane had its distinct flight characteristics and its pilot had to become familiar with them in order to form an effective "team" in air combat. After all, his life depended on his "crate."

Former fighter pilot Heinz Geisseler compared this familiarization to a couple's first dates: *"You check each other out, how you interact, how good a match you are, and how well you understand one another."*

Heinz Arntzen, a member of Jasta 15 and later Staffelführer of Jasta 50, observed: *"No one in his right mind would fly to the front in an untried machine."*[9]

Lastly, the machine was painted in accordance with the pilot's wishes. It took one or two days to paint the entire

10. Lt. Manfred Freiherr von Richthofen,
Lt. Lothar von Richthofen,
Albatros D III D.2006/16

11. Lt. Karl-Emil Schäfer,
Albatros D III D.2062/16

**11. Lt. Karl-Emil Schäfer,
Albatros D III D.2062/16**

**12. Lt. Manfred Freiherr von Richthofen,
Albatros D III**

13. Lt. Kurt Wolff,
Albatros D III

14. Lt. Eddy Lübbert,
Albatros D III D.1996/16

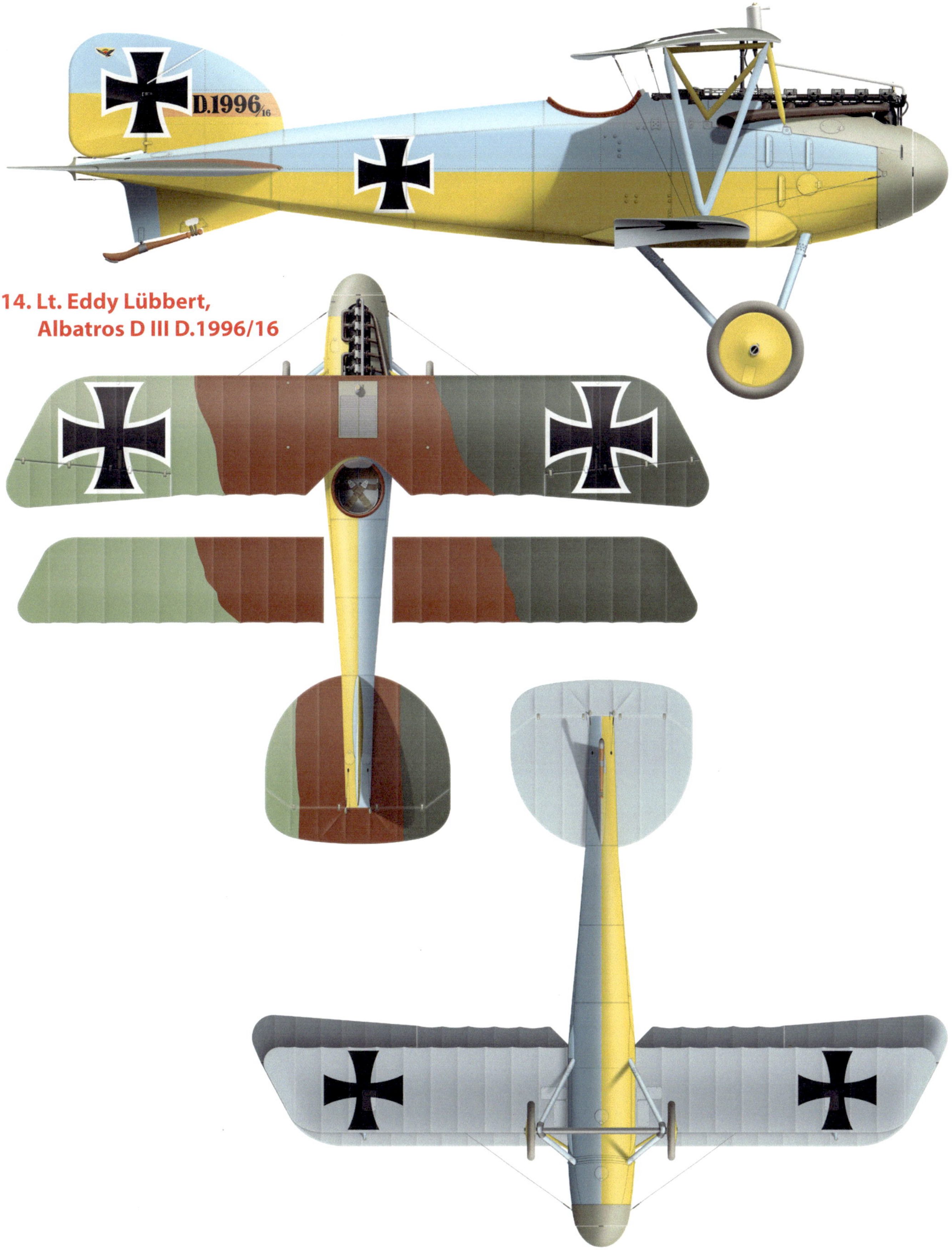

14. Lt. Eddy Lübbert,
Albatros D III D.1996/16

15. Lt. Carl Allmenröder,
Albatros D III (Reconstruction)

16. Lt. Carl Allmenröder,
Albatros D III (Reconstruction)

17. Lt. Georg Simon,
Albatros D III D.2015/16
(Reconstruction)

18. Vzfw. Sebastian Festner,
Albatros D III
(Partial reconstruction)

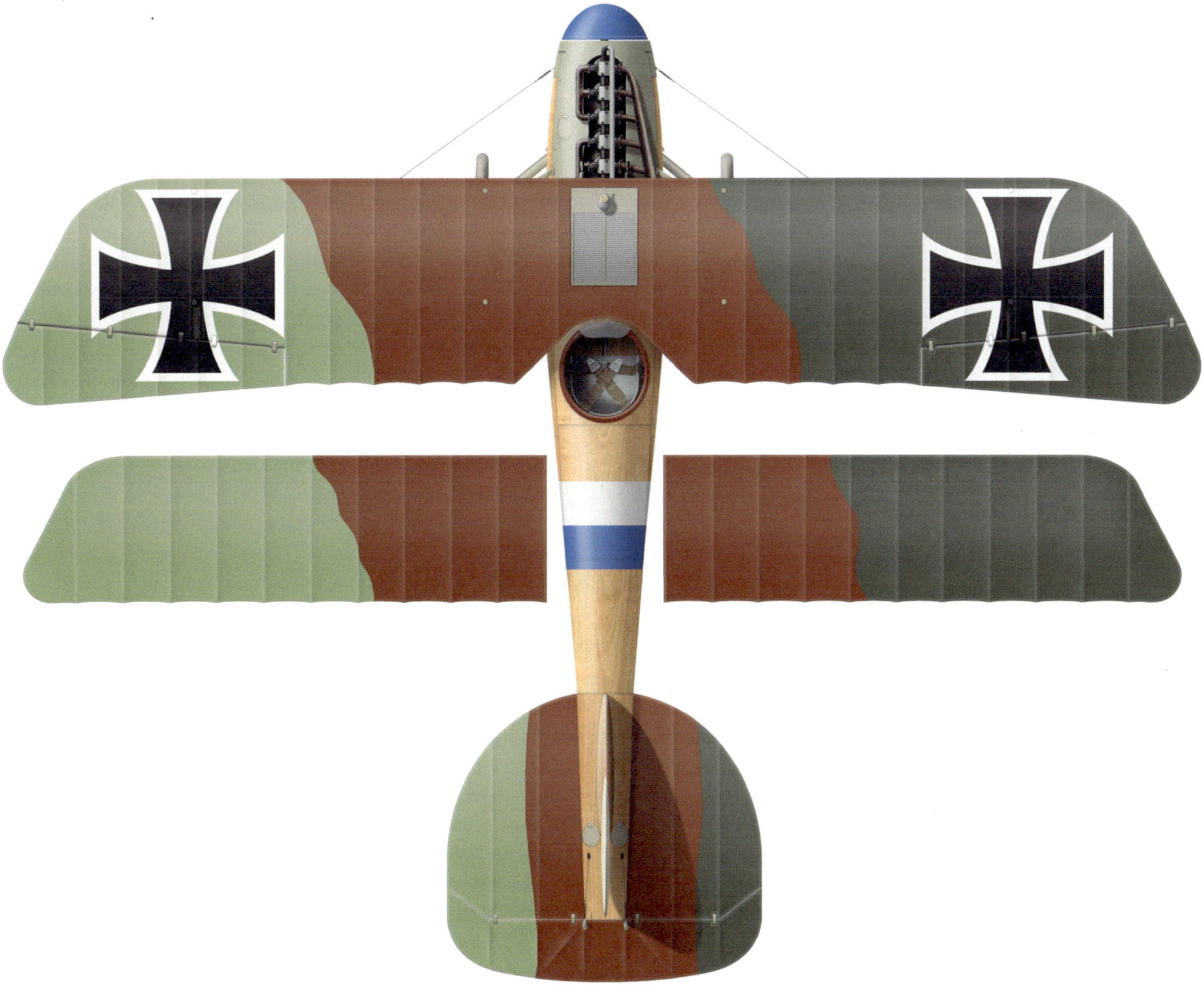

Above: Uniform Ulanen Regiment Kaiser Alexander III von Russland (Westpreußisches Nr. 1)

fuselage since the paint had to dry completely before the aircraft could be used; and drying time depended on the weather conditions. In warm, dry conditions, the paint would dry more quickly than in cold, humid conditions such as those found in January 1917.[10] According to mechanics and pilots, it normally took at least three or four days for a new machine to be judged as ready for frontline operation.[11] It could sometimes take even longer, as is demonstrated by the Halberstadt fighters delivered to Jasta 11 on 14 October 1916. Their first flight to the front occurred on 22 October– eight days after their arrival.

This means that the two Albatros D IIIs delivered to Jasta 11 on 21 February were by no means ready for use in combat on 23 or 24 January – especially if one of them had its entire fuselage painted red.

It remains to be added that the red color was a reference to his parent regiment, the Ulanen-Regiment Kaiser Alexander III von Russland (Westpreußisches Nr. 1) which had red trim stripes of the caps, collars, and cuffs.[12]

10. Lt. Manfred Freiherr von Richthofen, Lt. Lothar von Richthofen, Albatros D III D.2006/16

This aircraft was delivered to Jagdstaffel 11 via Armeeflugpark 6 on 24 January 1917.[13] This Albatros was the direct replacement for "le petit rouge", which was damaged beyond repair in the crash-landing on 24 January 1917. Unlike its predecessor, this aircraft enjoyed a comparatively long, and undoubtedly highly successful, frontline service. It is listed in the frontline inventory of Jasta 11 dated 31 March 1917.[14] It also appears in a lineup photo of Jagdstaffel 11 Albatros fighters taken at Roucourt airfield in late April 1917.

The aircraft was marked in a comparatively unspectacular manner, carrying just a red band around the fuselage as its personal identification, and the spinner was also painted red. These red markings represented the red trim stripe of the caps of his parent regiment. The reason why the machine was only marked with a red fuselage band and the red spinner as a personal identifier may have been that Manfred von Richthofen needed a replacement aircraft very quickly, and there was no time to apply an all-over red scheme to the fuselage. Or was it possible that the aircraft was not expected to enjoy a long period of frontline service anyway, just like the previous one?

After his brother Lothar joined the unit on 10 March 1917, Manfred von Richthofen provided him with his personal Albatros D III D.2006/16. In his typical, somewhat generalized way, Lothar recorded in relation to this:

"As I came to my brother's Staffel as a complete beginner, Manfred gave me one of his old machines, with which he had achieved ten aerial victories."[15]

However, Lothar von Richthofen's statement cannot be regarded as being entirely correct when cross-checking it against available evidence. During the period from 1 February 1917 to 10 March 1917, the date of Lothar´s arrival, Manfred von Richthofen had been credited with his 19th to 25th victories, amounting to a total of seven victories. This number includes two that were scored while flying a Halberstadt on 14 February when the Albatros D III was banned from frontline sorties due to recent wing failures. Thus, it is impossible for him to have scored ten kills with Albatros D III 2006/16 during this time span. However, it cannot be ruled out that Manfred continued to score a few more victories while flying this plane after 10 March, because he is known to have flown it on a visit to Jagdstaffel 12 in April 1917.

11. Lt. Karl-Emil Schäfer, Albatros D III D.2062/16

Albatros D III D.2062/16 was delivered to the Staffel on 28 February 1917.[16] The aircraft had a varnished fuselage, resulting in a yellowish appearance of the plywood, and Schäfer had the fuselage's rear section painted black as his personal marking. During the application of the black personal marking, the military number was left in place. The upper surfaces of the wings retained their light green/dark green/reddish brown three-colored factory scheme, the lower surfaces being painted light blue. Unless noted otherwise, this factory paint scheme was carried by all Jagdstaffel 11 Albatros D III fighters described in this book.

The letter from Eddy Lübbert dated 6 March 1917 confirms the coloring of this aircraft.[17] Photos of the machine taken after Emil Schäfer´s emergency landing on 9 March 1917 also document the described paint scheme of his aircraft.

A report by the 9th Bavarian Brigade filed on 6 April 1917 describes the aircraft as a ***"grey Abatros with black tail"***.[18] Another report describes his machine as a ***"yellow Albatros with black tail."*** The reason why Schäfer chose black as his personal color at that time is unknown.

12. Lt. Manfred Freiherr von Richthofen, Albatros D III

Eddy Lübbert wrote in his letter relating to the events of 6 March 1917 that Manfred von Richthofen flew a ***"bright red Albatros."***[19] This means that, in his position as Staffelführer, he already flew another personal machine with a red fuselage at this time, obviously alternating between this "largely red" aircraft and his Albatros D III D.2006/16 marked with just the red fuselage band and spinner. Since Jagdstaffel 11 was not assigned any new aircraft in March 1917, this particular machine must have been one of the six aircraft delivered to the Staffel between on 26 and 27 February 1917. After elimination of those aircraft that were known to have been assigned to other pilots, this aircraft must have been one of the following three machines: D.2014/16, D.2026/16, or D.2036/16.[20]

During the sortie which ended in his emergency landing on 6 March 1917, he flew the "bright red" Albatros D III. The same aircraft also appears in the report filed by the Flak unit at Drocourt on 17 March 1917, in which his machine is described as ***"red Albatros."***[21]

The fuselage, including the engine covers, propeller spinner, landing gear, hubcaps, tail skid and the fin and rudder were painted in red; only the company logo of the Albatros factory on the rudder was left out. The fuselage and tail crosses remained intact, but their white borders were thinly painted over in red, remaining slightly visible. The wings retained the factory finish described under Schäfer´s plane above.

13. Lt. Kurt Wolff, Albatros D III

Hermann Lohmeyer told the author that the entire fuselage of the machine was painted purple by him and the other groundcrew members. This indicates that the spinner, engine fairing, fuselage, fin, rudder and horizontal control surfaces were painted purple. The Iron Cross markings on the fuselage and rudder were not painted over. The metal panels on the nose and the spinner may or may not have been painted over, however, the photo of this airplane taking off is inconclusive in this respect due to the sunlight. The artwork in this book shows the metal panels in factory finish, while the profile shows these components painted purple. This color may have been obtained by a mix based on the "Mauve" upper wing camouflage color that was available in the depots of the Armeeflugparks. The wings of the aircraft retained their factory scheme.

Above: The Bavarian Flag, white and blue, was the model for the personal marking of the aircraft of Sebastian Festner.

Hermann Lohmeyer continued:

"In order to obtain the paint we needed, we drove to the Armeeflugpark and procured them in various colors. Sometimes we visited other depots in order to obtain a specific color. Often, the special duties officer accompanied us."[22]

This statement is confirmed by Eddy Lübbert´s letter dated 6 March 1917, which described the Albatros D III flown by Kurt Wolff as "violet plum."[23]

Hermann Lohmeyer also related that:

"Kurt Wolff, unlike other pilots, was indifferent to the paint scheme of his aircraft and did not have a special relationship with his parent unit, the transport troops. The 'violet plum' was simply created on a whim. Kurt Wolff simply found it funny to have such a strikingly painted machine". One reason may also have been that, according to Hermann Lohmeyer: *"... Wolff loved to eat plum cake throughout his life."*[24]

14. Lt. Eddy Lübbert, Albatros D III D.1996/16

In his letter, Hans-Georg Lübbert also described the coloring applied to his personal machine, calling it 'half blue and half yellow.'[25] The wording chosen by him raises the question of how exactly the 'half blue – half yellow' description of his Albatros D III should be interpreted.

During a visit, Eddy Lübbert told me that his father always had a framed picture of his brother with his plane on his desk, and I stayed over an extra day while he looked for it.

Unfortunately, he could not find it. I then showed him some photos of planes of Jagdstaffel 11 taken around February/ March 1917 which I had brought along. Eddy Lübbert identified his uncle's plane in a photo taken in April 1917 at Roucourt airfield with the words: "*This is his machine as in the photo.*" This was an Albatros D III in which the top of the fuselage had a whitish shade of gray. Accordingly, I assumed that the top of the fuselage was painted in blue, while the bottom was left in a yellowish plywood tone.[26]

However, this interpretation of mine turned out to be not quite right. My friend and colleague Jörn Leckscheid, after close evaluation of available photos, proved that the blue and yellow painted sections were applied in a "quartered lengthwise" manner to the fuselage. When seen from the left side, the darker (yellow) paint was applied to the upper fuselage quarter, while the lower one was lighter (light blue). On the right side, this pattern was reversed, the upper quarter being light-colored (light blue), while the lower quarter appears dark (yellow) in the photos. The photos also show that the yellow areas were not left in the clear-varnished plywood factory pattern, rather, they were indeed painted yellow.

When putting the available photos into context with the somewhat generalized description given in the letter, the statement 'half blue and half yellow' gets an entirely different meaning. On his Albatros D III D.1996/16, the left half of the fuselage side could be described as being "yellow above – blue below", while the right fuselage half might be termed as "blue above – yellow below". But no matter which terms might best describe the coloring of Lübbert´s Albatros, this aircraft was painted in a striking and unusual manner, which may have been unique in the Fliegertruppe at the time. The photos also show that the wings of this fighter retained the factory camouflage scheme described earlier, while the nose seems to have retained the factory grey-green color.

On 4 March 1917, 2Lt. W. B. Hills, from 40 Sqn. RFC noted in his combat report: *"I saw a Sopwith being attacked by a fast scout,* ***painted blue****…"*. The aircraft described here was likely Lübbert´s Albatros D III 1996/16.[27]

With the support of pilots of Jagdstaffel 11 and his brother's mechanics, Friedrich-Wilhelm Lübbert reconstructed his last days and the circumstances of his fatal crash. According to this information, Eddy Lübbert flew a comrade's plane on the day of his fatal crash, as his plane was not operational due to an engine failure.[28] An entry in his diary dated 24 March 1917 notes that his Albatros D III had "engine problems" and that he had to return to the airfield.[29] Since the entire engine had to be exchanged, the machine was still not yet ready to fly on 30 March 1917. This explains why the aircraft was still in use by Jagdstaffel 11 in late April 1917, but whoever took it over after Lübbert´s death is not known.

15. Lt. Carl Allmenröder, Albatros D III (Reconstruction)

Eddy Lübbert described Carl Allmenröder's machine in his letter dated 6 March 1917 as a "field-gray crate".[30] Unfortunately, further information relating to the appearance of this aircraft is unavailable. The color of the field-gray uniforms of the German troops has been used as the basis for the fuselage color depicted in the profile. However, this aircraft may also be a replacement machine, because on the same day he demonstrably also flew a "white Albatros".

16. Lt. Carl Allmenröder, Albatros D III (Reconstruction)

The combat report of the crew of BE.2d 5834, 16th Sqn. RFC, flown by F/Sergt. Quick and Capt. Claremont, dated 6 March 1917, notes that they were attacked by a "red" and a "white" enemy plane.[31] Since only Manfred von Richthofen and Carl Allmenröder from the German side were involved in this combat, the "white" machine must have been flown by Carl Allmenröder.[32] However, this would not have been a completely white machine, rather only the fuselage was painted white, perhaps only the wooden sections. Wilhelm Allmenröder stated that the planes flown by his brother and him carried white personal markings at Jagdstaffel 11.[33]

17. Lt. Georg Simon, Albatros D III D.2015/16 (Reconstruction)

Albatros D III D.2015/16 was delivered to the Staffel on 26 February 1917. According to statements made by Georg Simon to researcher and author Alex Imrie, the aircraft had a green painted fuselage as his personal identification.[34] This could have been the color of the cap band of his original unit, which is, however, unknown to the author. Unfortunately, the author never asked Alex Imrie about Georg Simon's parent unit.

In the combat report filed by Lt. Pescoe and Lt. Pearson, 29th Sqn. RFC, dated 9 March 1917, one of the attacking German aircraft is described as **"painted green"**.[35] This indicates that their attacker on this date was Lt. Simon, flying his green-painted aircraft. Albatros D III D.2015/16 served with Jasta 11 for a fairly long period of time. Simon became a prisoner-of-war after he made an emergency landing in British territory on 4 June 1917. At that time, the machine had a red-painted fuselage with a green fuselage band behind the cockpit, in line with the generous application of red paint to the aircraft of Jasta 11 in late April. During this process of repainting the aircraft, much of the previous green paint was covered by the new red color, leaving only a green fuselage band in place. Again, the wings retained the factory-applied camouflage scheme. This Albatros D III was in frontline service for more than three months of very intense aerial combat.

18. Vzfw. Sebastian Festner, Albatros D III (Partial reconstruction)

Vzfw. Sebastian Festner´s Albatros D III carried a white-blue fuselage band that represented the colors of his native Kingdom of Bavaria. For the reasons described previously, this also must have been one of the Albatros fighters delivered to Jagdstaffel 11 in January or February of 1917.

A report filed by the anti-aircraft group command at Denain, relating to an English plane being shot down by a German Albatros D III on 5 April 1917, describes the victorious fighter plane as a **"yellow Albatros with white-blue banding"**. The victory was claimed by and credited to Sebastian Festner, who brought down a Bristol F2a from 48th Sqn. RFC. On this occasion, three other aircraft from the same British unit were also brought down by Jasta 11.[36] More details relating to this combat will be described in Volume 2 of this series.

There are two photos of the Albatros D III flown by Sebastian Festner that were taken about one month later in May 1917. This was another case of the aircraft surviving its pilot, and at this time, the aircraft's fuselage was painted red, again in common with the repainting of several Jasta 11 Albatros D IIIs (see the previous description of Simon's aircraft), but still wore the white-blue fuselage band mentioned in the earlier anti-aircraft report. Before then, the white-blue band had been applied to the varnished wood fuselage, as depicted in the color profile.

Unit Marking

From the formation of the Staffel in October 1916 up to the end of March 1917, the aircraft of Jagdstaffel 11 did not carry any kind of uniform Staffel identification. Only various styles of personal markings of the respective pilots are documented for this timespan of roughly six months. These typically consisted of fuselage bands, painted tails, or painted fuselages, all of which allowed for easy identification in the air.

Above: Hope you enjoy the book, see you soon with Volume 2!

Endnotes

Establishment and First Front-Line Flights of Jagdstaffel 11, October 1916

1. The pilots and ground crewmen were in effect only on loan to their Kampfeinsitzer-Kommandos. Officially they remained regular members of their respective Feldflieger-Abteilung.
2. Wöchentlicher Tätigkeitsbericht des Stabsoffiziers der Flieger der 6. Armee (Stofl 6th Army weekly activity report (hereinafter referred to as "WTB Stofl 6") from 25.09.1916 – 01.10,1916. General der Pioniere 6 Band 58, Bayerisches Hauptstaatsarchiv München.
3. "WTB Stofl 6" from 25.09.1916 – 01.10,1916, ibid.
4. "WTB Stofl 6" from 25.09.1916 – 01.10,1916, ibid.
5. War diary excerpt of Jasta 11 by Erich Tornuß, transcribed in the 1930´s, revised and supplemented by Dr. Gustav Bock and Winfried Bock.
6. "WTB Stofl 6" from 02.10.1916 – 08.10,1916, ibid.
7. War diary excerpt of Jasta 11, ibid.
8. Flugzeuge Armeeflugpark 6, La Briquette, 31.10.1916 Iluft folder No. 206, Bayerisches Hautstaatsarchiv, Abt. 4, München. The aircraft numbers delivered to Jasta 11 are listed in: Iluft folder No. 36, 204, 205, and 206. Compiled and made available by Reinhard Zankl
9. Albert Oesterreicher OP (Personal record) file 1188, Bayerisches Hauptstaatsarchiv München. No.12 Squadron's B.E.2c 2100, piloted by 2Lt. R.M. Wilson-Browne, got into a fight with two Fokkers over Aubigny-au-Bac on 21 July and was forced to land nearby. Oesterreicher and Lt. Kurt Wintgens were in those Fokkers but only Wintgens was given full credit for the kill. Oesterreicher's 9 August victory may have been over No.2 Squadron's B.E.2d 5741, which was attacked by a Fokker and a Nieuport, and forced down in the vicinity of Lens (pilot Capt. E.W. Leggatt was taken prisoner).
10. For this book, the author has kept to the dates of the personal records (OP), Kriegsstammliste (KrStL.) [War Master Roll lists for officers].) or the Kriegsstammrolle (KrStR.) [War Master Roll lists for NCO and ranks] of Armeeflugpark 6 (Army Aviation supply depot), according to their availability. These may differ from dates quoted in other official documents by a few days. Lt. Oesterreicher's personal information provides an example of how certain dates can differ in various official documents. For example, his personnel file gives 16 May 1914 as the date of his transfer to Flieger-Bataillon Schleissheim, whereas the KrStL. notes it as 1 June 1914. His personnel file indicates 18 March 1915 as his transfer date to Feldflieger-Abteilung 2b, whereas a review written by Hptm. Oefele (FFA 2b) says it was 19 April 1915. His transfer date to FFA 4b is given as 28 July 1915 by his personnel file, yet a review written by the unit's CO, Hptm. Sorg, stated it was 1 August 1915.
11. The Czermak family's son, Johann, also became a fighter pilot and served as the Staffelführer of Jagdstaffel 32 for part of 1918. (See Schmäling and Bock, *Royal Bavarian Jagdstaffel 23*).
12. Sebstian Festner KrStR. 18154/659; 17961/48, Bayerisches Hauptstaatsarchiv München
13. Konstantin Krefft OP 11364, Kr.StR. 17956/114, Bayerisches Hauptstaatsarchiv München
14. Documents and photos by Lt. Heinz von Beaulieu-Marconnay, an observer with Feldflieger-Abteilung 2 attest to this, Bruno Schmäling collection
15. Konstantin Krefft OP 11364, Kr.StR. 17956/114, ibid.
16. Personal information by Carl-August von Schoenebeck, Jasta 11 to Bruno Schmäling in 1974–1978. It is sometimes stated that Krefft became the Technical Officer of Jasta 11. This is incorrect because no Jasta had the budgeted position of "Technical Officer" in their job structure. In Krefft's case, he voluntarily (with the approval of his Staffelführer) assumed that role as a secondary responsibility because of his technical expertise. Richthofen's and Jasta 11's positive experience with this led to the formal, budgeted position of "Technical Officer" being established when Jagdgeschwader Nr.1 was formed. According to his Kr.St.L., Krefft was assigned that role on 25 June 1917. JG I's war diary, however, recorded that he was transferred to Geschwader staff in that capacity on 2 July 1917.
17. Personal information by Carl-August von Schoenebeck, ibid.
18. Personal information by Carl-August von Schoenebeck, ibid.
19. Karl Kleinhenz OP 25963, Bayerisches Hauptstaatsarchiv Abt. 4, München.
20. Erwin Howe KrStR. 18038/454, Armee-Flug-Park 6, Bayerisches Haupstaatsarchiv, München.
21. Alfred Prehn, Rangliste KrStR. 18035/402, and 18040/1184, Armee-Flug-Park 6, Bayerisches Hauptstaatsarchiv, Abt.4, München.
22. OP Franz Anslinger, Badisches Generallandesarchiv, Karlsruhe. The Kastas of the Kagohls were equipped with two-seat aircraft to perform strategic bombing missions. Due to the German lack of fighter planes to oppose the allies during the Battles of the Somme and Verdun, the Kagohl were employed to protect the aircraft of the Artillerieflieger-Abteilungen and Feldflieger-Abteilungen as well as to engage enemy aircraft. At the end of 1916, four Kagohls were disbanded and their former Kasta units were converted into Schutzstaffels whose task was to protect their own artillery and reconnaissance aircraft. Many Kagohl/Kasta

pilots opted to join the new Jagdstaffeln.

23. Inspektion des Militär-Luftfahrwesens (Iluft) Akte (File) Nr. 204, 205, 206, ibid. Hereinafter referred to as "Iluft Akte".
24. OP 15292 Rudolf Lang, Bayerisches Hauptstaatsarchiv, München.
25. The former airfield ran west of today's Rue du Champ de Tir from a spot just below the D425 traffic circle to Chemin de la Brayelle to the south and is bisected by the D 621.
26. Ernst von Hoeppner, *Deutschlands Krieg in der Luft 1914/1918,* p. 52. Leipzig 1921.
27. Trevor Henshaw, *The Sky Their Battlefield II* (Second revised and supplemented edition), p. 57. Although the site of their landing is not given, we know that other No.3 Squadron aircraft engaged in actions near Le Transloy (on 25 September) and Maricourt (26 September 1916), which places the squadron well within Jasta 11's sphere of action.
28. Trevor Henshaw, *The Sky Their Battlefield II*, p. 58. 2Lt. G.G. Callender was the pilot and 2Lt. H.W. Soulby his observer.
29. Akte Iluft 204, ibid.
30. "WTB Stofl 6" from 22.10.1916 – 29.10.1916, ibid.

Front Line Operations of Jagdstaffel 11 November 1916 – December 1916

1. "WTB Stofl 6" from 30.10.1916 – 05.11.1916. Ibid.
2. Akte Iluft Nr 204, 205, 206, ibid.
3. For the numbers D.801/16, D.802/16 and D.805/16, there is no precise indication of when they were delivered from Armeeflugpark 6 to Jagdstaffel 11. However, it can be assumed that the four aircraft were delivered to the squadron at approximately the same time. The aircraft with the numbers D.801/16, D.802/16 and D.805/16 were still in the inventory of Jagdstaffel 11 on 31.12.1916.
4. Akte Iluft Nr 204, 205, 206, ibid.
5. Kurt Wolff, KrStL. 18031/208, Armee-Flug-Park 6, Bayerisches Hauptstaatsarchiv, Abt.4, München.
6. Zuerl, *Pour le Mérite-Flieger*, p.480. Zuerl described the trainer as an LVG, but the school only used aircraft produced by AEG in nearby Hennigsdorf.
7. Kurt Wolff, KrStL. 18031/208, Armee-Flug-Park 6, ibid.
8. Kurt Wolff's handwritten diary for the period 30 September 1915 to 13 May 1917, now in a private collection. Hereafter referred to as Wolff Diary.
9. Kurt Wolff, KrStL. 18031/208, Armee-Flug-Park 6, ibid.
10. Wolff Diary. The aircraft Wolff had flown previously used steering wheels and not "joysticks" for control.
11. Wolff Diary, ibid.
12. Bodenschatz, *Jagd in Flanderns Himmel*, p.17, München 1935.
13. Die Woche 28 (14 July 1917), p.941.
14. Personal information by Hermann Lohmeyer to Bruno Schmäling 1982. Wolff's fiancée was Marta Bömeleit.
15. Georg Simon KrStL. 18031/208, Armee-Flug-Park 6, Bayerisches Hauptstaatsarchiv, Abt.4, München.
16. "WTB Stofl 6" from 06.11.1916 – 12.11.1916. Ibid.
17. "WTB Stofl 6" from 06.11.1916 – 12.11.1916. Ibid.
18. "WTB Stofl 6" from 12.11.1916 – 18.11.1916. Ibid.
19. The information about Alfred Prehn's transfers is contradictory. According to his KrStR. 18035/402, and 18040/1184 he was commanded to the naval air station in Tondern on 28.02.1917. On 1 March 1917, the command was changed to a transfer. According to Erich Tornuß' war diary transcripts, he had already been ordered to Tondern on 13 November 1916. This is confirmed by the personnel list of Jagdstaffel 11 dated 31 December 1916. Vfzw Prehn is not included in the list of pilots. Instead, it reads: "Vzfw. Prehn commanded to Tondern air station according to daily order of Kofl 6 dated 13.11.16".
20. Wolff Diary, ibid.
21. Akte Iluft Nr. 204, 205, 206, ibid.
22. "WTB Stolf 6 from 12.11.1916 – 18.11.1916, ibid.
23. Personal Information by Frau Helene Allmenröder (née Kortenbach) to Bruno Schmäling in 1976. Helene Kortenbach lived next door to the Allmenröders and knew both brothers from their childhood days. She fell in love with Carl first and became engaged to him in April 1917. After Carl was killed in action, she was heartbroken and brother Wilhelm – no longer fit for frontline duty due to injury – comforted her. They eventually fell in love and married, and Helene later proclaimed that she had found the "best husband in the world."
24. Carl Allmenröder, KrStL. 18031/173, Armee-Flug-Park 6, Bayerisches Hauptstaatsarchiv, Abt.4, München. Lance Bronnenkant, *Blue Max Airmen*, Vol. 9.
25. Schnitzler, Carl Allmenröder, Der bergische Kampfflieger, p.6, Wald, 1927.
26. Personal Information by Wilhelm Allmenröder to Ludwig Zacharias about 1960.
27. Carl Allmenröder, KrStL., 18031 No. 173, ibid.
28. Personal Information by Frau Helene Allmenröder, ibid.
29. Personal Information by Otto Fuchs to Bruno Schmäling, 1978. At the time of Fuchs' short time with Jasta 11, Allmenröder was its acting Staffelführer while Manfred von Richthofen was away on leave. See also: Bruno Schmäling, Winfried Bock, *Royal Prussian Jagdstaffel 30*, Aeronaut Books 2014, ISBN 978-1-935881-25-4.
30. Josef Baierlein, KrStr. 18153/17; 18039/626; 18076/71, Bayerisches Hauptstaatsarchiv, Abt.4, München.
31. "WTB Stolf 6 from 19.11.1916 – 25.11.1916, ibid.
32. Wolff Diary, ibid.
33. Cole, Royal Flying Corps 1915-1916, pp.320-21.
34. "WTB Stolf 6 from 26.11.1916 – 02.12.1916, AOK 6 Band 288, ibid.

35. Wolff Diary, ibid.
36. Wöchentlicher Tätigkeitsbericht des Kommandeurs der Flieger der 6. Armee vom from 02.12.1916 – 10.12.1916, ibid. (Bayerisches Hauptstaatsarchiv, Abt. 4). On 29 November 1916, the position of the "Stabsoffizier der Flieger" was changed to that of a "Kommandeur der Flieger" (Chief of all flying units assigned to an Army), who thus also had the corresponding command authority. Hereinafter referred to as WTB Kofl 6".
37. Wolff Diary, ibid.
38. "WTB Kofl 6"from 11.12.1916 – 16.12.1916, ibid. War diary excerpt of Jagdstaffel 12, by Erich Tornuß, revised and supplemented by Dr. Gustav Bock and Winfried Bock.
39. Volder Iluft Nr. 204, 205, 206, ibid.
40. "WTB Kofl 6" from 17.12.1916 – 24.12.1916, ibid.
41. "WTB Kofl 6" from 24.12.1916 – 31.12.1916, ibid.
42. Letter by Kurt Wolff to his sister Elisabeth, dated 21 December 1916. The letters of Kurt Wolff provided by Obst.Lt. Reinhard Schröder, Taktisches Luftkampfgeschwader 71 "Richthofen", Wittmund, Germany.
43. Letter by Kurt Wolff to his sister Elisabeth, dated 21 December 1916, ibid.
44. Wolff Diary. "Liebesgabe" or "Gifts to Dear Ones" was a program where packages containing gifts (e.g., woolen socks, cookies, etc.) were sent to soldiers serving at the front by German relatives and citizens.
45. War diary excerpt of Jagdstaffel 12, by Erich Tornuß, revised and completed by Dr. Gustav Bock and Winfried Bock. See also: Bruno Schmäling, Winfried Bock, *Royal Bavarian Jagdstaffel 30*, Aeronaut Books 2014.
46. "WTB Kofl 6" from 01.01.1917 – 07.01.1917, ibid.
47. "WTB Kofl 6" from 14.01.1917 – 21.01.1917, ibid.
48. Georg Braumüller, KrStR_31421_BH18040-00029, Armee-Flug-Park 6, (Bayerisches Hauptstaatsarchiv, Abt.4, München). Some publications refer to a Georg Braumiller, but this is a confusion of names.
49. "WTB Kofl 6" from 14.01.1917 – 21.01.1917, ibid.

Der neue Staffelführer Lt. Manfred Freiherr von Richthofen

3.1 The Book Der rote Kampfflieger

1. Erich von Salzmann, born 1876 in Stettin, died in Shanghai in 1941, was an artillery officer who had served with the German colonial force in German Southwest Africa. By 1917, he worked as a journalist and author for the Vossische Zeitung, which was taken over by Ullstein Verlag in 1913. His books mainly dealt with China, and, by his own account, he spoke Mandarin. His best-known book was his personal adventure report "*In the Saddle through Central Asia: 6000 Kilometers in 176 days*", which is very reminiscent of Karl May's adventure books. Other books include: "*Gelb gegen Weiß - aus dem Freiheitskampf Asiens*" (Yellow against white - from Asia's fight for freedom), the novel "*Yü Fong - Ein chinesischer Revolutionsroman aus dem Beginn des 20. Jahrhunderts* (A Chinese revolutionary novel from the beginning of the 20th century)", the novelistic account "*Über die Weltmeere zur deutschen Front in Flandern, eine Kriegsodyssee* (Across the oceans to the German front in Flanders, a war odyssey)" and the book "*Im Kampfe gegen die Herero* (Fighting against the Herero)". The contemporary readership found his books exciting, novelistic and imaginatively written. In his writings, he refrains from the typical arrogant attitude of whites towards peoples of other colors. A particular admiration for Chinese culture can be noted in his books. However, he was a layman when it came to the air force and air warfare, which explains the military and aeronautical inconsistencies in his book on von Richthofen. https://www.deutsche-biographie.de/pnd116777206.html, https://www.pressreader.com/uk/iron-cross/20210331/281775631945724
2. Daheim, *Richthofen – Erich von Salzmann*, p. 10–11, 18.05.1918.
3. Combat report Manfred von Richthofen, 23.11.1916, Public Records Office, London, via Alex Imrie collection. The English translations of von Richthofen´s original combat reports are kept in the United Kingdom's Public Record Office. Alex Imrie copied them and provided copies to the author.
4. All former members of Jagdstaffel 11 who were interviewed denied that von Richthofen had kept a diary or anything similar. The adjutant of Jagdgeschwader I, Karl Bodenschatz, also expressly denied this in conversation with Bruno Schmäling and Neal O'Connor in 1978.
5. Journal Daheim, *Richthofen – Erich von Salzmann*, 18.05.1918, ibid.
6. Iron Cross, Dr. Immanuel Voigt, Ghostwriting "*Der rote Kampfflieger*", p. 8–9, Windsor. o.A.
7. Iron Cross, Dr. Immanuel Voigt, ibid.
8. Personal Information by Carl August von Schoenebeck, confirmed by Hans-Georg von der Osten to Bruno Schmäling,1974–1975.
9. As example: Joachim Castan, Der rote Baron, Germany 2016, as well as Friedrich Wilhelm Korff, *Der rote Kampfflieger von Rittmeister Manfred Freiherrn von Richthofen*, Germany 1977.

3.2 Manfred von Richthofen's Military Career Before his Appointment as Staffelführer

1. *Der rote Kampfflieger* (1917), pp.106-07, Berlin-Wien 1917.
2. Walter Zuerl, "*Pour le Mérite Flieger*" p. 327, Munich 1938. Walter Zuerl gives the birthplace as Kleinburg. Other sources give his birthplace as Breslau (now Wroclaw, Poland). His father was stationed at Breslau, but the home of the family was in Kleinburg. There are two reasons why Kleinburg was the real birthplace: First, the author

spoke with Walter Zuerl in 1974. He told him, that he had the full support from the Reichsluftfahrtministerium (Göring) to work on this book and therefore was able to access the original personal records at the Reichsarchiv in Potsdam. In the personal record, Kleinburg was named as the birthplace. Second, during this time, home births were common. A personnel file of Manfred von Richtofen is not available in the military archives in Freiburg, which makes it difficult to give an exact account of his military career. Finally, the author also wanted to know from Herrn Zuerl why the Jewish Pour le Mérite Pilot Wilhelm Frankl was not mentioned in his book. No answer was given and the conversation was stopped.

3. Personal Information by Eddy Lübbert (son of Friedrich-Wilhelm Lübbert) in 1983 to Bruno Schmäling. After Freifrau Kunigunde von Richthofen fled from Silesia in 1945, she and her daughter Ilse lived for a long time with the Lübberts as guests.
4. *Der rote Kampfflieger* (1917), p. 12, ibid. Kadettenanstalten (Cadet schools) were secondary schools that led up to the Abitur (High school degree) and usually served as preparation for a military career and were very beneficial for a future military career. The Wahlstatt cadet school was known for its iron discipline and strict drill.
5. *Der rote Kampfflieger* (1917), p. 13, ibid.
6. Walter Zuerl, *Pour le Mérite Flieger* p. 372, ibid.
7. There is no personal file on Manfred von Richthofen in the Freiburg Military Archives. The biographical data on Manfred von Richthofen is taken from: Lance Bronnenkant, *Blue Max Airmen*, Vol. 5, Manfred von Richthofen 2014, Aeronaut Books, Walter Zuerl, *Pour le Mérite Flieger*, ibid. Compilation of personal data from various sources by the author.
8. *Der rote Kampfflieger* (1917), pp. 61 - 62.
9. *Der rote Kampfflieger* (1917), pp. 65 - 66.
10. War diary excerpt of Jagdstaffel Boelcke, by Erich Tornuß ca. 1930–1933, revised and completed by Dr. Gustav Bock and Winfried Bock.
11. Combat report Manfred von Richthofen, 23.11.1916, Public Records Office, London, via Alex Imrie collection. The English translations of Richthofen original combat reports are stored in the United Kingdom's Public Record Office in London. Alex Imrie copied them and gave copies to the author. These reports were used for the book. They differ from the combat reports published by Floyd Gibbons in his book *The Red Knight of Germany*. It seems that Gibbons made his own translation of original German reports. For this book, the author used the wording of the report from the Public Record Office. As far as we know, the original reports were destroyed in Germany during World War II.
12. War diary excerpt of Jagdstaffel Boelcke, ibid.

3.3 Manfred von Richthofen in Contemporary Reports

1. Die Woche 34 (25 August 1917), p.1150.
2. Letter from Carl Allmenröder to his fiancée Helene Allmenröder née Kortenbach, handwritten transcription by Bruno Schmäling 1976.
3. Letter from Otto Brauneck dated 20 April 1917, Original private property of Frau Papendiek, sister of Otto Brauneck, copy Michael Schmeelke collection.
4. Letter from Busso von Alvernsleben dated 21 April 1917, published by his family as a book, copy Schmäling Collection.
5. Personal Information by Carl-August von Schoenebeck, ibid.
6. Richthofen, *Ein Heldenleben*, pp. 321-24.
7. Wilhelm Reinhard was killed during a test flight on 3 July 1918.
8. Schwester Käte Oltersdorf, *Tagebuch aus dem Kriegslazarett 76 Kortrijk* (Courtrai) Publisher: Oberstleutnant Reinhard "Kanzler" Schröder, tradition officer of Taktisches Luftwaffengeschwader 71 "Richthofen".
9. Schwester Käte Oltersdorf, *Tagebuch aus dem Kriegslazarett 76 Kortrijk,* ibid.
10. Richthofen, *Ein Heldenleben*, ibid., pp. 312-15.
11. Handwritten notes by Alex Imrie according to the interview with Wilhelm Gisbert Groos 1968.
12. Personal information by Johann Janzen, Jasta 6 to Alex Imrie, 1960.
13. Personal Information by Franz Hemer, Jasta 6 to Bruno Schmäling,1976.
14. Letter of Walter Böning to his parents from 14 April 1917, provat property, Copy Bruno Schmäling Collection. Se also: Bruno Schmäling, Reinhard Kastner, Jörn Leckscheid, *Royal Bavarian Jagdstaffel 76*, p. 75, Aeronaut Books 2022.
15. Personal Information by Walter Böning, Staffelführer Jasta 76 to Bruno Schmäling, 1975.
16. Personal Information by Hans-Georg von der Osten, ibid. He served with Jagdgeschwader I's Jasta 11 from August 1917 to March 1918 and then for a short time as the acting Staffelführer of Jagdstaffel 4.
17. Personal information by Otto Fuchs to Bruno Schmäling, 1978, See also *Royal Prussian Jagdstaffel 30*, Bruno Schmäling, Winifred Bock, Aeronaut Books 2014.
18. Personal information by Alois Heldmann, Jasta 10, to Bruno Schmäling, 1975.
19. Personal Information by Karl Bodenschatz to Bruno Schmäling & Neal O'Connor, 1975. Manfred von Richthofen must have been equally saddened by the loss of Carl Allmenröder and Emil Schäfer. Bodenschatz is unlikely to have mentioned them because they had fallen before he joined Jagdgeschwader I and, unlike Voss and Wolff, he was unlikely to have met them personally.
20. Personal Information by Hans-Georg von der Osten, ibid.

21. Letter from Carl Deilmann to Dr. Heinz Penzlin, newspaper "*Die Welt*", dated 3 May 1978, copy Bruno Schmäling Collection. Carl Deimann was credited with a total of 5 aerial victories, as well as an aircraft forced to land on the Entente side of the lines.
22. Personal Information by Carl-August von Schoenebeck, ibid.
23. Manfred von Richthofen, *Geschwaderflüge*, Copy of the original. Alex Imrie obtained the original from Willi Gabriel, a former member of Jasta 11.
24. Manfred von Richthofen, *Geschwaderflüge*, ibid.
25. Letter from Kurt Delang, Jasta 10 to Bruno Schmäling, dated 7 April 1980.
26. In the 1970´s, fighter pilot Josef Jacobs gave the author a piece of fabric with the serial number of a British aircraft he had forced to land. Aviation historians such as Alex Imrie, Ed Ferko and Bill Puglisi also received such "souvenirs" from former fighter pilots. Photos of the casinos or quarters of both German and Entente fighter pilots often show such "trophies" of downed enemy aircraft decorating the walls.
27. *Cross & Cockade* Vol. 1, Nb. 3. P.61, ibid.
28. Personal information by Carl-August von Schoenebeck, ibid.
29. *Cross & Cockade* Vol. 1, Nb. 3. P.61, USA 1960.
30. *Der rote Kampfflieger* (1917) p. 130.
31. Personal Information by Hermann Lohmeyer, mechanic, Jasta 11 to Bruno Schmäling, 1981/1982.
32. Kölner Zeitung, 22 April 1938, Interview with Josef Holzapfel.
33. Kölner Zeitung, ibid.
34. Kölner Zeitung, ibid.
35. Personal Information Carl-August von Schoenebeck, ibid.
36. Personal Information by by Hans-Georg von der Osten, ibid.
37. Personal Information by Hans-Georg von der Osten, ibid.
38. Personal Information by Carl-August von Schoenebeck, ibid.
39. Combat report dated 18 March 1918 in Bodenschatz, Jagd in Flanderns Himmel, ibid.
40. *Cross & Cockade* Vol. 1, No. 3., p. 61, ibid.
41. Chef des Generalstabes der Feldheeres, Weisungen über den Einsatz von Jagdstaffeln (Directives on the deployment of of Jagdstaffeln), October 1917, Archive Bruno Schmäling.
42. Manfred von Richthofen, Geschwaderflüge, ibid.

3.4 Approaching the Personality of Manfred von Richthofen

1. The portrayal of his personality is based on the sources analyzed, as well as on interviews with former members of Jagdstaffel 11 and Jagdgeschwader I, who knew Manfred von Richthofen personally:
   Alex Imrie: Johann Janzen (6), Carl Bodenschatz (JG I), Justus Grassmann (10), Wilhelm-Gisbert Groos (11), Hans-Georg von der Osten (11 und 4), Alois Heldmann (10), Georg Simon (10), Oskar Rouselle (11), Wilhelm Allmenröder (11), Lutz-Otto Förster, Franz Hemer (6).
   Bruno Schmäling: Carl-August von Schoenebeck (11), Otto Fuchs (FA (A) 292, 11), Hans-Georg von der Osten (11 und 4), Hermann Lohmeyer (11), Alois Heldmann (10), Kurt Delang (10), Wilhelm-Gisbert Groos (11), Xaver Leinemüller (10), Oskar Rouselle (4), Franz Hemer (6), Walter Böning (77),
   Michael Schmeelke: Frau Papendiek, sister of Otto Brauneck.
2. Personal Information by Josef Jacobs, Carl August von Schoenebeck, Otto Fuchs and Theodor Rumpel to Bruno Schmäling. According to their statements, Otto Fuchs and Theodor Rumpel served under Staffelführers who also possessed this ability, namely Oblt. Hans Bethge and Lt. Otto Kissenberth.
3. Schwester Käte Oltersdorf, *Tagebuch aus dem Kriegslazarett 76 Kortrijk,* ibid.
4. Ernst Udet, Herausgegeben von Ernst Eichler "*Kreuz wider Kokarde - Jagdflüge des Leutnants Ernst Udet*", p. 132, Berlin 1918.
5. Karl Bodenschatz, *Jagd in Flanderns Himmel*, p. 38, München 1935.
6. Joachim Castan, *Der rote Baron*, p. 230, Deutschland 2016.
7. War diary excerpt Jagdstaffel 37 by Erich Tornuß ca. 1930, copy Schmäling collection.
8. Lance Bronnenkant *Blue Max Airmen* 15, Udet, p. 43–45, Aeronaut Books, USA 2020. War diary excerpt Jagdstaffel 11, ibid.
9. Combat reports Manfred von Richthofen, Public Records Office, London, copy archive Bruno Schmäling.
10. Questions regarding the influence of Göring in relation to the creation of the book, *Jagd in Flanderns Himmel*, were studiously evaded by Karl Bodenschatz in conversation with Neal O'Connor and the author.
11. Combat reports Manfred von Richthofen, July 1917–April 1918 ibid.
12. Combat reports Manfred von Richthofen, September 1917–April 1918 ibid.
13. The author was in personal or written contact with 74 former fighter pilots between 1972 and 1983. For a list of the pilots see: Bruno Schmäling, Jörn Leckscheid, *Jasta Colors Volume 1*, pp. 262–263. The shooting down of an enemy aircraft had to be confirmed by witnesses who did not belong to one's own unit. As a rule, these were frontline troops, airspace observers, anti-aircraft units or balloon observers. The testimony of fellow squadron members

alone was not enough to recognize the air victory. Bruno Schmäling, Jörn Leckscheid, *Jasta Colors Volume 2*, pp. 118–121, Aeronaut Books, USA 2023.
14. Combat reports Manfred von Richthofen, ibid.
15. Werner von Langsdorff, *Flieger am Feind,* p. 145–146, Gütersloh 1934.
16. Movie made by Anthony Fokker, personal information by Wilhelm Gisbert Groos to Alex Imrie, ibid.
17. Manfred von Richthofen wrote this report to Anthony Fokker. The latter seemed to consider the report so important that he sent it to the MAG - Magyar Általános Gépgyár, which, among other aircraft types, produced Fokker airplanes under license for the Austro-Hungarian Empire.
18. Archive of the MAG-Magyar Általános Gépgyár, Peter Grosz collection.
19. Letter written by Manfred von Richthofen to Oblt. Fritz von Falkenhayn, 18 July 1917, personal information from Wilhelm-Gisbert Groos to Aelx Imrie, ca. 1960.
20. Manfred von Richthofen, *Geschwaderflüge* and further written documents, e. g. in Floyd Gibbons, *The Red Knight of Germany,* USA 1927.
21. Personal Information by Josef Jacobs, Jasta 22 and Staffelführer Jasta 7, to Bruno Schmäling, 1976. Of his 47 aerial victories, 43 were officially confirmed. The aerial victories 44 to 47 were submitted to the Kogenluft for confirmation towards the end of the war, but it is not clear whether these were officially confirmed.

4. The First Successes for Jagdstaffel 11, January 1917

1. 1. Iluft Akte Nr. 205. Flugzeuge Armeeflugpark 6, Pont a Chin, 28.2.1917 / BHStA/KAMünchen Akt Iluft 205.
2. Personal Information by Hermann Lohmeyer, ibid.
3. Letter of Hermann Lohmeyer to Bruno Schmäling dated 11 November 1982..
4. Personal Information by Hermann Lohmeyer, ibid.
5. Combat report Manfred von Richthofen, 23 January 1917, ibid.
6. Bowyer, *Royal Flying Corps Communiqués 1917–1918*, pp. 11–12, 1998.
7. Combat report 40 Sqn. RFC, 23 January 1917,by Lt. E.L. Benbow, summarized in the daily intelligence summary of the RFC "War diary", Winfried Bock collection.
8. To avoid repetition in the text, the pilot is always named first and the observer second.
9. Combat report Manfred von Richthofen, 24 January 1917, ibid.
10. Floyd Gibbons, *The Red Knight of Germany*, p. 117, New York, 1927.
11. Combat report 25 Sqn. RFC, 24 January 1917, summarized in the daily intelligence summary of the RFC War diary, Winfried Bock collection.
12. Oberkommando der 6. Armee, Kraftfahr- und Fliegertruppe (WK) Bund 2150, Capt. Greig, 2/Lt. Mac Lennan, Bayerisches Hauptstaatsarchiv, Abt. 4 München.
13. Franks et al., *Under the Guns of the Red Baron*, p. 56, 2000.
14. Malcolm Brown, Ed., *The Imperial War Museum Book of the First World War* , pp. 123–125, London, 2002.
15. Norman Macmillan, *Wings of Fate*, p. 81, London, 1967. Unfortunately, Macmillan did not give a word by word quotation of the interview.
16. Based on the content of Norman Macmillan, *Wings of Fate*, pp. 82–83, ibid.
17. Wolff Diary.
18. Richthofen, *Der rote Kampfflieger,* p. 119.
19. Iluft Nr. 205, ibid.
20. Richthofen, *Der rote Kampfflieger,* p. 119.
21. According to Mrs. Helene Allmenröder, Carl Allmenröder reported this experience during his vacation. Handwritten note by the author based on the conversation with Mrs. Allmenröder.
22. Letter of Carl Allmenröder to Helene Kortenbach, owned by Frau Helene Allmenröder. Handwritten transcription by Bruno Schmäling, 1976.
23. Letter by Kurt Wolff to is his aunt, dated 8 February 1917, ibid.
24. Personal Information by Carl-August von Schoenebeck, ibid.
25. Floyd Gibbons, *The Red Knight of Germany,* p. 126. The German original document is no longer in existence.
26. Personal information by Kurt Delang, ibid.
27. Wöchentlicher Tätigkeitsbericht des Kommandeurs der Flieger of the 6th Army from 21 January 1917 – 28 January 1917 (hereinafter referred to as "WTB Kofl 6"), General der Pioniere 6 Band 58, Bayerisches Hauptstaatsarchiv, Abt. 4 München.
28. Bowyer, *Royal Flying Corps Communiqués 1917–1918*, RFC Communiqué No. 73 pp. 14–15.
29. "WTB Kofl 6 from 29 January 1917 - 04 February 1917, ibid.
30. Wolff Diary, ibid.

5. Jagdstaffel 11 during Operation "Alberich", February 1917

1. Reichsarchiv Band XII. Beilage 7 - Die Siegfriedbewegung, Lageskizze: late March 1917.
2. Jagdstaffel 4 arrived only on 23 February 1917 to reinforce the German 6th Army sector; Jagdstaffel 30 at Phalempin was ready for action only from 28 February 1917.
3. WTB Kofl 6 from 29 January 1917–08 February 1917, ibid.
4. Combat report Manfred von Richthofen, 1 February

1917, ibid.

5. James F. Miller, *Inside the Victories of Manfred von Richthofen*, Vol. 1, p. 101. Aeronaut Books 2016.
6. WTB Kofl 6 from 29 January 1917–08 February 1917, ibid.
7. Wolff Diary, ibid.
8. A.O.K. der 6. Armee, Flieger-Nachrichten und Verfolgungsstelle Douai, Flugmeldebuch XV, 2. 2. – 3. 3. 1917. Bestand AOK 6/318, Bayerisches Hauptstaatsarchiv, Abt. 4 München.
9. Letter by Kurt Wolff to his aunt, dated 18 February 1917, ibid.
10. Wolff Diary, ibid.
11. WTB Kofl 6 from 29 January 1917–08 February 1917, ibid.
12. Iluft Akte Nr. 205. Flugzeuge Armeeflugpark 6, BHStA/KA ILUFT205.
13. Combat report Manfred von Richthofen, 14 April 1917, ibid.
14. A.O.K. der 6. Armee, Flieger-Nachrichten und Verfolgungsstelle Douai, Flugmeldebuch XV, 2. 2. – 3. 3. 1917, ibid.
15. Oberkommando der 6. Armee, Kraftfahr- und Fliegertruppe (WK) Bund 2150 und 2170 Lt. C.D. Bennet, Bayerisches Hauptstaatsarchiv, Abt. 4, München.
16. Oberkommando der 6. Armee, Kraftfahr- und Fliegertruppe (WK) Bund 2150, medical report of Lt. C.D. Bennet, ibid.
17. Combat report Manfred von Richthofen, 14 February 1917, ibid.
18. Combat report 2. Sqn. RFC, 14 February 1917, summarized in the daily intelligence summary of the RFC War diary, Winfried Bock collection.
19. The designation LB stands for Lichtbild; the aircraft of the Fliegerabteilungen LB were equipped with special cameras for photographing enemy positions.
20. The "Nachrichtenblatt der Luftstreitkräfte" erroneously notes the date of the downing as 16 February 1917 (not the only mistake in the "Nachrichtenblatt"). According to the transcript of the war diary of Jasta 11 and the "WTB Kofl 6", the downing took place on 15 February 1917.
21. A.O.K. der 6. Armee, Flieger-Nachrichten und Verfolgungsstelle Douai Flugmeldebuch XV, 2. February – 3. 3. 1917, ibid.
22. A.O.K. der 6. Armee, Flieger-Nachrichten und Verfolgungsstelle Douai Flugmeldebuch XV, 2. February – 3. 3. 1917, ibid.
23. Herris, *Siemens-Schuckert Aircraft of World War I*, p.62. Personal information by Rudolf Nebel, Jasta 5 to Bruno Schmäling, 1975.
24. WTB Kofl 6 from 15.02.1917–22.02.1917, ibid.
25. Beate-Christine Fiedler, *Verfolgung und Ausgrenzung sogenannter "Mischlinge". Der Fall Friedrich Wilhelm Lübbert* (Persecution and Exclusion of So-Called "Mixed Breeds". The Case of Friedrich Wilhelm Lübbert) in: Hamburger Schlüsseldokumente zur deutsch-jüdischen Geschichte, 22 September 2016.
26. After Alma del Banco received the deportation order to Theresienstadt, she took her own life with morphine on 8 March 1943. Some of her paintings were destroyed by the National Socialists.
27. In 1963, two streets, Lübbertkai and Lübbertstraße in Cuxhaven, were named after Hans Julius Lübbert.
28. Personal Information by Eddy Lübbert, son of Friedrich-Wilhelm Lübbert and nephew of Eddy Lübbert, ibid. Beate-Christine Fiedler, ibid.
29. Personal Information by Eddy Lübbert, ibid. Beate-Christine Fiedler, ibid.
30. Wartime letter by Hans-Georg Lübbert 19 August 1914. His original war letters were published in a booklet compiled for the family. Privately owned by the family. Copies are maintained in the Bruno Schmäling collection. Eddy and Fritz Lübbert's father placed himself in the service of the German military administration and was director of the military administration of occupied Romania in 1917.
31. Eddy Lübbert wartime letter, mid-November 1914. ibid.
32. Eddy Lübbert war letter, 12 November 1915, ibid.
33. Hans-Georg, Eduard Lübbert, KrStL. 32301/211, Armee-Flug-Park 6, Bayerisches Hauptstaatsarchiv, Abt.4, München.
34. Eddy Lübbert war letter 21 May 1916, ibid.
35. Eddy Lübbert war letter 27 May 1916, ibid.
36. Hans-Georg Lübbert, KrStL. 32301/211, ibid.
37. Eddy Lübbert war letter 7 January 1917.
38. WTB Kofl 6 from 01 January 1917 – 07 January 1917, ibid.
39. Eddy Lübbert war letter 29 January 1917.
40. Eddy Lübbert war letter 20 February 1917. War diary excerpt Jasta 11, ibid.
41. Carl-Emil Schäfer, KrStL. 18031/264, Armee-Flug-Park 6, Bayerisches Hauptstaatsarchiv, Abt. 4 München.
42. Carl-Emil Schäfer, KrStL. 18031/264, ibid.
43. Emil Schäfer, *Vom Jäger zum Flieger,* p. 65, Berlin 1918.
44. Carl-Emil Schäfer, KrStL. 18031/264, ibid. Other sources named the 28 July as date of transfer to Kagohl II.
45. Emil Schäfer, *Vom Jäger zum Flieger* p. 77. Schaefer's comrades were: Lt. Lothar von Hausen, Lt. Werner Albert, Lt. Franz Christian von Scheele, Lt. Paul Hermann, and Lt. Stern.
46. Emil Schäfer, *Vom Jäger zum Flieger*, pp. 77-79.
47. Emil Schäfer, *Vom Jäger zum Flieger*, p. 80.
48. Carl-Emil Schäfer, KrStL. 18031/264, ibid.
49. Die Woche 23 (9 June 1917) Report of Georg von

Ompteda, pp. 767–768.
50. Iluft Akte Nr. 36, 204, 205, 206, ibid.
51. Jasta 4 war diary excerpt compiled by Erich Tornuß in the 1930s, copy Bruno Schmäling collection.
52. Personal Information by Freifrau von Döring to Bruno Schmäling, 1980.
53. Bruno Schmäling, Winfried Bock, Royal Prussian Jagdstaffel 30, Aeronaut Books 2014.
54. Personal information by Otto Fuchs, ibid.
55. WTB Kofl 6 from 23 February 1917–01 March 1917, ibid. From this "WTB" onwards, only the total number of wartime flights is given, and these are no longer allocated to the individual Jagdstaffeln.
56. WTB Kofl 6, February 1917, Zusammenstellung Dr. Gustav Bock.

6. The Birth of the Richthofen Myth, March 1917

1. Personal Information by Hermann Lohmeyer, ibid.
2. WTB Kofl 6 from 02 March 1917 – 08 March 1917, ibid.

6.1 1st March 1917 – 6th March 1917

1. Schäfer, *Vom Jäger zum Flieger,* p. 81, ibid.
2. War Diary excerpt Jasta 11, and Jasta 35 by Erich Tornuß, ibid.
3. Letter by Kurt Wolff to his sister Elisabeth, dated 3 March 1917, ibid.
4. WTB Kofl 6 from 02 March 1917 – 08 March 1917, ibid.
5. At this time, Jagdstaffel 11 was divided into two groups, referred to as swarms.
6. Emil Schäfer, *Vom Jäger zum Flieger*, p. 81ff.
7. Combat report 40 Sqn. RFC, 4 March 1917, summarized in the daily intelligence summary of the RFC War diary, Winfried Bock collection.
8. War Diary excerpt Jasta 11, by Erich Tornuß, ibid.
9. Wolff Diary.
10. AIR1 Files at The National Archives UK, Combat report 40 Sqn. RFC, 4 March 1917.
11. Combat report Manfred von Richthofen, 4 March 1917, 23rd aerial victory, ibid.
12. Combat report 2 Sqn. RFC, 4 March 1917, summarized in the daily intelligence summary of the RFC War diary, Winfried Bock collection.
13. A.O.K. der 6. Armee, Flieger-Nachrichten und Verfolgungsstelle Douai Flugmeldebuch XVI, 4 March – 24 March 1917, ibid.
14. Combat report Manfred von Richthofen, 4 March 1917, (22nd aerial victory), ibid.
15. Emil Schäfer, *Vom Jäger zum Flieger,* p. 82, ibid.
16. Emil Schäfer, *Vom Jäger zum Flieger* p. 82, ibid. For Lt. Anslinger, the flight on 4 March 1917 may have been his last mission with Jagdstaffel 11.
17. Emil Schäfer, *Vom Jäger zum Flieger,* p. 83, ibid.
18. Combat report 40 Sqn. RFC, 4 March 1917, summarized in the daily intelligence summary of the RFC War diary, Winfried Bock collection.
19. Wolff Diary, ibid.
20. A.O.K. der 6. Armee, Flieger-Nachrichten und Verfolgungsstelle Douai Flugmeldebuch XVI, 4.3. – 24. 3. 1917, ibid.
21. Personal Information by Hermann Lohmeyer, ibid.
22. Emil Schäfer, *Vom Jäger zum Flieger*, p. 93–96, ibid.
23. AIR1 Files at The National Archives UK, Combat report 43 Sqn. RFC, 06 March 1917.
24. AIR1 Files at The National Archives UK, Combat report 43 Sqn. RFC, 06 March 1917.
25. WTB Kofl 6 from 02 March 1917–08 March 1917, ibid.
26. Manfred von Richthofen, *Der rote Kampfflieger*, p. 128–131, ibid.
27. AIR1 Files at The National Archives UK, Combat report 43 Sqn. RFC, 06 March 1917.
28. Eddy Lübbert war letter, 13 March 1917, ibid.
29. Combat report 40 Sqn. RFC, 4.March 1917, summarized in the daily intelligence summary of the RFC War diary, Winfried Bock collection.
30. AIR1 Files at The National Archives UK, Combat report 43 Sqn. RFC, 09 March 1917.
31. Manfred von Richthofen, *Der rote Kampfflieger*, p. 131, ibid.
32. Wolff Diary, ibid.
33. Bowyer, *RFC Communiqués 1917–1918*, p. 24. Trevor Henshaw, p. 70, ibid.
34. AIR1 Files at The National Archives UK, Report on casualties to personal and machines when flying, 16 Sqn. RFC, 06 March.1917.
35. AIR1 Files at The National Archives UK, Combat report 16 Sqn. RFC, 6 March 1917.
36. Personal Information by Hermann Lohmeyer. ibid.
37. Letter by Kurt Wolff to his sister Elisabeth, dated 3 March 1917, ibid. The pleasure of downing an enemy plane in aerial combat may sound brutal, but one must not forget that this was a fight to the death and that it was and still is the task of fighter pilots to shoot down enemy planes.
38. Combat report Manfred von Richthofen, 6 March 1917, ibid.
39. AIR1 Files at The National Archives UK, Combat report 16 Sqn. RFC, 06 March 1917.
40. Personal Information by Hermann Lohmeyer, ibid.
41. War letter Lübbert, 07 March 1917.
42. Personal Information by Eddy Lübbert, the nephew of the pilot Hans-Georg Lübbert, ibid.

6.2 7th March 1917 – 9th March 1917

1. Personal Information by Xaver Leinemüller, Jasta 10 to Bruno Schmäling, confirmed by Hermann Lohmeyer.
2. Personal Information by Carl-August von Schoenebeck, ibid.
3. WTB Kofl 6 from 02 March 1917 – 08 March 1917, ibid.
4. WTB Kofl 6 from 09 March 1917 – 15 March 1917, ibid.
5. Wolff Diary.

6. AIR1 Files at The National Archives UK. The Secretary, Air Ministry: Repatriated or escaped Officer of war.
7. AIR1 Files at The National Archives UK. The Secretary, Air Ministry: Repatriated or escaped Officer of war, ibid.
8. AIR1 Files at The National Archives UK. The Secretary, Air Ministry: Repatriated or escaped Officer of war, ibid.
9. AIR1 Files at The National Archives UK, Combat report 40 Sqn. RFC, 09 March 1917.
10. Combat report 40 Sqn. RFC, 9 March 1917, summarized in the daily intelligence summary of the RFC "War diary), Winfried Bock collection.
11. Emil Schäfer, *Vom Jäger zum Flieger* p. 103.
12. Letter by Kurt Wolff to his sister Elisabeth, dated 3 March 1917, ibid.
13. Oberkommando der 6. Armee, Kraftfahr- und Fliegertruppe (WK) Bund 2150, interrogation report of 2/Lt. A Shepard, 2/Lt. G.J. Haseler and 2/Lt. W.B. Hill, Bayerisches Hauptstaatsarchiv, Abt. 4, München.
14. Combat report Manfred von Richthofen, 09 March 1917, 25th aerial victory, ibid.
15. AIR1 Files at The National Archives UK, Combat report 29 Sqn. RFC, 09 March 1917.

6.3 10th March 1917 – 17th March 1917

1. Lothar von Richthofen KrStL. 18031/267, Armee-Flug-Park 6, Bayerisches Hauptstaatsarchiv, Abt. 4 München.
2. Information from Wolf-Manfred von Richthofen, the son of Lothar von Richthofen, to the Swedish aviation historian Joachim Tinnerholm. Wolf-Manfred von Richthofen lived until his death in Sweden and Joachim Tinnerholm was in close contact with him. Joachim Tinnerholm thankfully supported the author of this book with information he got from Wolf von Richthofen.
3. Lothar von Richthofen KrStL. 18031/267, ibid.
4. Excerpt from Lieutenant Joachim von Alt-Stütterheim's report on Lothar von Richthofen, Joachim Tinnerholm collection.
5. *Ein Heldenleben*, p. 128 – 129, a.a.O. Flieger-Ring-Nachrichtenblatt, Herausgeber Otto Fuchs 2. Jahrgang Nr. 8, 1. August 1922, Bruno Schmäling collection.
6. Kunigunde Freifrau von Richthofen, *Mein Kriegstagebuch*, p. 78, Berlin 1937
7. Lothar von Richthofen KrStL. 18031/267, ibid.
8. Flieger-Ring-Nachrichtenblatt, Herausgeber Otto Fuchs 2. Jahrgang Nr. 8, 1. August 1922.
9. Personal information from Wolf-Manfred von Richthofen, towards Joachim Tinnerholm.
10. Report from Alex Imrie according to his handwritten notes of the conversation with Carl Bodenschatz.
11. Personal Information by Carl-August von Schoenebeck, to Bruno Schmäling.
12. Personal Information by Hans-Georg von der Osten, to Bruno Schmäling. As far as Mr. von der Osten can remember, his rescue by Lothar von Richthofen took place in November 1917.
13. Personal information by Wolf-Manfred von Richthofen, towards Joachim Tinnerholm.
14. Journal Daheim, Erich von Salzmann, *Richthofen*, p. 10–11, 18. May 1918.
15. Personal Information by Otto Fuchs, ibid.
16. Summary of the description of the personality of Lothar von Richthofen by former pilots of Jasta 11 as related to Bruno Schmäling and Alex Imrie.
17. Personal Information by Otto Fuchs, ibid. Strangely enough, despite his resentment towards Prussia, Hptm. Franz Hailer wore the Prussian airman's badge instead of the Bavarian one. It is possible that his resentment towards the "Saupreussen (swine-Prussians)", as he called them, stemmed from his time training as an airplane pilot in Döberitz, Prussia.
18. Personal Information by Otto Fuchs, ibid.
19. "Island of Genius", also known as savant syndrome, means that these people have a very high, often far above-average talent in some areas, but are below average in other areas. A number of computer experts, for example, probably fall into this personality.
20. Knötel d.J., Herbert, Pietsch, Paul, Baron Collas, *Das deutsche Heer, Friedensuniformen bei Ausbruch des Weltkrieges*, Band I und II, Stuttgart 1982.
21. *Ein Heldenleben*, p. 224 – 225, a.a.O.
22. War letter Lübbert, 10 March 1917.
23. Emil Schäfer, a.a.O., p. 97–99.
24. A.O.K. der 6. Armee, Flieger-Nachrichten und Verfolgungsstelle Douai Flugmeldebuch XVI, 4.3. – 24. 3. 1917, Ibid.
25. The Nachrichtenblatt report was probably incorrect because a crash between Givenchy-en-Gohelle and Vimy would have been in German territory and easily confirmed. If Krefft's victim had been observed to go down across the lines near Souchez, it probably caught itself and returned to base. Errors do occur in the Nachrichtenblatt and whenever possible, its information should be verified by another source.
26. Combat report Manfred von Richthofen, 11 March 1917 (26th aerial victory), ibid.
27. Mentioned in the Wöchentlicher Tätigkeitsbericht des Stabsoffiziers der Flieger der 6. Armee Kofl 6 vom 16.03.1917 point 9. (Bayerisches Hauptstaatsarchiv, Abt. 4, München). Of the 5 claimed aerial victories from 4. March 1917 only 3 were officially credited.
28. Albert Oesterreicher OP (Personal record) file 1188, Bayerisches Hauptstaatsarchiv München.
29. Wolff Diary, ibid.
30. Combat report Manfred von Richthofen, 17 March 1917 (27 aerial victory), ibid.
31. Wolff Diary, ibid.
32. Postcard by Carl Allmenröder from the 21 March 1917, Stadtarchiv Solingen.

33. AIR1 Files at The National Archives UK, Combat report 25 Sqn. RFC 17.03.1917. There were no German rotary engine fighters involved.
34. AIR1 Files at The National Archives UK, Combat report 43 Sqn. RFC 17.03.1917.
35. Combat report 43 Sqn. RFC, 17 March 1917, summarized in the daily intelligence summary of the RFC "War diary", Winfried Bock collection.
36. WTB Kofl 6 from 23. März 1917
37. The human brain can only perceive and process parts of what it perceives. This perception is interpreted by the brain on the basis of opinions, experiences, attitudes, wishes and ideas. Therefore, every report, especially under stress, is a personal interpretation of what is seen and can never be objective. Reference: Christian Callo, Bruno Schmäling "Perception of dangerous situation", Training material, Germany 2000.
38. Kraftfahr und Fliegertruppe (WK) 2341/I, Flugzeuge Armeeflugpark 6, La Briquette, 31.03.1917. (Bayerisches Hauptstaatsarchiv Abt. 4, München)
39. Combat report Manfred von Richthofen, 17 March 1917 (28 aerial victory), ibid.
40. A.O.K. der 6. Armee, Flieger-Nachrichten und Verfolgungsstelle Douai Flugmeldebuch XVI, 4.3. – 24. 3. 1917, Ibid.
41. Andrew White, *Fire-Step to Fokker Fodder, From the Trenches to the Red Baron, the first world War Diaries of William 'Jack' Lidsey*, p. 148, Fighting High, Ltd., 2019.

6.4 18th March 1917 – 31st March 1917

1. Flight log of Hans-Georg Lübbert, Entries on the flights on 18 March 1917, ibid. The entry shows that at this time one or more Halberstadts remained in Jasta 11's inventory as reserve or "touring" aircraft.
2. Flight log Hans-Georg Lübbert, entry relating to the flights on 18 March, ibid.
3. Flight log Hans-Georg Lübbert, entry relating to the flights on 19 March, ibid.
4. Flight log Hans-Georg Lübbert, entry relating to the flights on 21 March, ibid.
5. Flight log Hans-Georg Lübbert, entry relating to the flights on 21 March, ibid.
6. Flight log Hans-Georg Lübbert, entry relating to the flights on 21 March, ibid.
7. A.O.K. der 6. Armee, Flieger-Nachrichten und Verfolgungsstelle Douai Flugmeldebuch XVI, 4.3. – 24. 3. 1917, ibid.
8. A.O.K. der 6. Armee, Flieger-Nachrichten und Verfolgungsstelle Douai Flugmeldebuch XVI, 4.3. – 24. 3. 1917, ibid.
9. AIR1 Files at The National Archives UK, Casualty report 43 Sqn. RFC 21.03.1917.
10. Postcard from Carl Allmenröder from 21.03.1917, Stadtarchiv Solingen
11. Combat report Manfred von Richthofen, 21 March 1917 (29th aerial victory), ibid.
12. AIR1 Files at The National Archives UK, Casualty report 16 Sqn. RFC 21.03.1917.
13. Flight log Hans-Georg Lübbert, entry relating to the flights on 22 March 1917, a.a.O.
14. Flight log Hans-Georg Lübbert, entry relating to the flights on 22 March 1917, a.a.O.
15. War diary excerpt Jagdstaffel 11, ibid.
16. The author Ed Ferko states in his book *Richthofen*, without naming a source, that this meeting took place on 15 April 1917. According to the caption in the photo album of Lt. Oskar Müller, Officer z. b. V. of Jagdstaffel 12, the photos were taken on 23 March 1917. Photo album Oskar Müller, B. Schmäling and M. Schmeelke Collection.
17. Flight log Hans-Georg Lübbert, entry relating to the flights on 23 March 1917.
18. Letter by Kurt Wolff, dated 23 March 1917, ibid.
19. Thorsten Pietsch, *Mein Lieber Hans, Feldbriefe einer Mutter*, p. 144, Münster 2006.
20. Thorsten Pietsch, *Mein Lieber Hans*, ibid.
21. Flight log Hans-Georg Lübbert, entry relating to the flights on 23 March 1917, a.a.O.
22. Combat report Manfred von Richthofen, 24 March 1917 (30th aerial victory), ibid.
23. Oberkommando der 6. Armee, Kraftfahr- und Fliegertruppe (WK) Bund 2150, Lt. R.P. Backer, Bayerisches Hauptstaatsarchiv, Abt. 4, München.
24. Combat report 19 Sqn. RFC, 24 March 1917, summarized in the daily intelligence summary of the RFC War diary, Winfried Bock collection.
25. AIR1 Files at The National Archives UK, Combat report 19 Sqn. RFC 25.03.1917.
26. Flight log Hans-Georg Lübbert, entry relating to the flights on 24 March 1917, a.a.O.
27. *Ein Heldenleben*, p. 225
28. Combat report Manfred von Richthofen, 25 March 1917 (31st aerial victory), ibid.
29. Oberkommando der 6. Armee, Kraftfahr- und Fliegertruppe (WK) Bund,2150, Lt. C.G. Gilbert, Bayerisches Hauptstaatsarchiv, Abt. 4, München.
30. AIR1 Files at The National Archives UK. The Secretary, Air Ministry: Repatriated or escaped Officer of war.
31. *Ein Heldenleben*, p. 226.
32. Flightlog Hans-Georg Lübbert, entry relating to the flights on 25 March 1917, a.a.O.
33. *Ein Heldenleben*, p. 195.
34. Flight log Hans-Georg Lübbert, 27 March 1917.
35. AIR1 Files at The National Archives UK, Combat report 25 Sqn. RFC, 28 March 1917. Although by this time all German fighter units in the German 6th Army area were equipped with Albatros D IIIs and some D IIs, the designation "Halberstadt" still appears in British air combat reports. Misidentification of enemy aircraft was the rule rather than the exception on both sides of the front.
36. AIR1 Files at The National Archives UK, Combat report 16 Sqn. RFC 28.03.1917.

37. Oberkommando der 6. Armee, Kraftfahr- und Fliegertruppe (WK) Bund 2150, 2/Lt. Leslie Knight, Bayerisches Hauptstaatsarchiv, Abt. 4, München.
38. War letter of Eddy Lübbert dated 29 March1917, ibid.
39. Wolff Diary, ibid.
40. Combat report 19 Sqn. RFC, 24 March 1917, summarized in the daily intelligence summary of the RFC War diary, Winfried Bock collection.
41. *Cross & Cockade USA*, 11, Vol. 2, 1980, p. 56.
42. Personal Information by Eddy Lübbert, ibid. Friedrich-Wilhelm Lübbert researched his brother's last flight. According to him, his brother flew another pilot's airplane that day because his 'half-blue and half-yellow' airplane was still not ready for service.
43. Combat report 40 Sqn. RFC, 30 March 1917, summarized in the daily intelligence summary of the RFC War diary, Winfried Bock collection.
44. A.O.K. der 6. Armee, Flieger-Nachrichten und Verfolgungsstelle Douai Flugmeldebuch XVII, 25.3. – 6.4.1917, ibid.
45. Wolff diary, ibid.
46. Telegram dated 30. March 917, private property Eddy Lübbert.
47. Conversation with Richthofen as related by Helene Allmenröder to Bruno Schmäling, ibid.
48. Behaviour of Manfred von Richthofen as related by the pilots Jasta 11 and Jagdgeschwader I towards Bruno Schmäling and Alex Imrie, ibid.
49. Eddy's brother, Friedrich-Wilhelm Lübbert, was a fighter pilot in Fighter Squadron 11 from 26 December 1917 to 17 February 1918. He narrowly escaped deportation and to be murdered in a concentration camp. Beate-Christine Fiedler, ibid.
50. Combat report 60. Sqn. RFC, 30 March 1917, summarized in the daily intelligence summary of the RFC War diary, Winfried Bock collection.
51. Combat report 60 Sqn. RFC, 30 March 1917, summarized in the daily intelligence of the RFC War diary, Winfried Bock collection.
52. WTB Kofl 6 from 7 April 1917.
53. Wolff Diary, ibid.
54. Oberkommando der 6. Armee, Kraftfahr- und Fliegertruppe (WK) Bund 2150, 2/Lt. A. Strange, Bayerisches Hauptstaatsarchiv, Abt. 4, München.
55. AIR1 Files at The National Archives UK. The Secretary, Air Ministry: Repatriated or escaped Officer of war, ibid
56. AIR1 Files at The National Archives UK, Combat report 11 Sqn. RFC, 31 March 1917.
57. *Cross & Cockade Great Britain* Journal, Vol. 14 1983, Presentation Aircraft 1914–1918 by Raymon Vann, Colin Waugh.
58. Kriegstagebuchabschriften Jasta Boelcke, ibid.
59. WTB Kofl 6 from 1 March 1917 to 7 April 1917, ibid.
60. Personal Information by Hermann Lohmeyer, ibid.

7. Kennung und Bemalung der Flugzeuge

7.1 The Halberstadt Era

1. Heinz Nowarra, *Eisernes Kreuz und Balkenkreuz*, p. 130, Mainz 1968.
2. Personal Information by Rudolf Nebel, Jasta 5, Staffelführer KeSt. 1a, 1b, Jasta 90 to Bruno Schmäling.
3. Hans-Joachim Buddecke, *El Schahin – Der Jagdfalke*, p. 103, Berlin 1918,
4. Bruno Schmäling, *Jasta Colors Volume 2*, The book contains a detailed treatise on the various fabric coverings of the Halberstadt fighter planes.
5. Flugzeuge Armeeflugpark 6, La Briquette, Iluft folder Nr. 36, 204, 205 and 206, 207 Bayerisches Hauptstaatsarchiv, Abt. 4 München.
6. Flugzeuge Armeeflugpark 6, La Briquette, ibid.
7. Flugzeuge Armeeflugpark 6, La Briquette, ibid.
8. Fotoalbum Carl Allmenröder.

7.2 The Albatros D III Era

1. Personal Information by Hermann Lohmeyer, ibid.
2. *Ein Heldenleben*, pp. 302–307.
3. British G-Report 42 captured Albatros D III D.2015/16, Alex Imrie collection.
4. Photo of the emergency-landed Albatros D III of Manfred von Richthofen, taken on 24 January 1917, R. Kastner Collection.
5. Combat report 40 Sqn. RFC, 23 January 1917, summarized in the daily intelligence summary of the RFC War diary, Winfried Bock collection.
6. Combat report Manfred von Richthofen, 24 January 1917, ibid.
7. Photo of the emergency-landed Albatros D III of Manfred von Richthofen, , taken on 24 January 1917, R. Kastner Collection.
8. Armeeflugpark 6, La Briquette, Iluft folder Nr. 36, 204, 205 and 206, ibid. The Albatros D III D.1996/16 later became the 'half-blue and half-yellow' aircraft of Lt. Hans-Georg Lübbert.
9. Personal information by by Heinz Geisseler, Jasta 20, 33, 46, and Kest 4b and Heinz Arntzen, Jasta 15 and Staffelführer Jasta 50 to Bruno Schmäling, ibid.
10. Personal Information by ground crew members Hermann Lohmeyer, Xaver Leinemüller, Jasta 10 and Ernst Schnabel, Jasta 79.
11. Personal information by Heinz Arntzen, Otto Fuchs, Josef Jacobs, Carl-August von Schoenebeck a. o., as well as the mechanics Xaver Leinemüller, Hermann Lohmeyer, and Erich Schnabel, Jasta 79.
12. Knötel d.J., Herbert, Pietsch, Paul, Baron Collas, *Das deutsche Heer, Friedensuniformen bei Ausbruch des Weltkrieges*, Band I, II and III, Stuttgart 1982.
13. Armeeflugpark 6, La Briquette, ibid.
14. Armeeflugpark 6, La Briquette, ibid.
15. Ein Heldenleben, S. 221, ibid.
16. Armeeflugpark 6, La Briquette, ibid.
17. War letter Eddy Lübbert dated 6. March 1917, ibid.

18. „A.O.K. der 6. Armee, Flieger-Nachrichten und Verfolgungsstelle Douai, Flugmeldebuch XVII, 25 March – 6 April 1917, ibid.
19. War letter Eddy Lübbert dated 6 March 1917, ibid.
20. Armeeflugpark 6, La Briquette, ibid.
21. A.O.K. der 6. Armee, Flieger-Nachrichten und Verfolgungsstelle Douai Flugmeldebuch XVI, 4 March – 24 March 1917, ibid.
22. Personal Information by Hermann Lohmeyer ibid.
23. War letter Lübbert, 13 March 1917, ibid.
24. Personal Information by Hermann Lohmeyer, ibid.
25. War letter Eddy Lübbert of 6 March 1917, ibid.
26. Personal Information Eddy Lübbert, nephew of Hans-Georg Lübbert based on the information he got from his father Friedrich-Wilhelm Lübbert. ibid.
27. AIR1 Files at The National Archives UK, Combat report 40 Sqn. RFC, 4.March 1917.
28. Personal Information by Eddy Lübbert,
29. Flight log Hans-Georg Lübbert, entry relating to the flights on 24 March 1917, a.a.O.
30. War letter Lübbert, 6 March 1917, ibid.
31. AIR1 Files at The National Archives UK, Combat report 16 Sqn. RFC 06 March 1917 via T. Henshaw collection.
32. Combat report Lt. Manfred von Richthofen, 6 March 1917, ibid.
33. Personal Information Ludwig Zacharias based on his interview with Wilhelm Allmenröder, ibid.
34. Personal Information by Georg Simon to Alex Imrie. Unfortunately, the author never asked Alex Imrie if he knew the reason why Georg Simon painted his aircraft green.
35. AIR1 Files at The National Archives UK, Combat report 29 Sqn. RFC 09.03.1917 via T. Henshaw collection.
36. Report of the Gruppenführer der Flieger Gruppe Vimy vom 5 April 1917, Bayerisches Hauptstaatsarchiv, Abt, 4 München Abschrift durch Dr. Gustav Bock, Sammlung Winfried Bock.
37. A.O.K. der 6. Armee, Flieger-Nachrichten und Verfolgungsstelle Douai, Bericht der Flakgruppe Denain, dated 3 April 1917, Bayerisches Hauptstaatsarchiv München, Abt. 4.
38. *Ein Heldenleben*, p. 205, ibid. Knötel d. J., Herbert, Pietsch, Paul, Baron Collas, *Das deutsche Heer*, ibid.
39. Flugmeldebuch des A.O.K. 6 dated 7 April 1917, Report of Flak Vitry, dated 07 April 1917, Bayerische Hauptstaatsarchiv, Abt. 4 München.
40. *Ein Heldenleben*, p. 221, ibid.

Excerpts from Royal Prussian Jagdstaffel 11's War Diary

Date	Event
28.09.16	In accordance with an order from the Chief of the General Staff of the Field Army, Kofl 6th Army ordered the immediate formation of a Staffel at Douai airfield consisting of the Bavarian portion of Kampfeinsitzer-Kommando III.
05.10.16	Formation of the Staffel at Douai (La Brayelle) airfield began.
11.10.16	The Staffels' first day of mobilization. Kampfeinsitzer-Kommando III is considered disbanded as of today. The relevant ground personnel and pilots were transferred to the Staffel: Lt. Albert Oesterreicher and Lt. Karl Kleinhenz, who will assume the duties of Offizier z.b.V. for the time being, and Gefr. Sebastian Festner, all from Bavarian Feldflieger-Abteilung 5, and Sergt. Erwin Howe from Flieger-Abteilung 24. Lt. Konstantin Krefft is transferred from Bavarian Flieger-Ersatz-Abteilung 1 in Schleissheim to the Staffel. Vzfw. Alfred Prehn, formerly with Jagdstaffel 10, is transferred to the Staffel.
14.10.16	Lt. Franz Anslinger, formerly with Kagohl 3's Kampfstaffel 14, is transferred to the Staffel via Armee-Flug-Park 6.
20.10.16	Oblt. Rudolf Lang, formerly the leader of Kagohl 6's Kampfstaffel 31, is appointed Staffelführer in accordance with Feldflugchef Order No.53826. Lt. Konstantin Krefft, on a solo flight, shoots down an enemy airplane far over enemy territory. The downing was not confirmed due to a lack of air and ground witnesses.
21.10.16	Gefr. Sebastian Festner is promoted to Unteroffizier.
22.10.16	The Staffel's first combat flight under the leadership of its Staffelführer.
05.11.16	Lt. Kurt Wolff, formerly with Kagohl 7's Kampfstaffel 40, is transferred to the Staffel via Armee-Flug-Park 6.
07.11.16	Lt. Georg Simon, formerly with Kagohl 2's Kampfstaffel 12, is transferred to the Staffel via Armee-Flug-Park 6.
14.11.16	Vzfw. Alfred Prehn is ordered to Seeflugstation Tondern.
20.11.16	Lt. Carl Allmenröder, formerly with Feldflieger-Abteilung 18, is transferred to the Staffel.
26.11.16	Vzfw. Josef Baierlein, formerly with Bavarian Feldflieger-Abteilung 2, is ordered to the Staffel.
22.12.16	Vzfw. Josef Baierlein is now considered as transferred to the Staffel.
14.01.17	In accordance with Kogenluft order I Flieg. No.57439 of 14.01.17, Oblt. Rudolf Lang is to lead Jagdstaffel 28, currently situated with Flieger-Ersatz-Abteilung 10 at Böblingen, and is transferred there as Staffelführer. In the same order, Lt. Manfred Frhr. von Richthofen, formerly with Jagdstaffel 2, is appointed as the new Staffelführer.
16.01.17	Vzfw. Georg Braunmiller from Jagdstaffelschule I in Valenciennes is transferred to the Staffel via Armee-Flug-Park 6.
20.01.17	Lt. Manfred Frhr. von Richthofen arrives at the Staffel and takes command; Oblt. Rudolf Lang sets off for Jagdstaffel 28 after handing over command.
21.01.17	Lt. Manfred Frhr. von Richthofen is awarded the Orden Pour le Mérite.
23.01.17	Lt. Manfred Frhr. von Richthofen shoots down a Vickers single-seater in flames over enemy territory southwest of Lens at around 4:10 p.m. Confirmed by Kogenluft as his 17th air victory. According to available documents, his opponent was F.E.8 6388 from 40 Squadron RFC. Its pilot, 2/Lt. J. Hay, was killed.
24.01.17	Lt. Manfred Frhr. von Richthofen forces F.E. 6937 to land in German territory west of Vimy, but then had to make an emergency landing himself near his opponent's landing place due to a breakage in his Albatros D.III's lower wing. Confirmed by Kogenluft as his 18th air victory. According to available documents, his opponent was F.E.2b 6997 from 25 Squadron RFC. The crewmen, Capt. O. Greig (pilot), who was wounded,and 2/Lt. J.E. MacLennan (observer), were both taken prisoner.

Date	Event
29.01.17	Lt. Georg Simon is slightly wounded by a glancing shot during a dogfight with an F.E. at around 10:50 a.m. east of Lens. He landed safely at his airfield and remains with the Staffel.
	His opponent was one of two F.E.2b crews from 25 Squadron RFC who reported that they had shot down a Halberstadt fighter during the dogfight: either 2/Lt. W. Shirtcliffe and 2/Lt. A.V. Blenkiron (WIA) in F.E.2b A784 or 2/Lt. J.L. Leith and 2Lt. D.C. Woolen in F.E.2b 7693.
01.02.17	Lt. Manfred Frhr. von Richthofen shoots down B.E. 6742 around 4:00 p.m. over the front lines where it came down in a barbed wire barrier about 1 kilometer southwest of Thélus. Confirmed by Kogenluft as his 19th air victory.
	His opponent, according to available documents, was B.E.2d 6742 from 16 Squadron RFC. Its crew, Lt. P.W. Murray (pilot) and Lt. D.J. McRae (observer), were seriously wounded and taken prisoner, but died of their injuries the next day.
05.02.17	Vzfw. Sebastian Festner shoots down a B.E. that trailed smoke around 5:00 p.m. over enemy territory near Neuville (north of Arras). Confirmed by Kogenluft as his 1st air victory.
14.02.17	Lt. Manfred Frhr. von Richthofen shoots down B.E. 6231 around 12:00 p.m. over German territory west of Loos on the Lens-Hulluch road. Confirmed by Kogenluft as his 20th air victory. Lt. Manfred Frhr. von Richthofen shoots down a B.E. in pieces around 4:45 p.m. over enemy territory about 1.5 kilometers southwest of Mazingarbe. Confirmed by Kogenluft as his 21st air victory.
	According to available documents, his two opponents were B.E.2d 6231 from 2 Squadron RFC, manned by 2/Lt. C.D. Bennett (pilot, WIA, POW) and 2/Lt. H.A. Croft (observer, KIA), followed by B.E.2c 2543 from the same squadron with Capt. G.C. Bailey (pilot, WIA) and 2/Lt. G.W.B. Hampton (observer) as its crew.
15.02.17	Following a long pursuit, Vzfw. Sebastian Festner shoots down an F.E. at 12:10 p.m. over enemy territory southwest of Grénay. Confirmed by Kogenluft as his 2nd air victory.
16.02.17	Lt. Carl Allmenröder forces B.E. 4179 to land in German territory around 1:00 p.m. about 1.5 kilometers south of Roeux. Confirmed by Kogenluft as his 1st air victory.
	According to available documents, his opponent was B.E.2c 4179 from 16 Squadron RFC, manned by 2/Lt. E.W. Lindley (pilot) and 2/Lt. L.V. Munn (observer). The former was taken prisoner and died of his injuries two days later, the latter was killed in action.
17.02.17	Sergt. Erwin Howe is transferred to Armee-Flug-Park 6.
18.02.17	Lt. Hans-Georg Lübbert, formerly with Flieger-Abteilung 18, is ordered to the Staffel.
21.02.17	Lt. Karl-Emil Schäfer, formerly with Kagohl 2's Kampfstaffel 11, is ordered to the Staffel.
26.02.17	Lt. Hans-Georg Lübbert is now considered as transferred to the Staffel.
28.02.17	Vzfw. Alfred Prehn, previously ordered to Seeflugstation Tondern, is now considered as having been transferred there.
01.03.17	Vzfw. Georg Braumüller is transferred to Flieger-Ersatz-Abteilung 4 in Posen via Armee-Flug-Park 6. Lt. Friedrich von Hartmann als Offizier zur besonderen Verwendung zur Staffel versetzt.
02.03.17	Lt. Franz Anslinger is transferred to Jagdstaffel 35 via Armee-Flug-Park 6.
04.03.17	Lt. Karl-Emil Schäfer shoots a Sopwith down in flames around 11:50 a.m. that broke apart over German territory southwest of Haisnes. Upon his return to the airfield, he crashed while landing but was uninjured; machine totaled. Confirmed by Kogenluft as his 2nd air victory. Lt. Manfred Frhr. von Richthofen shoots down a B.E. around 12:45 p.m. over enemy territory in the Loosbogen northeast of Loos. Confirmed by Kogenkuft as his 23rd air victory. Lt. Manfred Frhr. von Richthofen shoots down Sopwith A1108 around 4:20 p.m. that broke apart over German territory near Acheville. Confirmed by Kogenluft as his 22nd victory. Lt. Karl-Emil Schäfer shoots down two F.E. single-seaters, one after the other, around 5:20 p.m. far over enemy territory near Béthune. Both were not confirmed due to a lack of ground witnesses.

Date	Event
	According to available documents, Schäfer's first opponent was Sopwith Strutter A1109 from 43 Squadron RFC. Both of its crewmen, 2/Lt. P.L. Wood (pilot) and 2/Lt. A.H. Fenton (observer), were killed. Richthofen's first opponent was B.E.2d 5785 from 2 Squadron RFC with Lt. J.B.E. Crosbee (pilot) and Sgt. J.E. Prance (observer, WIA) aboard. Richthofen's second opponent was Sopwith Strutter A1108 from 43 Squadron RFC, manned by 2/Lt. H.J. Green (pilot) and 2/Lt. A.W Reid (observer). Both men were killed. Schäfer's unconfirmed opponents were F.E.8s from 40 Squadron RFC: Lt. C.O. Usborne in A4872 and 2/Lt. K.S. Henderson in 6419. Both pilots had to make emergency landings in their damaged aircraft.
06.03.17	Lt. Karl-Emil Schäfer shoots down a Sopwith in flames around 11:45 a.m. over German territory between Lens and Cité St. Pierre as well as a second Sopwith around 11:55 a.m. over enemy territory west of Lens. Confirmed by Kogenluft as his 3rd and 4th air victories. During this dogfight, Lt. Manfred Frhr. von Richthofen received several hits to his engine and fuel tank from a Sopwith's observer and had to make an emergency landing with a dead engine on a meadow along the Lens-Hénin-Liétard road. Lt. Hans-Georg Lübbert was lightly wounded by a bullet to the shoulder during a dogfight with a Sopwith and two F.E. single-seaters around 12:00 p.m. and made a safe emergency landing near Lens. Lt. Kurt Wolff shoots down B.E. 5856 at around 12:30 p.m. over German territory on the southeastern edge of Givenchy. Confirmed by Kogenluft as his 1st air victory. Lt. Manfred Frhr. von Richthofen shoots a B.E. down in pieces around 5:00 p.m. over enemy territory near Souchez. Confirmed by Kogenluft as his 24th air victory.
	According to available documents, Schäfer's first opponent was Sopwith Strutter A978 from 43 Squadron RFC, with Lt. S.J. Pepler (pilot) and Capt. J.D. Stuart (observer) as its crew. Pepler was killed in action and the wounded Stuart died the next day. Schäfer's second opponent was the same unit's A1097, manned by Capt. H. Balfour (pilot) and 2/Lt. A. Roberts (observer), who landed unhurt. Wolff's opponent, B.E.2d 5856 from 16 Squadron RFC, was flown by 2/Lt. G.M. Underwood (pilot) and 2/Lt. A.E. Watts (observer), both of whom were killed. Richthofen's opponent was the same unit's B.E.2e A2785. Both of its crewmen, 2/Lt. G.M. Gosset-Bibby (pilot) and Lt. G.J.O. Brichta (observer), were also killed. 2/Lt. C.P. Thornton (pilot) and 2/Lt. H.D. Blackburn (observer) in 43 Squadron's Sopwith Strutter A961, who reported that they shot down a Halberstadt that hit the ground south of Lens, were probably responsible for Richthofen's emergency landing. Capt. R. Gregory, flying in 40 Squadron's F.E.8 6384, caused an Albatros scout to crash out of control near Givenchy, and therefore probably was Lübbert's victor.
09.03.17	Lt. Karl-Emil Schäfer is now considered as transferred to the Staffel. Lt. Kurt Wolff forces F.E. single-seater 6456 to land in German territory north of Lens near Annay around 10:20 a.m. Confirmed by Kogenluft as his 2nd air victory. Lt. Karl-Emil Schäfer also shot down F.E. single-seater 6397 over German territory near Cité Fachoda near Meurchin around 10:20 a.m., and two minutes later shot down Vickers single-seater 4874 over German territory near Pont-à-Vendin. He then had to make an emergency landing in a shot-up machine in German territory but was uninjured. Confirmed by Kogenluft as his 5th and 6th air victories. Lt. Carl Allmenröder shoots an F.E. single-seater down in flames over enemy territory west of Hulluch. Confirmed by Kogenluft as his 6th air victory. Lt. Manfred Frhr. von Richthofen shoots a Vickers single-seater down in flames over the front lines near Roclincourt, southwest of Bailleul-Sir-Berthoult, north of Arras, around 11:55 a.m. Confirmed by Kogenluft as his 25th air victory.
	According to available documents, Wolff's opponent was F.E.8 6456 from 40 Squadron RFC, manned by 2/Lt. T.A. Shepard (POW), and Schäfer's initial victim was F.E.8 6397 from the same unit with 2/Lt. W.B. Hills (WIA, POW) as its pilot. Schäfer's second opponent was F.E.8 A4874, also from 40 Squadron, piloted by 2/Lt. G.F. Haseler (POW). Allmenröder's victim was yet another 40 Squadron fighter, F.E.8 6399, with 2/Lt. R.E. Neve (WIA) at its controls. Richthofen's opponent was D.H.2 A2571 from 29 Squadron, flown by Lt. A.J. Pearson (KIA). The pilot who was responsible for Schäefer's emergency landing probably was 2/Lt. H.C. Todd who, flying in 40 Squadron's F.E.8 6425, caused an Albatros to crash out of control east of Lens.

Date	Event
10.03.17	Generaloberst Ludwig von Falkenhausen visits Jagdstaffel 11 at La Brayelle. Lt. Lothar Frhr. von Richthofen is transferred to the Staffel via Armee-Flug-Park 1 and then Armee-Flug-Park 6.
11.03.17	Lt. Konstantin Krefft shoots down an F.E. single-seater over German territory between Givenchy and Vimy around 11:00 a.m. Confirmed by Kogenluft as his 1st air victory. Lt. Karl-Emil Schäfer shoots a B.E. down in flames at around 11:20 a.m. over enemy territory southwest of Hulluch. Confirmed by Kogenluft as his 7th air victory. Lt. Manfred Frhr. von Richthofen shoots B.E. 6232 down in flames at around 11:50 a.m. over German territory near La Folie Grove southwest of Vimy. Confirmed by Kogenluft as his 26th air victory. Vzfw. Josef Baierlein shoots down a B.E. over enemy territory near Villers-au-Bois around 12:05 p.m. Not confirmed, probably due to a lack of witnesses.
	According to available documents, Schäfer's opponent was B.E.2c 4541 from 2 Squadron RFC, crewed by 2/Lt. G.C. Hoskins (pilot) and 2/Lt. G.W.B. Hampton (observer), both of whom were killed. Richthofen's victim was B.E.2d 6232 from the same unit, manned by 2/Lt. J. Smyth (pilot) and 2/Lt. E.G. Byrne (observer), who were also killed.
16.03.17	Lt. Albert Oesterreicher is transferred to Bavarian Flieger-Ersatz-Abteilung 1 in Schleissheim via Armee-Flug-Park 6.
17.03.17	Lt. Manfred Frhr. von Richthofen shoots down Vickers A5439 in flames at around 11:40 a.m. that broke apart over German territory near Oppy Grove. Confirmed by Kogenluft as his 27th air victory. Lt. Kurt Wolff shoots a Sopwith down in flames over German territory southwest of Athies at around 11:43 a.m. Confirmed by Kogenluft as his 3rd victory. Lt. Carl Allmenröder shoots down Sopwith A1111 over German territory near Point-du-Jour north of Athies. Confirmed by Kogenluft as his 2nd air victory. Lt. Manfred Frhr. von Richthofen shoots down a B.E. that broke apart between the lines south of Givenchy around 5:00 p.m. Confirmed by Kogenluft as his 28th air victory.
	According to available documents, Richthofen's first opponent was 25 Squadron's F.E.2b A5439 with Lt. A.E. Boultbee (pilot) and 2/AM F. King (observer) as its crew. His second victim was B.E.2e A2814 from 16 Squadron, flown by 2/Lt. G.M. Watt (pilot) and Sgt. E.A. Howlett (observer). Wolff's opponent was Sopwith Strutter A1097 from 43 Squadron RFC, crewed by 2/Lt. A.L. Constable (pilot) and 2/Lt. C.D. Knox (observer). Allmenröder's success was over Sopwith A1111 from the same unit, manned by 2/Lt. J.C. Rimer (pilot) and 2/Lt. R.H. Lownds (observer). All eight crewmen were killed in action.
21.03.17	Lt. Carl Allmenröder shoots a B.E. down in flames over enemy territory south of Vermelles at around 3:40 p.m. Confirmed by Kogenluft as his 3rd air victory. Lt. Manfred Frhr. von Richthofen shoots a B.E. down over enemy territory near Neuville St. Vaast around 5:40 p.m. Confirmed by Kogenluft as his 29th air victory.
	According to available documents, Allmenröder's victim was 43 Squadron's Sopwith Strutter A2390, occupied by Lt. H.W.L. Poole (pilot, WIA) and 2/AM A.J. Ball (observer, WIA, later died of wounds). Richthofen's opponent was 16 Squadron's B.E.2e A3154, with Sgt. S.H. Quicke (pilot, KIA) and 2/Lt. W.J. Lidsey (observer, WIA, died of wounds next day).
22.03.17	Lt. Hans Hinsch, formerly with Schutzstaffel 7, is transferred to the Staffel. Vzfw. Josef Baierlein is transferred to Armee-Flug-Park 6.
23.03.17	Lt. Manfred Frhr. von Richthofen is promoted ahead of normal protocol to Oberleutnant.
24.03.17	Lt. Karl-Emil Schäfer shoots down a Sopwith in flames over enemy territory near Anzin, north of Arras, at around 11:00 a.m. Confirmed by Kogenluft as his 8th air victory. Oblt. Manfred Frhr. von Richthofen shoots down Spad 6706 over German territory near Givenchy at around 11:55 a.m. Confirmed by Kogenluft as his 30th air victory.
	According to available documents, Schäfer's opponent was B.E.2e 7248 from 8 Squadron RFC, with 2/Lt.H.W. Tagent (pilot, KIA) and 2/Lt. G.T. Gray (observer, KIA) as its crew. Richthofen's victim was 19 Squadron's Spad VII A6706, piloted by Lt. R.P. Baker, who was wounded and taken prisoner.

Date	Event
25.03.17	Oblt. Manfred Frhr. von Richthofen shoots a Nieuport down in flames at around 8:20 a.m. over German territory near Tilloy-lès-Mofflaines. Confirmed by Kogenluft as his 31st air victory.
	According to available documents, his opponent was Nieuport 17 A6689 from 29 Squadron RFC. Its pilot, 2/Lt. C.G. Gilbert, was taken prisoner.
28.03.17	Lt. Lothar Frhr. von Richthofen sets a Vickers on fire at around 5:20 p.m. and forces it to land in German territory east of Vimy's train station. Confirmed by Kogenluft as his 1st air victory.
	According to available documents, his opponent was F.E.2b 7715 from 25 Squadron RFC, manned by 2/Lt. N.L. Knight (pilot, WIA, POW) and 2/Lt. A.G. Severs (observer, KIA).
30.03.17	Lt. Kurt Wolff shoots down a Nieuport over German territory east of Gavrelle at around 11:45 a.m. Confirmed by Kogenluft as his 4th air victory. Lt. Carl Allmenröder shoots down a Nieuport over German territory southeast of Bailleul-Sir-Berthoult at around 2:15 p.m. Confirmed by Kogenluft as his 4th air victory. Lt. Hans-Georg Lübbert was shot down during a dogfight with a Nieuport and crashed in flames at around 2:15 p.m. between Bailleul-Sir-Berhoult and Thélus.
	According to available documents, Wolff's opponent was Nieuport 17 A273 from 60 Squadron RFC, piloted by Lt. W.P. Garnett. Allmenröder's victim was Nieuport 23 A6780 from the same unit, flown by Lt. D.M.F. Sinclair. Both men were killed. Hans-Georg Lübbert was apparently brought down by 40 Squadron's Capt. R. Gregory, who in Nieuport 17 A6680 caused a Halberstadt to crash out of control south of Vimy.
31.03.17	Lt. Kurt Wolff forces an F.E. to land in German territory near the sugar factory southeast of Bailleul-Sir-Berthoult at around 7:50 a.m. Confirmed by Kogenluft as his 5th air victory.
	According to available documents, his opponent was F.E.2b 7691 from 11 Squadron RFC, with Lt. L.A. Strange (pilot, POW) and 2/Lt. W.G.T. Clifton (observer, KIA) as its crew.

Personnel Roster Royal Prussian Jagdstaffel 11: Oktober 1916 – März 1917					
Formation					
Order for deployment	28/09/16				
Start of the deployment	05/10/16				
Day of mobilization	11/10/16				
Airfields	**Timeframe**	**Army**			
La Brayelle by Douai	05.10.16 - 15.04.17	6. Armee			
Staffelführer	**Timeframe**	**from**	**Remain**	**Additional Information**	**Source**
Rudolf Lang	20.10.16 - 14.01.17	Kasta 31	Jasta 28 Staffelführer		OP 15292 BHSt.A
Manfred von Richthofen	14.01.17 - 25.06.17	Jasta Boelcke	Leader Jagdgeschwader I		Zuerl, War diary excerpt Jasta 11
Deputy Staffelführer	**Timeframe**	**from**	**Remain**	**Additional Information**	**Source**
Unknown	02.02.17 - 07.(09).02.17				War diary excerpt Jasta 11
Offizier z.b.V	**Timeframe**	**from**	**Remain**	**Additional Information**	**Source**
Karl Kleinhenz	11.10.16 - 01.03.17	Aflup 6			KrStL. Aflug 6, BHSt.A.KTB-Tornuß
von Hartmann	01.03.17 -06.07.17				
Pilots	**Timeframe**	**from**	**Remain**	**Additional Information**	**Source**
Albert Oesterreicher	11.10.16 - 16.03.17	KEK III at the Feld.Flg.Abt. 5	bay. FEA 1		KrStL. Aflup 6, Bay. HSt.A./ OPAkte 1188
Sebastian Festner	11.10.16 - 25.04.17	KEK III at the Feld.Flg.Abt. 5	Killed in action (Flak?/ 43. Sqn.)	Albatros D III D.2251/16	Rangl. Aflup 6, Bay. HSt.A.
Karl Kleinhenz	11.10.16 - 05.04.17	KEK III at the Feld.Flg.Abt. 5	Aflup 6		KrStR. Aflup 6, Bay. HSt.A.
Konstantin Krefft	11.10.16 - 25.06.17	Bavarian FEA I	Technical Officer JG I		KrStL. Aflup 6, Bay. HSt.A.
Erwin Howe	11.10.16 - 17.02.17	Feld.Flg.Abt. 24	Aflup 6		KrStR. Aflup 6, Bay. HSt.A.
Alfred Prehn	11.10.16 - 14.11.16	Jasta 10	Naval Station Ton-dern commanded posted 28.02.17		KrStR. Aflup 6, Bay. HSt.A.
Franz Anslinger	14.10.16 - 02.03.17	Kagohl III/14	Jasta 35		KrStL. Aflup 6, Bay. HSt.A.
Kurt Wolff	05.11.16 - 06.05.17	Kagohl VII/40	Jasta 29 Staffel-führer		KrStL. Aflup 6, Bay. HSt.A.
Georg Simon	07.11.16 - 04.06.17	Kagohl II/12	Emergency land-ing on enemy territory, P.O.W.	Albatros D III D.2015/16	KrStL. Aflup 6, Bay. HSt.A.

Royal Prussian Jagdstaffel 11: Oktober 1916 – März 1917					
Pilots	**Timeframe**	**from**	**Remain**	**Additional Information**	**Source**
Carl Allmenroeder	20.11.16 - 27.06.17	Feld.Flg.Abt. 18	fatally crashed after ground fire		Rangl. Aflup 6, Bay. HSt.A.
Josef Baierlein	26.11.16 - 22.03.17	Bavarian Feld. Flg.Abt. 2	Aflup 6		KrStR. Aflup 6, Bay. HSt.A.
Georg Braumüller	16.01.17 - 01.03.17	Jagdstaffel-schule I	Aflup 6 zur FEA 4 Posen		KrStR. Aflup 6, Bay. HSt.A.
Eduard Lübbert	26.02.17 - 06.03.17	Feld.Flg.Abt. 18	Wounded in action, hospital		KrStL. Aflup 6, Bay. HSt.A.
	18.03.17 - 30.03.17		Killed in action		
Karl-Emil Schäfer	21.02.17 - 26.04.17	Kagohl II/11	Jasta 28 Staffel-führer		KrStL. Aflup 6, Bay. HSt.A.
Lothar von Richthofen	10.03.17 - 13.05.17	Kaghol IV/23	Wounded in ac-tion		KrStL. Aflup 6, Bay. HSt.A.
Hans Hinsch	22.03.17 - 24.05.17	Schutzstaffel 7	Fatally wounded in air combat died 25.05.17	Albatros D III D.2016/16	War diary excerpt Jasta 11, Thorsten Pietsch, Lieber Hans.

References for Aircraft Inventory (Facing Page)

Source Bayerisches Hauptstaatsarchiv, Abt. 4, München:
Flugzeuge Armeeflugpark 6, La Briquette, BHStA/KA ILUFT 203,
Flugzeuge Armeeflugpark 6, La Briquette, 31.10.1916 / BHStA/KA München Akt Iluft 206
Flugzeugbestand der Jagstaffel 11 in Douai 31.12.1916 / BHStA/KA München Akt Iluft 204
Flugzeuge Armeeflugpark 6, La Briquette, 31.12.1916 / BHStA/KA München Akt Iluft 205
Flugzeuge Armeeflugpark 6, Pont a Chin, 28.2.1917 / BHStA/KA München Akt Iluft 205

Jagdstaffel 11 Aircraft Inventory: October 1916 – March 1917			
Military Number	**Type of aircraft**	**Date**	**Comment**
Aircraft Inventory Record October 1916 **Aircraft delivered to or handed over by Jagdstaffel 11 during this period.**			
D.425/16	Alb D.I	11.10.1916	Armeeflugpark 6, delivery to Jagdstaffel 11.
D.410/16	Halb D.V	14.10.1916	Armeeflugpark 6, delivery to Jagdstaffel 11.
D.412/16	Halb D.V	14.10.1916	Armeeflugpark 6, delivery to Jagdstaffel 11.
D.412/16	Halb D.V	25.10.1916	Jasta 11, departer to Armeeflugpark 6 insufficient deflection in altitude control.
D.416/16	Halb D.V	25.10.1916	Armeeflugpark 6, delivery to Jagdstaffel 11.
D.412/16	Halb D.V	28.10.1916	Armeeflugpark 6, delivery back to Jagdstaffel 11.
D.417/16	Halb D.V	26.10.1916	Armeeflugpark 6, delivery to Jagdstaffel 11.
D.418/16	Halb D.V	26.10.1916	Armeeflugpark 6, delivery to Jagdstaffel 11.
D.107/16	Halb D.II	31.10.1916	Armeeflugpark 6, delivery to Jagdstaffel 11.
Aircraft inventory on 31 October 1916 **Aircraft in the inventory of Jagdstaffel 11 on this date.**			
D.107/16	Halb D.II	31.10.1916	Jasta 11 Douai
D.410/16	Halb D.V	31.10.1916	Jasta 11 Douai
D.412/16	Halb D.V	31.10.1916	Jasta 11 Douai
D.416/16	Halb D.V	31.10.1916	Jasta 11 Douai
D.417/16	Halb D.V	31.10.1916	Jasta 11 Douai
D.418/16	Halb D.V	31.10.1916	Jasta 11 Douai
D.425/16	Alb D.I	31.10.1916	Jasta 11 Douai
Aircraft Inventory Record December 1916 **Aircraft delivered to or handed over by Jagdstaffel 11 during this period.**			
D.420/16	Halb D.V	01.11.1916	Armeeflugpark 6, delivery to Jagdstaffel 11
D.421/16	Halb D.V	01.11.1916	Armeeflugpark 6, delivery to Jagdstaffel 11
D.413/16	Halb D.V	01.11.1916	Armeeflugpark 6, delivery to Jagdstaffel 11.
D.803/16	Halb D.II(Han)	Nov.	Armeeflugpark 6, delivery to Jagdstaffel 11.
D.801/16	Halb D.II(Han)	Nov.	Armeeflugpark 6, delivery to Jagdstaffel 11.
D.802/16	Halb D.II(Han)	Nov.	Armeeflugpark 6, delivery to Jagdstaffel 11.
D.805/16	Halb D.II(Han)	Nov.	Armeeflugpark 6, delivery to Jagdstaffel 11.
D.412/16	Halb D.V	05.11.1916	Jasta 11, damaged by emergency landing.
D.803/16	Halb D.II(Han)	05.11.1916	Jasta 11, departure to Armeeflugpark 6, heavily damaged in evening landing.
D.473/16	Alb D.II	15.11.1916	Armeeflugpark 6, delivery to Jagdstaffel 11.
D.425/16	Alb D.I	15.11.1916	Jasta 11 Alb. D I D.425 departure to Armeeflugpark
D.1099/16	Halb D.V	16.11.1916	Armeeflugpark 6, delivery to Jagdstaffel 11.
D.1100/16	Halb D.V	27.11.1916	Armeeflugpark 6, delivery to Jagdstaffel 11.
D.116/16	Halb D.III	04.12.1916	Armeeflugpark 6, delivery to Jagdstaffel 11 after repair.
D.473/16	Alb D.II	11.12.1916	Alb. D II D.473 departure to Armeeflugpark 6
D.413/16	Halb D.V	11.12.1916	Jasta 11, departure to Armeeflugpark 6, damaged by bombing, return to Jasta 11 after repair.

Jagdstaffel 11 Aircraft Inventory: October 1916 – March 1917 (cont.)			
Military Number	**Type of aircraft**	**Date**	**Comment**
Aircraft inventory on 31. December 1916 **Aircraft in the inventory of Jagdstaffel 11 on this date.**			
D.107/16	Halb D.II	31.12.1916	Jasta 11 Douai
D.410/16	Halb D.V	31.12.1916	Jasta 11 Douai
D.412/16	Halb D.V	31.12.1916	Jasta 11 Douai
D.413/16	Halb D.V	31.12.1916	Jasta 11 Douai
D.414/16	Halb D.V	31.12.1916	Jasta 11 Douai
D.415/16	Halb D.V	31.12.1916	Jasta 11 Douai
D.416/16	Halb D.V	31.12.1916	Jasta 11 Douai
D.417/16	Halb D.V	31.12.1916	Jasta 11 Douai
D.418/16	Halb D.V	31.12.1916	Jasta 11 Douai
D.420/16	Halb D.V	31.12.1916	Jasta 11 Douai
D.421/16	Halb D.V	31.12.1916	Jasta 11 Douai
D.801/16	Halb D.II(Han)	31.12.1916	Jasta 11 Douai
D.802/16	Halb D.II(Han)	31.12.1916	Jasta 11 Douai
D.805/16	Halb D.II(Han)	31.12.1916	Jasta 11 Douai
D.1099/16	Halb D.V	31.12.1916	Jasta 11 Douai
D.1100/16	Halb D.V	31.12.1916	Jasta 11 Douai
Aircraft Inventory Record February 1917 **Aircraft delivered to or handed over by Jagdstaffel 11 during this period.**			
D.415/16	Halb D.V	04.01.1917	Jasta 11, departure to Armeeflugpark 6, unusable, exchanged for Alb. D III.
D.1118/16	Halb D.V	04.01.1917	Armeeflugpark 6, delivery to Jagdstaffel 11
D.413/16	Halb D.V	18.01.1917	departure to Armeeflugpark 6, damaged by bombing
		18.02.1917	departure to Armeeflugpark 6, unusable, exchanged for Alb. D III.
D.107/16	Halb D.II	20.01.1917	departure to Armeeflugpark 6, unusable, exchanged for Alb. D III.
D.116/16	Halb D.III	21.01.1917	departure to Armeeflugpark 6, damaged by landing, exchanged for Alb. D III.
D.1993/16	Alb D.III	21.01.1917	Armeeflugpark 6, delivery to Jagdstaffel 11
D.1996/16	Alb D.III	21.01.1917	Armeeflugpark 6, delivery to Jagdstaffel 11
D.412/16	Halb D.V	22.01.1917	departure to Armeeflugpark 6, unusable, exchanged for Alb. D III.
D.418/16	Halb D.V	22.01.1917	departure to Armeeflugpark 6, unusable, exchanged for Alb. D III.
D.1980/16	Alb D.III	22.01.1917	Armeeflugpark 6, delivery to Jagdstaffel 11
D.1991/16	Alb D.III	22.01.1917	Armeeflugpark 6, 11.10.17 to Jagdstaffel 11
D.420/16	Halb D.V	24.01.1917	departure to Armeeflugpark 6, unusable, exchanged for Alb. D III.
D.2006/16	Alb D.III	24.01.1917	Armeeflugpark 6, delivery to Jagdstaffel 11
D.1118/16	Halb D.V	25.02.1917	departure to Armeeflugpark 6, unusable, exchanged for Alb. D III.

Jagdstaffel 11 Aircraft Inventory: October 1916 – March 1917 (cont.)			
Military Number	**Type of aircraft**	**Date**	**Comment**
D.1985/16	Alb D.III	09.02.1917	Armeeflugpark 6, delivery to Jagdstaffel 11
D.3507/16	SSW D.I	17.02.1917	Armeeflugpark 6, delivery to Jagdstaffel 11
D.413/16	Halb D.V	18.02.1917	departure to Armeeflugpark 6, unusable, exchanged for Alb. D III.
D.2026/16	Alb D.III	25.02.1917	Armeeflugpark 6, delivery to Jagdstaffel 11
D.421/16	Halb D.V	26.02.1917	departure to Armeeflugpark 6, unusable, exchanged for Alb. D III.
D.801/16	Halb D.II(Han)	26.02.1917	departure to Armeeflugpark 6, unusable, exchanged for Alb. D III.
D.805/16	Halb D.II(Han)	26.02.1917	departure to Armeeflugpark 6, unusable, exchanged for Alb. D III.
D.2015/16	Alb D.III	26.02.1917	Armeeflugpark 6, delivery to Jagdstaffel 11
D.2016/16	Alb D.III	26.02.1917	Armeeflugpark 6, delivery to Jagdstaffel 11
D.2036/16	Alb D.III	26.02.1917	Armeeflugpark 6, delivery to Jagdstaffel 11
D.414/16	Halb D.V	27.02.1917	departure to Armeeflugpark 6, unusable, exchanged for Alb. D III.
D.1100/16	Halb D.V	27.02.1917	departure to Armeeflugpark 6, unusable, exchanged for Alb. D III.
D.2014/16	Alb D.III	27.02.1917	Armeeflugpark 6, delivery to Jagdstaffel 11
D.2026/16	Alb D.III	27.02.1917	Armeeflugpark 6, delivery to Jagdstaffel 11
Aircraft inventory on 28 February 1916 **Aircraft in the inventory of Jagdstaffel 11 on this date**			
D.410/16	Halb D.V	28.02.1917	Jasta 11 Douai
D.416/16	Halb D.V	28.02.1917	Jasta 11 Douai
D.417/16	Halb D.V	28.02.1917	Jasta 11 Douai
D.802/16	Halb D.II(Han)	28.02.1917	Jasta 11 Douai
D.1099/16	Halb D.V	28.02.1917	Jasta 11 Douai
D.1980/16	Alb D.III	28.02.1917	Jasta 11 Douai
D.1985/16	Alb D.III	28.02.1917	Jasta 11 Douai
D.1991/16	Alb D.III	28.02.1917	Jasta 11 Douai
D.1993/16	Alb D.III	28.02.1917	Jasta 11 Douai
D.1996/16	Alb D.III	28.02.1917	Jasta 11 Douai
D.2006/16	Alb D.III	28.02.1917	Jasta 11 Douai
D.2014/16	Alb D.III	28.02.1917	Jasta 11 Douai
D.2015/16	Alb D.III	28.02.1917	Jasta 11 Douai
D.2016/16	Alb D.III	28.02.1917	Jasta 11 Douai
D.2026/16	Alb D.III	28.02.1917	Jasta 11 Douai
D.2036/16	Alb D.III	28.02.1917	Jasta 11 Douai
D.2062/16	Alb D.III	28.02.1917	Jasta 11 Douai
March 1917			
No new aircraft were assigned to Jagdstaffel 11 in March 1917.			

The Research for the Book

Above: Signed photo by Carl-August von Schoenebeck for the author.

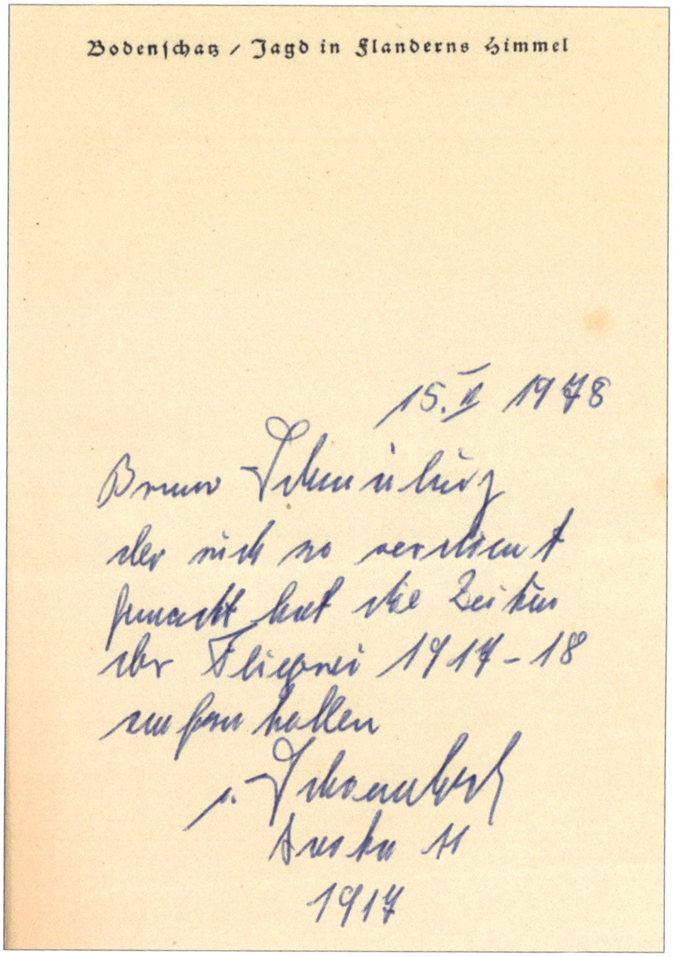

Above: Dedication by Carl-August von Schoenebeck for the author, in the book *"Jagd in Flanderns Himmel"*.

"Would you like another piece of cake?" Before I could answer, *Frau* von Schoenebeck had already put another piece on my plate. It was a beautiful, sunny day in July 1975 when I, for the fourth time, was the guest of retired General Carl-August von Schoenebeck. We sat on the terrace of his large, spacious home in Harlaching, an upscale district of Munich. Already, on three previous occasions, he had patiently told me in detail about his time as a fighter pilot with Jagdstaffeln 11 and 59 and as the Staffelführer of Jasta 33. His memories of Jastas 11 and 33 were excellent, but I had the feeling that he had little interest in his time with Jasta 59. As he told me, he was quite irritated when Oblt. von Boddien took him from his beloved Jasta 11 to serve with him as a Kette leader there. He had even approached Manfred von Richthofen to ask if he could stay with Jasta 11. But Richthofen had explained that a newly-appointed Staffelführer had the right to take one or two seasoned, proficient fighter pilots with him from his former unit for Kette leadership at his new one. Richthofen would not break that tradition, even if he regretted the departures because as Richthofen said, *"They'll make something of themselves."*

I visited Herr von Schoenebeck a good ten times in all. Every meeting was an experience, especially since he was one of the persons who had personally known Manfred von Richthofen. He generously made all of his photos and documents available to me so that I could make copies of them.

Excerpts of the Jasta 11 War Diary that were transcribed by Erich Tornuß and carefully revised by Dr. Gustav Bock, as well as documents and information supplied by former Jasta 11 members to Alex Imrie, served as the basis for our discussions.

I had found the retired General's address in the directory of the *Alten Adler* (Old Eagles), the association of pre-war and World War I airmen. When I first appeared before him, Herr von Schoenebeck was visibly astonished that someone in Germany was interested in the history of German Jagdstaffeln. Previously, such visits had been conducted only by aviation enthusiasts from the United States and Great Britain.

But on that day in July 1975, he had a special gift for me. He gave me the addresses and telephone numbers of all of the still-living members of *Der Eisgrauen*, the association of airmen from Manfred von Richthofen's Jagdsgeschwader

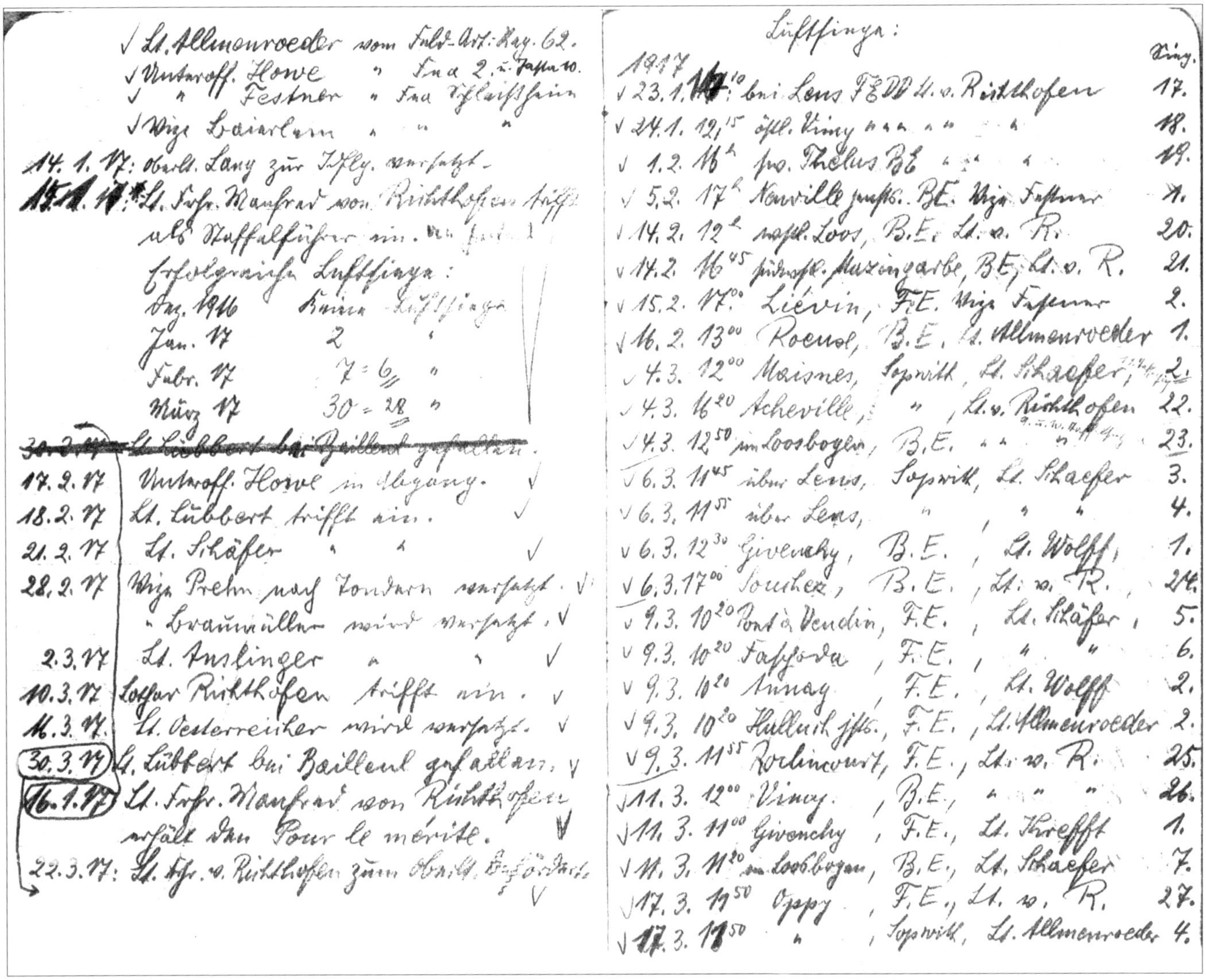
√ Lt. Allmenroeder vom Feld-Art.-Reg. 62.
√ Unteroff. Howe " " 2. [illegible]
√ " Festner " [illegible]
√ Vize Baierlein " " "
14. 1. 17: Oblt. Lang zur [illegible] versetzt.
[illegible]: Lt. Frhr. Manfred von Richthofen trifft als Staffelführer ein.
Erfolgreiche Luftsiege:
Okt. 1916 keine Luftsiege
Jan. 17 2
Febr. 17 7 = 6
März 17 30 = 28
~~30.3.17 Lt. Lübbert bei Bailleul gefallen.~~
17.2.17 Unteroff. Howe in Abgang. √
18.2.17 Lt. Lübbert trifft ein. √
21.2.17 Lt. Schäfer " " √
28.2.17 Vize Prehn nach Tondern versetzt. √
" Braunmüller wird versetzt. √
2.3.17 Lt. Auslinger " " √
10.3.17 Lothar Richthofen trifft ein. √
11.3.17 Lt. Oesterreicher wird versetzt. √
30.3.17 Lt. Lübbert bei Bailleul gefallen. √
16.1.17 Lt. Frhr. Manfred von Richthofen erhält den Pour le mérite. √
22.3.17: Lt. Frhr. v. Richthofen zum Oblt. befördert. √

Luftsiege:
1917
√ 23.1. [illegible] bei Lens F.E. Lt. v. Richthofen 17.
√ 24.1. 12.15 östl. Vimy " " " " 18.
√ 1.2. 16 [illegible] Thelus B.E. " " " 19.
√ 5.2. 17 Neuville [illegible] B.E. Vize Festner 1.
√ 14.2. 12 [illegible] Loos, B.E. Lt. v. R. 20.
√ 14.2. 16.45 [illegible] Mazingarbe, B.E. Lt. v. R. 21.
√ 15.2. 17.00 Liévin, F.E. Vize Festner 2.
√ 16.2. 13.00 Roeux, B.E. Lt. Allmenroeder 1.
√ 4.3. 12.00 Maisnes, Sopwith, Lt. Schaefer 2.
√ 4.3. 16.20 Acheville, " Lt. v. Richthofen 22.
√ 4.3. 12.50 im Loosbogen, B.E. " " " 23.
√ 6.3. 11.45 über Lens, Sopwith, Lt. Schaefer 3.
√ 6.3. 11.55 über Lens, " " " 4.
√ 6.3. 12.30 Givenchy, B.E. Lt. Wolff 1.
√ 6.3. 17.00 Souchez, B.E. Lt. v. R. 24.
√ 9.3. 10.20 Pont à Vendin, F.E. Lt. Schäfer 5.
√ 9.3. 10.20 [illegible] F.E. " " 6.
√ 9.3. 10.20 Annay F.E. Lt. Wolff 2.
√ 9.3. 10.20 Hulluch [illegible], F.E. Lt. Allmenroeder 2.
√ 9.3. 11.55 Roclincourt, F.E. Lt. v. R. 25.
√ 11.3. 12.00 Vimy, B.E. " " " 26.
√ 11.3. 11.00 Givenchy, F.E. Lt. Krefft 1.
√ 11.3. 11.20 im Loosbogen, B.E. Lt. Schaefer 7.
√ 17.3. 11.50 Oppy, F.E. Lt. v. R. 27.
√ 17.3. 11.50 " Sopwith, Lt. Allmenroeder 4.

Above: Erich Tornuß' original notations from Jagdstaffel 11's war diary that was available at the Reichsarchiv in Potsdam.

I. He also handed me a letter asking his JG I comrades to support my research with their photos and documents. I included a copy of this letter in my initial contact with den "*Eisgrauen*", which proved to be worth its weight in gold and opened the door for me to contact them. There was one exception, however: Herr Maushake, who strangely enough, was not a member of the association .

When I said goodbye to the Schoenebecks that day, Herr von Schoenebeck uttered a sentence that he continued to remind me about during my subsequent visits: "*Herr Schmäling, you are welcome at any time because I am counting on you to write the history of Jagdstaffel 11.*"

A little over a year later, Herr von Schoenebeck informed me that the BBC was planning a radio documentary about World War I air warfare and that he was going to be interviewed for it. At his request, I accompanied him to the interview and took numerous notes of what he said. Then I too was interviewed about my conversations with former airmen and my assessment of Manfred von Richthofen as a person. The documentary, titled *The History of Flight*, was broadcast shortly afterwards.

In 1975, aviation historian Gero von Langsdorff provided me with the address of Herr Ludwig Zacharias, who belonged to a group of aviation enthusiasts who had already been collecting documents with regard to the First World War's Fliegertruppe in the 1930's and had been in contact with former combat pilots. Herr Zacharias gave me his list of their addresses and let me review his correspondence and interviews with them. One of the combat pilots who Herr Zacharias had been in contact with for some time was Wilhelm Allmenröder, the brother of Pour le Mérite pilot Carl Allmenröder. Herr Zacharias had visited and interviewed him several times in Wald near Solingen during the 1960s. Herr Zacharias also gave me the address of Wilhelm Allmenröder's widow, Helene, who lived in Prien am Chiemsee, about 60 km east of Munich.

Die Eisgrauen 1972

Bodenschatz, Carl, Gen.d.Fl.	852 Erlangen, Döderleinstr.2, Tel.09131/22502
v.Conta, Gottlieb, Dr.med.	675 Kaiserslautern, Auf dem Bännjerrück 85, Tel.3779
Eisfeldt, Hans	799 Friedrichshafen, Tannenweg 53, Tel. ?
Förster, Otto-Lutz	3388 Bad Harzburg, Wilh.Raabestr.7, Tel.05322/3107
v.Fabrice, Victor	479 Paderborn, Neuhauser Str.49, Tel.05251/22860
Graf Gluszewski, Heinz	81 Garmisch-Partenkirchen, Triftstr.49, Tel. 4253
Groos, Gisbert-Wilhelm	5 Köln 41, Landgrafenstr.104, Tel. 0221/444106
Grosch, Walter, Gen.	8 München 90, Reichenhaller Str.7, App.106 Tel.0811/6215617
Heldmann, Alois	8201 Au, Bad Aibling, Gottschalling 4 a, Tel. ?
Hemer, Franz	6 Frankfurt, Humperdinckstr.17, Tel.633629
Klamt, Rudolf	8 München 21, Joergstraße 39
Frhr.v.Köckritz, Fr.-August	7082 Oberkochen/Württ., Apotheke, Tel. ?
Koepsch, Egon	7407 Mössingen/Tübingen, Belsener Weg 36, Tel.474
Kraut, Richard	1 Berlin 31, Babelsberger Str.48, Tel. ?
Lehmann, Walter	8182 Wiessee, Adamstraße 3, Tel.08022/81473
Maushake, Heinrich	3339 Ingeleben-Schöningen, Tel. ?
Niemz, Alfred	6380 Bad Homburg, v.d.Höh, Fröhlingstr.5, Tel.20403
Nöldeke, Werner	4470 Meppen, Nodeker Str.45, Tel.05931/13896
v.d.Osten, Hans-Georg	5 Köln 41, Franz-Marc-Str.11, Tel.0221/426874
Otto, Heinz	8971 Ofterschwang/Immenstedt, Allgäu, Tel.08321/2316
Petersen-List, Hans	2 Hamburg 70, Jenfelder Tannenweg 1D, Tel.662162
Riehm, Karl	3 Hannover, Emdenstraße 15, Tel.0511/15635
Rousselle, Oscar	Lissabon/Portugal, Praca de Alegria 58
v.Schoenebeck, Karl-August	8 München 90, Griechenplatz 22, Tel.0811/647437
Schibilsky, Kurt	4048 Grevenbroich-Oeken, Goethestr.32, Tel.02181/5182
Suck, Hans	2940 Wilhelmshaven, Grenzstraße 91, Tel.?
Dasher, Josef	8 München 27, Denninger Str.100, Tel.0811/914323
Horten, Walter	3105 Fassberg, Meisenweg 37, Tel. ?
Geiges, Karl	7700 Singen, Rieslinger Straße 134
Gebhard, Hans	7 Stuttgart, Kerner Straße 26
Leinmüller, Xaver	7030 Böblingen, Wielandstraße 3

Above: List of members of the "Eisgrauen", the association of former members of Jagdgeschwader I Freiherr von Richthofen from 1972. This list was given to the author by Carl-August von Schoenebeck for contact purposes.

CARL AUGUST VON SCHOENEBECK

D 8 MÜNCHEN 90
GRIECHENPLATZ 22
☎ 089/64 74 37

15.XII.81

Lieber Herr Schmäling !

In meinem Brief vom 14.11. habe ich Ihnen versprochen, über das Treffen in Paris zu berichten. Es war einzigartig, voll Harminie und Freundschäft, wie man es nicht für möglich gehalten hätte.

Sie werden Näheres in dem beiliegenden Bericht finden und gleichzeitig lege ich Ihnen noch einen Artikel aus einer engl. Zeitung bei.

Mit den besten Wünschen zum Weihnachtsfest und alles Gute für das Jahr 1982 !

Above: The personal contact with Mr. Carl-August von Schioenebeck went on for many years, this is a card from 1981.

When I contacted Frau Allmenröder by telephone, she too was quite surprised that a young aviation enthusiast from Germany was interested in World War I combat pilots and invited me to visit her. When I told her that my girlfriend would be accompanying me, coffee, tea, and homemade cakes were waiting for us – so too the Allmenröder brothers' photo albums and records. It was a very nice afternoon.

Frau Allmenröder told us about her childhood as the daughter of the neighbor of Pastor Wilhelm Allmenröder's family. She explained that she fell in love with Carl as a teenager and eventually became his fiancée. Whenever Carl came home on leave, she was always one of the first persons he would contact. During their time together, he told her about his life as an aviator and his time with Jagdstaffel 11. It was really quite moving when she spoke of his death and how the entire city sympathized with her and shared her great sadness. Carl's older brother, Wilhelm, took loving care of her following his brother's death, which eventually led to her falling in love with and marrying him. She also told how the memory of Carl was erased in Wald after the Second World War. A street that had been named in his honor was renamed by some local politicians who were looking to make their own personal mark.

At the close of our meeting, she entrusted me with the brothers' photo albums so that I could copy their contents. She did not want to surrender Carl's personal letters, however, and instead allowed me to copy those sections that related to Jagdstaffel 11. Two more visits followed and each time we were deeply impressed by her charm and friendliness.

My next Jasta 11 contact was with Herr Hans-Georg

Above: Photo of Hans-Georg von der Osten with signature. The author received the photo during his visit.

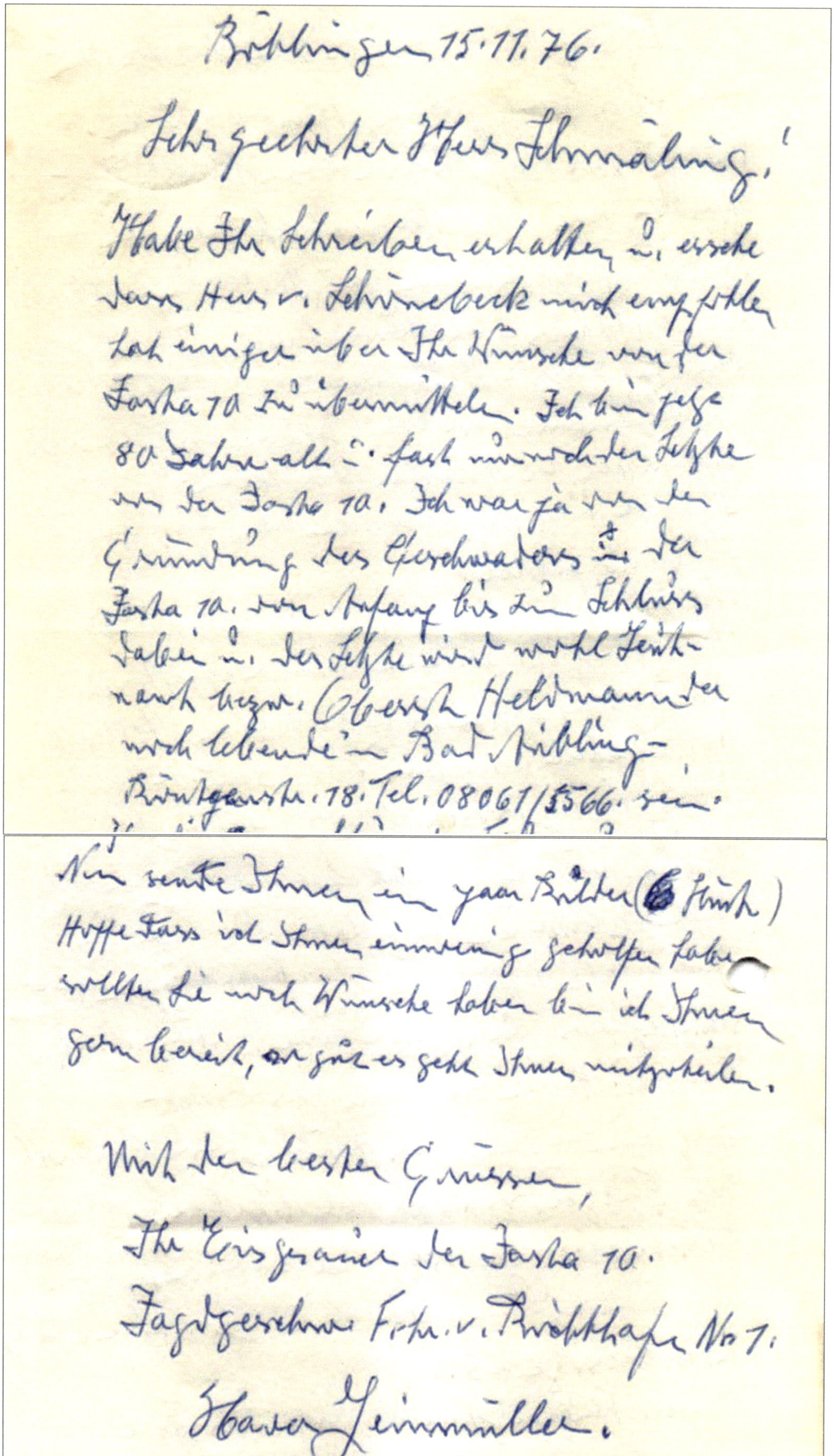
Böblingen 15.11.76.

Sehr geehrter Herr Schmäling!

Habe Ihr Schreiben erhalten u. ersehe daß Herr v. Schönebeck mich empfohlen hat einiges über Ihr Wünsche von der Jasta 10 zu übermitteln. Ich bin jetzt 80 Jahre alt u. fast nur noch der Letzte von der Jasta 10. Ich war ja von der Gründung des Geschwaders u. der Jasta 10. von Anfang bis zum Schluss dabei u. der Letzte wird wohl Leutnant bezw. Oberst Heldmann der noch lebende in Bad Aibling - Röntgenstr. 18. Tel. 08061/3566. sein.

Nun sende Ihnen ein paar Bilder (6 Stück) Hoffe dass ich Ihnen ein wenig geholfen habe, sollten Sie noch Wünsche haben bin ich Ihnen gern bereit, was ich weiss es gerne Ihnen mitzuteilen.

Mit den besten Grüssen,

Ihr Eisgrauer der Jasta 10.

Jagdgeschw. Frhr. v. Richthofen No 1.

Xaver Leinemüller.

Left & Below Right: Letter from Mr. Xaver Leinemüller to the author dated November 15, 1976.

Above: The author as a guest of Otto Fuchs, who was with Jagdstaffel 11 for a few days and later a member of Jagdstaffel 30, as well as squadron leader of Jagdstaffel 35 and 77.

von der Osten. Because Herr von Schoenebeck had already introduced me through his letter, Herr von der Osten dedicated an entire day to my visit. Although, due to the ravages of World War II, he had only been able to save a very small portion of his photos from his time with Jasta 11, he did share many memories with me. What was particularly interesting to me was that Herr von der Osten also knew Manfred von Richthofen personally. What he told me helped form my impression of Richthofen as a Staffel leader and person.

A short time later, Herr Xaver Leinemüller, who was a mechanic with Jasta 10 from June 1917 until the war's end, invited me to visit. Once again, Herr von Schoenebeck's letter had opened the door for me. Herr Leinemüller had a quite nice photo album featuring Jasta 10 from the summer to the autumn of 1917. He shared his memories of his time with the unit during three visits, and it was particularly interesting to hear about the life and activities of a Jagdstaffel from a mechanic's perspective.

Besides Herr von Schoenebeck, one of my earliest and most extended contacts was Otto Fuchs, formerly with Jasta 30 and Staffelführer of Jastas 35 and 77, whom I visited frequently over a period of eight years. In addition to fighter pilot history, we shared an interest in painting since Herr Fuchs was a well-known and respected painter of landscapes. He had been with *Jasta* 11 for only a few days, but nonetheless had precise and thoughtful recollections regarding the atmosphere at Jasta 11 and what Carl Allmenröder was like as a person.

It took some time before I was able to get in touch with

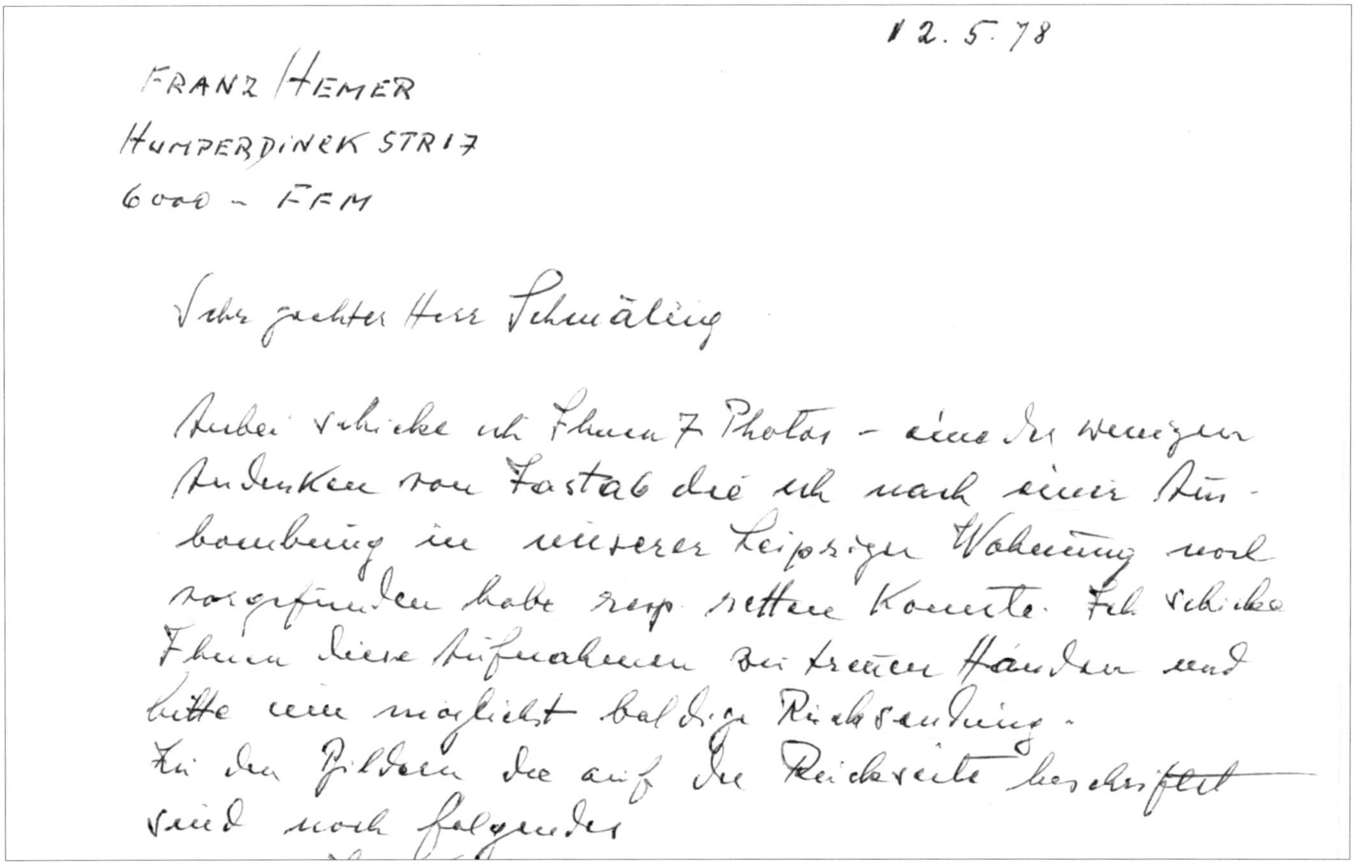

12.5.78

FRANZ HEMER
HUMPERDINCK STR 17
6000 - FFM

Sehr geehrter Herr Schmäling

Anbei schicke ich Ihnen 7 Photos – eine der wenigen Andenken von Jasta 6 die ich nach einer Ausbombung in unserer Leipziger Wohnung noch vorgefunden habe resp. retten konnte. Ich schicke Ihnen diese Aufnahmen zu treuen Händen und bitte um möglichst baldige Rücksendung.
Zu den Bildern die auf der Rückseite beschriftet sind noch folgendes

Above: Excerpt of the letter from the 2 May 1978 written by Franz Hemer, Jagdstaffel 6 to the author.

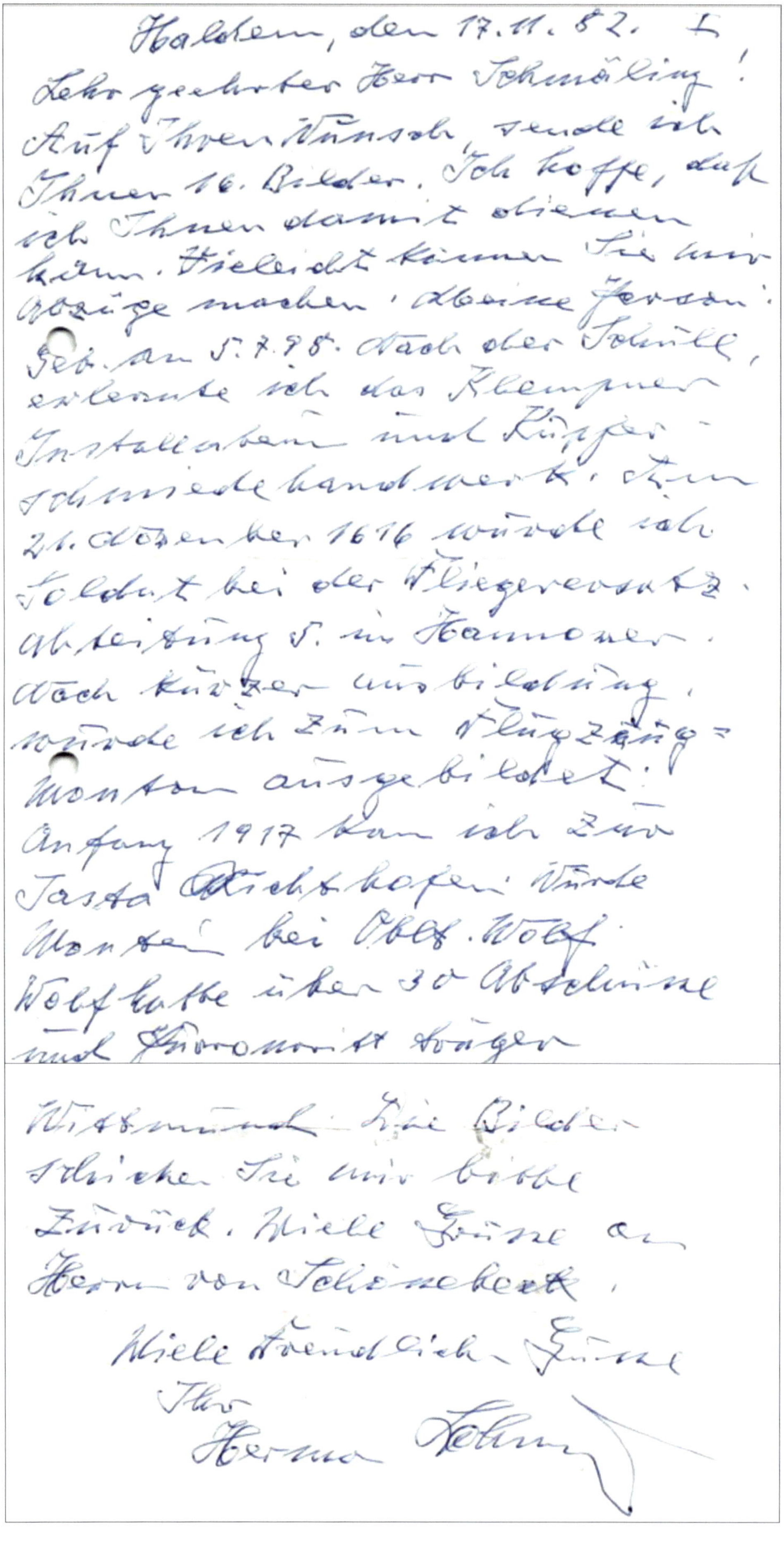
Haltern, den 17.11.82. I
Sehr geehrter Herr Schmäling!
Auf Ihren Wunsch, sende ich Ihnen 16. Bilder. Ich hoffe, daß ich Ihnen damit dienen kann. Vielleicht können Sie mir Abzüge machen. Meine Person: Geb. am 5.7.98. Nach der Schule, erlernte ich das Klempner Installateur und Kupferschmiedehandwerk. Am 21. Dezember 1916 wurde ich Soldat bei der Fliegerersatz-abteilung 5. in Hannover. Nach kurzer Ausbildung, wurde ich zum Flugzeug=monteur ausgebildet. Anfang 1917 kam ich zur Jasta Richthofen. Wurde Monteur bei Oblt. Wolff. Wolff hatte über 30 Abschüsse und Chronometer Träger

Wittmund. Die Bilder schicken Sie mir bitte zurück. Viele Grüße an Herrn von Schönebeck.
Viele freundliche Grüße
Ihr
Herman Lohm

Left: Excerpt of the long letter from 17 November 1982 written by Hermann Lohmeyer, mechanic of Jasta 11 to the author.

Above: Photo of Kurt Delang as pilot of Jagdstaffel 10 with dedication to the author.

Herr Wilhelm-Gisbert Groos. A personal visit was not possible for me because he lived in Switzerland, but I did enjoy a lengthy telephone conversation with him. It was quite interesting, although he informed me that he had unfortunately lost all of his photos and records of Jasta 11 during the Second World War. Memories of him were published in the *Cross and Cockade (U.S.)* journal and were also of interest. Years later, to my surprise, I received a photo album from his son. Unfortunately, it consisted only of reproductions of well-known photos he had been given by other former Staffel members.

I tried to contact Karl Bodenschatz, the former adjutant of Jagdgeschwader I, who lived in Erlangen, but none of my letters were answered. A telephone call to arrange a visit was also refused. When aviation historian Neal O'Connor visited me, he said that when he had met with Bodenschatz, the retired general had complained bitterly that no aviation historian from Germany had ever contacted him with regard to the Richthofen Geschwader. I then showed Neal my unanswered letters. Neal subsequently explained what had occurred to General Bodenschatz, and while Neal was in Germany I was able to visit with the General. He told me about Manfred von Richthofen and shared a few experiences but was generally quite reserved with his information.

I had several conversations with Herr Alois Heldmann (Jasta 10), who lived in Bad Aibling. He too knew Richthofen personally and shared his memories of the "Chief." I also contacted Franz Hemer (Jasta 6), who lived in Frankfurt, but our few conversations were limited to the telephone since his wife had strictly forbidden any visits from "aviation people." As a result, our conversations were conducted while she was out shopping. Since there were no hands-free systems available at the time, I had to pin the telephone receiver between my shoulder and ear and write down whatever he told me. He also sent me – in a "conspiratorial manner" – several photos from his time with Jasta 6. His son-in-law in the United States possessed his photo album, so I gave his address to an American aviation historian who promised to send me copies from the album. Although that historian indeed visited with the son-in-law and relatives, I never received a single photo copy.

On a rainy day in 1983, I visited Herr Eddy Lübbert at his wholesale fish company in Bremerhaven in northern Germany. He was the son of Friedrich-Wilhelm Lübbert and therefore the nephew of Hans-Georg "Eddy" Lübbert. His mother was the sister of Jasta 11's Carlos Meyer-Baldó, who returned to Venezuela after the war but always stayed in touch with and frequently visited his relatives back in Germany.

- 1 -

Ing.Kurt DELANG
Hochfeldstrasse 28
8900 AUGSBURG 1

Augsburg,d.7.April 1980.

Sehr geehrter Herr S c h m ä l i n g.

Herzlichen Dank für die Übersendung des ausgearbeiteten Tagebuches der Jasta 54. Ich entnahme daraus,da Sie die Daten wohl einem vorhandenen"Kriegstagebuch" entnommen haben. Sehen Sie - darüber stolpere ich! Möglicherweise wurde dieses Kriegstagebuch - am Ende des WW I - etwas frisiert! Damit Sie die Dinge richtig sehen,muß ich etwas umfangreicher berichten.

Ich habe nur zwei Luftsiege! Nummer eins war der Abschuß eines engl. Jagdfliegers an der engl. Ostküste,über See,brennend, am 12. August 1917,mit einer zweimotorigen Gotha Go.G. III. -
Der zweite Luftsieg war der Abschuß eines franz. Fesselballons bei der Jasta 54,am 26.Juli 1918.
Von einem dritten Abschuß weiß ich nichts!!!
Den ersten Luftsieg hatte ich bei meinem zweiten Englandflug. Bei meinem vierten Englandflug hat mich die englische Bodenabwehr so stark beschossen,daß ich -nachts - über Belgien,aus 1800 Meter Höhe, abstürzte. Ich konnte noch die Motore ausschalten,damit sie im Sturz erkalteten und das Flugzeug beim Aufschlag nicht brannte. Ich habe einige Zeit in einem Feldlazarett in Belgien gelegen und konnte nach meiner Entlassung die langen,sehr anstrengenden Flüge nicht mehr machen.Ich wurde anschließend,als Lehrer,bei der Geschwaderschule Paderborn eingesetzt von wo der Nachschub für das Englandgeschwader kam und hatte hier die neuen Besatzungen auf ihre Aufgaben beim Englandgeschwader vorzubereiten! Paderborn war aber ganz und gar keine Genesungs-Kompanie!

Am Weihnachtsabend 1917 wurde ich zum Kommandeur gerufen,der mir mitteilte,daß ich,auf Anforderung des Jagdgeschwaders Rittmeister Freiherr v. Richthofen,als Jagdflieger zum Jageri I versetzt sei,unter vorheriger Kommandierung zur Jagdstaffelschule Valenciennes!!
Am 1. Weihnachtsfeiertag war ich bei der Jagdstaffelschule,aber gesundheitlich leider noch schwer angeschlagen. Eine Woche später stand ich bereits,als fertiger Jagdflieger,vor dem Kommandeur des Geschwaders. Der Rittmeister Frhr.v.Richthofen war sehr aufgeschlossen.Er interessierte sich für alles,was ich in den vorangegangenen Kriegsjahren getan hatte und hatte Verständnis für meine gesundheitliche Lage. Ich wurde der Jasta 10 zugeteilt. -

Kurt Delang wrote a long typewritten report of his time with Jagdstaffel 10 and 54 for the author. The photo presents the first page.

Herr Lübbert was only able to provide a few photos to me, but the highlight of my visit were the letters that Hans-Georg sent home from November 1914 until his death in March 1917. Herr Lübbert told me that his father and Hans-Georg had a very close relationship and that his father had arranged for Hans-Georg's letters to be printed in a small book for the family after the war. I returned the next day and during our conversations Herr Lübbert indicated that his father had been persecuted during the time of the Third Reich. I also learned that Manfred von Richthofen's mother and his sister Ilse were guests of the Lübbert family for quite some time after they had fled Silesia in 1945.

During the 1960s, Alex Imrie was in contact with all of the surviving members of Jagdstaffel 11, including Willi Gabriel, who provided Alex with their addresses. Georg Simon, who served with the Staffel from November 1916 to June 1917, was among them. Alex's handwritten notes about their conversations were important for this book, especially since Herr Simon quite clearly stated that Manfred von Richthofen had arrived at Jasta 11 in a red Albatros from Jasta 2 'Boelcke'. Considering the amount of personal knowledge he had about the history of Jasta 11, it is very regrettable that he never incorporated it in a book.

In 1982, I visited the German Federal Air Force's Taktische Luftwaffengeschwader 71 Richthofen at Wittmund and was warmly welcomed there by Oberstabsfeldwebel Falk Hallensleben. He was very interested in my work and allowed me to visit the Geschwader's archives room. He also told me that a former Jasta 11 mechanic, Herr Hermann Lohmeyer, had recently contacted the Geschwader to provide it with copies of his photographs. Herr Lohmeyer's address was kindly given to me and I wrote to him shortly after returning home. I received a letter from him soon afterwards that included 15 beautiful photos. It also stated that he had joined the Jasta in January 1917, had served as Pour le Mérite pilot Kurt Wolff's mechanic for some time, and had remained with the Jasta until the end of the war. I quickly arranged to visit with him in Haldem, north of Osnabrück. There he told me that right after his arrival he had been assigned to Wolff as his third mechanic. When Wolff was transferred to take command of Jasta 29, Herr Lohmeyer became Otto Maashoff's second mechanic and then Eberhardt Mohnicke's first mechanic. After Mohnicke's transfer to the Idflieg in September 1918, Herr Lohmeyer served for the rest of the war as Rüdiger von Wedel's first mechanic. Herr Lohmeyer's recollections gave me interesting insights into the life of a Jagdstaffel mechanic, and his information regarding how the unit's aircraft were painted turned out to be somewhat of a gold mine. In him, I had finally found a contemporary witness to the Staffel's period before the formation of Jagdgeschwader I!

Personal contacts with other former members of Jagdstaffel 11 included Herren Otto-Lutz Förster, Alfred Niemz, Friedrich-August von Kockeritz, and Gottlieb von Conta, whose flight log I was able to copy; but those contacts will be discussed in a later volume. I have been in personal or written contact with a total of 19 former members of Jagdgeschwader I, and many of them such as Herren Alois Heldmann, Xaver Leinemüller, Kurt Delang, Franz Hemer, Oskar Rouselle, Heinz Graf von Gluszewski, and Egon Koepsch had personally known Manfred von Richthofen. The depiction of Manfred von Richthofen in this book is not only based on their direct testimony but also additional information provided by Alex Imrie's interviews.

During my interviews with these men, I always focused on three major topics:

1. What were the personalities of the individual Jasta members? In particular, what was Manfred von Richthofen actually like as a person, comrade, officer, and leader?
2. What was daily life in the Jasta like, including its missions and air combats?
3. How were the aircraft identified by Staffel and personal markings?

I was never interested in aircraft technology and am therefore grateful to Herr Reinhard Zankl and my co-author Jörn Leckscheid for sharing their in-depth knowledge in that field.

In addition to my direct contacts with former members of Jasta 11 and Jagdgeschwader I, several of their relatives provided documents and photos to me – in particular the von Döring and Döhl families. The latter gave me access to Emil Schäfer's small photo album.

During the course of my research, I learned that Kurt Wolff's handwritten diary was in a private collection. The owner, who wishes to remain anonymous, gave me complete access to it – and it turned out to be tremendously valuable in that it contained a great deal of previously unknown information that was recorded soon after the actual events.

My special thanks go to the owner because the Wolff diary, together with Eddy Lübbert's letters, those from Carl Allmenröder, and Emil Schäfer's book, form the living framework of this publication.

In addition, I have had the good fortune of being in contact with Herr Oskar Tinnerholm of Sweden. He knew Lothar von Richthofen's son, Wolf-Manfred, for many years and is an authority on Lothar. He has generously supported this and future Jasta 11 volumes with his documents and other information.

As with all of my books, I have drawn on the fundamental work of my mentor, Dr. Gustav Bock, who possessed the war diary information recorded by Erich Tornuß in the 1930s. Dr. Bock and his son, Winfried, painstakingly transcribed all of the Tornuß documentation and supplemented it with additional sources. Winfried Bock, through intense research and an evaluation of British sources, was able to identify many of Jasta 11's Allied opponents and shared that information with me. Without his work, this book would have been considerably less complete in its information.

Another valuable source for this book's research was

the Bayerischen Hauptstaatsarchiv (Abt. 4), in Munich. Its former director, Dr. Heil, informed Dr. Bock that the archives contained the so-called "Potsdam Inventory" of files from Army Group Crown Prince Rupprecht that were sent to Bavaria during the final turbulent days of World War II. They contain numerous documents concerning the aviation units and pilots deployed by the German 6th Army during World War I. The "Wöchentlicher Tätigkeitsbericht des Stabsoffiziers der Flieger der 6. Armee, 1916" ("6th Army Aviation Staff Officer's Weekly Activity Report, 1916") and the "Wöchentlicher Tätigkeitsbericht des Kommandeurs der Flieger der 6. Armee, 1917-1918" ("6th Army Aviation Commander's Weekly Activity Report, 1917-1918") were of particular value. The "Flugmeldebücher des A.O.K. der 6. Armee, Flieger-Nachrichten und Verfolgungsstelle Douai" ("6th Army High Command Aviation Report Books, Pilot Information and Tracking Center, Douai"), which reported on the observations of enemy aircraft and air combats, were also quite helpful, and contained information on the paint schemes of Jasta 11 airplanes. The wartime personnel rosters and records kept at Armee-Flug-Park 6, which contain biographical data on the pilots and observers transferred there, were equally valuable. Documents from the Bavarian Inspectorate of Military Aviation Aircraft, which included files from AFP 6, La Briquette (No. 36, 204, 205, and 206), were also discovered that provided lists of the aircraft delivered from AFP 6 to fighter units between October 1916 and June 1917. My friend and colleague, Reinhard Zankl, systematically reviewed their records and compiled a complete list of the aircraft supplied to Jasta 11 during that period.

Dr. Bock and I spent a lot of time at the Bayerischen Hauptstaatsarchiv during the 1970s. At that time, only so-called "Nasskopien" (wet process photocopies) were available that were very expensive. As a young student, I could only afford a few such copies and therefore had to laboriously copy many documents by hand.

Another of this book's interesting building blocks are the interrogation reports of British airmen who were forced down alive by Jasta 11 pilots. My friend and colleague, Reinhard Kastner, discovered these previously unpublished documents in the Bayerischen Hauptstaatsarchiv, copied them, and graciously shared them with me.

The German 6th Army documents in the Bayerischen Hauptstaatsarchiv probably offer the most extensive information on the deployment of the aviation units of a German Army during World War I. We were sometimes so overwhelmed by the wealth of information that we could not transcribe all of it into our records. Nevertheless, the period from October 1916 to the beginning of June 1917 is by far the best-documented period of Jasta 11's history thanks to that information.

What is most surprising is that the information available in the Hauptstaatsarchiv has barely been taken into account by previous publications focused on Jasta 11. Only the American aviation historian Peter Grosz transcribed some of the AFP 6 documents and made them available to Ed Ferko, who published their contents in his booklet titled *Richthofen*, but without citing the source.

Unfortunately, the Jagdstaffeln combat reports stored in the Reichs Archives in Potsdam were destroyed toward the end of World War II. Individual reports can now only be found among private records. Accordingly, to get a more complete picture of Jasta 11's combats, I turned to British records whenever possible. In this regard, my special thanks go to Trevor Henshaw, who was most helpful in locating British combat reports, casualty reports, and repatriated prisoner statements among the AIR1 files in London's National Archives.

In this context, I would like to refer the reader to Trevor Henshaw's *The Sky Their Battlefield II* (2nd revised edition) in which British air losses are presented in daily order. This 2nd edition with its new information is an irreplaceable contribution to the history of World War I air warfare and is highly recommended to every aviation enthusiast interested in the subject.

I have also received very important materials from my American friends and colleagues, Dr. Lance Bronnenkant and Greg VanWyngarden. In addition to extremely interesting photographs, they have also provided me with their own research results. Writing a book on your own in a quiet little room can be boring and frustrating, but the collaboration and exchange of information and thoughts that we have enjoyed has made the creation of this book a great deal of fun. I also owe Lance my gratitude for the excellent translation of my German manuscript into English.

Colin Huston and Colin Owers of the Great War Aviation Society (formerly known as *Cross & Cockade International*) have likewise provided me with photographs and documents.

In addition to books published in Germany between 1917 and the 1930s, I have consulted with Dr. Lance Bronnenkant's more recent biography of Oswald Boelcke (which is rightly considered to be the authoritative work on Boelcke) and his biographies of Manfred von Richthofen, Emil Schäfer, Kurt Wolff, Carl Allmenröder, and Lothar von Richthofen presented in his series, *The Blue Max Airmen* – all published by Aeronaut Books.

My friend and colleague, Jörn Leckscheid has been my irreplaceable research and discussion partner, just as he was with the *Jasta Colors* publications. In particular, his in-depth photo analyses and technical knowledge about the aircrafts added new dimensions to this work.

While working on the book, I received great support from Herr Oberstleutnant Reinhard "Kanzler" Schröder, Traditionsoffizier (tradition officer) of the Taktischen Luftwaffengeschwader 71 Richthofen of the German Air Force. The Geschwader had received the estate of Käte Oltersdorf, the nurse who took care of Manfred von Richthofen after he was wounded in July 1917. Her estate contained not only two very interesting albums with previously unknown photos, but also a diary that provides further insight into the personality of Manfred von Richthofen. My heartfelt thanks for the support!

The book was already in layout when I got the exiting news from Oberstlt. Schröder that copies of letters written by Kurt Wolff to his sisters and relatives had just come into his possession. I immediately asked Jack Herris to stop the layout. The copies were in a very hard-to-read condition, but with the support of Jörn Leckscheid they could be transcribed and used for the book, offering another unknown and unpublished insight for the history of Jasta 11.

I also received friendly support from Dr. Niklas Napp, who researched personnel documents in the Federal Archives in Freiburg for me. His book "*Die deutschen Luftstreitkräfte im ersten Weltkrieg*", published in 2017, is the first comprehensive academic publication in Germany on this topic and is recommended to any interested reader.

The intensive research that I undertook as a young man between 1972 and 1984 was an extraordinarily enjoyable, exciting, and highly interesting experience. It was more than worth all of the time, financial expenditures, and travel that was involved.

The most enjoyable and lasting memories of my research are the many personal conversations I had with former members of Jasta 11. It was fascinating to vicariously experience how my conversational partners mentally climbed once more into their Albatros, Pfalz, and Fokker fighters, soared up into the skies of western France or Belgium, flew over the Somme or Flanders, and encountered – during a senseless war's dogfights – young, enthusiastic pilots from "the other field post number" for whom they always had the greatest respect. Passing those experiences along and thus saving them from being forgotten has been my motivation for writing books about them in my free time.

I am now happy that, after so many years, I have been able to fulfill my promise to Herr Carl-August von Schoenebeck and Frau Helene Allmenröder to write the history of *Jagdstaffel* 11. This work has been more than just a book. It has been a personal and emotional obligation.

Bruno Schmäling, Spring 2024

Transcripts of Statements from Captured British Airmen

It would be beyond the scope of this work to publish in full all of the statements made by captured Allied airmen who, after being forced to land by Jasta 11 pilots, were interrogated by German intelligence officers. Therefore, only the six below will serve as examples.

Statements Made by the British Crewmen (Capt. O. Greig and Lt. J.E. MacLennan) Brought Down by Lt. Manfred von Richthofen on 25 January 1917:

"***1. Personal:***
1. Capt. Greig, 26 years old, born in London, career officer.
2. Lt. MacLennan, 20 years old, also a career officer who formerly belonged to the Scottish Rifles (8th Division) and has been with the RFC for about one year.
2. Machine: F.E.2 160 P.S. Motor.
3. Orders:
Reconnaissance flight with special task of taking photographs. Accompanied by six other aircraft which for some reason did not fly over the German lines.
4. Landing Cause: The engine had no fuel flow because the fuel line was destroyed by gunfire and gasoline flowed out of the tank which was also shot through.
5. Squadron:
Both prisoners belong to 25 Squadron, Auchel airfield; CO Major Cherry. 25 and 40 Squadron belong to the 10th Wing, whose leader until recently was Col. Herbert. His successor has not yet arrived – name unknown.
6. Air Combat:
The observer (MacLennan) started taking photographs and did not notice our [German] airplane's approach because it approached the enemy in a very clever manner. Intense fire immediately destroyed the fuel line and container. Greig was shot in the right ankle. He [MacLennan] couldn't fire because his machine gun failed to work.
7. Deutsche Defense:
Both prisoners praised our defensive measures. The English flyers particularly dread the defensive guns at Fampoux (east of Arras) and by Lille.
MacLennan claims to know that the defensive gun at Fampoux was on the Fampoux-Athies Road, but that it recently has not been at the same spot. He claims to have seen it on his most recent flight.
8. Ammunition:
P-ammunition (Phosphorus-ammunition) was included in a 1:4 ratio.
9. General:
Lt. MacLennan, who was with the Scottish Rifles (8th Division) on the Somme, admits that his battalion's losses were enormous – only two of his comrades are still alive.

An order has recently been issued to all airmen that in the event of capture, statements about squadrons, leaders, and airfields are strictly forbidden. A message recently dropped by us about a captured English pilot was said to have given the full address, squadron leader, etc.

Our airmen bombed a munitions depot at St. Omer and caused great damage. The camp is said to have burned for four days.

A German flyer landed at Tincques, east of St.Pol, around 1:00 p.m. on Sunday, 21 January. The observer inquired about the name of the place. When the pilot realized he was on enemy soil, he immediately ascended again without the observer, but then had to land intact again further to the east and was taken prisoner.

Very young captains are often found among aviation officers. This rank is awarded to particularly skilled officers, but only for the duration of the war.

Recently, one of our airmen is said to have given a signal to the English at night that the enemy was said to have responded to in kind. So our airman was deceived and landed among the enemy.

Zeppelins have caused significantly more damage in England than has been reported in the newspapers.

Captain Greig set his apparatus on fire after landing.

He described the maneuvering of our airplane in front of his apparatus before landing as careless because he certainly could have fired at him if his machine gun had been in order.

Both prisoners believe the war will end this year. No advance of any significance is possible for either party.

General war weariness is noticeable everywhere.

Tattenburg
Hauptmann and Intelligence Officer

Statements made by the English Aviation Officers Shot Down by Jasta 11 on 9 March 1917

Lt. T. Shepard
1. Personal:
Lt. T. Shepard, architect by profession, son of an active English Major, 5 brothers in the field of which 1 was killed, two wounded; joined the infantry when the war began, in the field for the first time on 15 August at Maricourt north of the Somme, took part in the Battle of the Somme in July 1916, entered RFC in August 1916, trained in England until January 1917, was then assigned to Army Aviation Depot Candas for three weeks to ferry aircraft there, has been with 40 Squadron for three weeks.
2. Aviation Associations:

Nachrichtenoffizier
A. O. K. 6.
B. Nr. 8139.

FE 2b 6997
25. Squadron
M. v. Richthofen

A.H.Qu., den 29.1.17.

6 bayr. R.D. 10 b
Ic № 2572.

This Page & 2 Following Pages: The original interrogation report of Capt. O. Greig and Lt. J.E. MacLennan. (Bayerisches Hauptstaatsarchiv, Abt. 4, München)

A u s s a g e n

von zwei am 25.1.17 (24.) bei Brebières südwestl. Douai abgeschossenen englischen Fliegeroffizieren.

1.) Persönliches : 1.) Capt. Graig, 26 Jahre alt, in London geboren, Berufsoffizier.

2.) Lt. Mac Lennan, 20 Jahre alt, ebenfalls Berufsoffizier, gehörte früher zum Scottish Rifles (8.Div.) und ist seit etwa einem Jahre beim R.F.C.

2.) Maschine : F.E.2 160 P.S.Motor.

3.) Auftrag : Erkundungsflug, mit dem besonderen Auftrage, Photographien anzufertigen. Er war von 6 anderen Flugzeugen begleitet, die jedoch aus unbekannten Gründen die deutsche Linie nicht überflogen.

4.) Ursache der Landung : Der Motor hatte keinen Benzinzufluss, weil die Leitung durch einen Schuss zerstört war und das Benzin aus dem gleichfalls zerschossenen Behälter floss.

5.) Squadron : Beide Gefangenen gehören dem 25.Squadron an.

Flugplatz : Auchel.

Führer der 25.Squadron : Major Cherry.- 25.Squadron gehört mit 40.Squadron zu 10.Wing, dessen Führer bis vor kurzem Col.Herbert. Sein Nachfolger ist noch nicht eingetroffen - Name unbekannt.

6.) Luftkampf : Der Beobachter (Mac Lennan) habe begonnen, Aufnahmen zu machen, und will das Herannahen unseres Fliegers nicht bemerkt haben, da er sich in sehr geschickter Weise dem Feinde näherte. Heftiges Feuer habe sofort die Benzinleitung und Behälter zerstört. Graig erhielt einen Schuss am rechten Fussgelenk. Da sein M.G.versagte, konnte er nicht feuern.

7.) Deutsche Abwehr : Beide Gefangene loben unsere Abwehrmassnahmen. Besonders sei das Abwehrgeschütz bei Fampoux (östl.Arras) und das bei Lille von den engl.Fliegern gefürchtet.

Mac Lennan will wissen, dass das Abwehrgeschütz bei Fampoux auf der Strasse Fampoux-Athies stand, aber

Ib

- 2 -

seit kurzem nicht mehr an der gleichen Stelle sei. Er will es gelegentlich seines jüngsten Fluges genau gesehen haben.-

8.) Munition : P-Munition (Phosphor-Munition) wird im Verhältnis 1 : 4 mitgeführt.-

9.) Allgemeines : Lt.Mac Lennan, der beim Scottish Rifles (8.Div.) an der Somme war, gibt zu, dass die Verluste seines Bataillons enorm gewesen seien.- Nur zwei seiner Kameraden wären noch am Leben.

Neuerdings ist ein Befehl an alle Flieger gegangen, dass bei Gefangennahme Aussagen über Squadron, Führer, Flugplätze auf das Strengste verboten sind. Vor kurzem soll eine Nachricht, dass ein englischer Flieger in Gefangenschaft geraten sei, mit der vollen Adresse unter Angabe von Squadron,Führer usw. von uns bei den Engländern abgeworfen worden sein.

Bei St.Omer habe einer unserer Flieger in der Nacht ein Munitionsdepot mit Bomben belegt und grossen Schaden angerichtet. Vier Tage hindurch soll das Lager gebrannt haben.

Am Sonntag,dem 21.1.gegen 1 Uhr Nachmittags sei bei Tinques östl. St.Pol ein deutscher Flieger gelandet. Der Beobachter habe sich nach dem Namen des Platzes erkundigt. Als der Führer bemerkte, dass er sich auf feindlichem Boden befände, sei er ohne den Beobachter sogleich wieder aufgestiegen, hätte aber weiter östl. wieder unversehrt landen müssen und sei in Gefangenschaft geraten.

Häufig findet man unter den Fliegeroffizieren sehr junge Hauptleute. Dieser Dienstgrad wird besonders gewandten Offizieren verliehen, aber nur für die Dauer des Krieges.-

Kürzlich soll einer unserer Flieger bei den Engländern nachts ein Signal abgegeben haben, was von dem Feinde in gleicher Weise erwidert worden sein soll, so dass unser Flieger getäuscht wurde und beim Gegner landete.-

Zeppeline hätten in England bedeutend grössere Schäden angerichtet als in den Zeitungen angegeben wurde.

Den Apparat habe Captain Graig nach der Landung angezündet.-

Das Manövrieren unseres Fliegers vor seinem Apparat kurz vor der Landung bezeichnet er als leichtsinnig, da er ihn mit Sicherheit hätte beschiessen können, wenn sein M.G. in Ordnung gewesen wäre.

Beide Gefangene sind der Ansicht, dass der Krieg noch in diesem Jahre zu Ende gehen werde.- Ein Vordringen von irgend welcher Bedeutung sei für keine der Parteien möglich.

Allgemeine Kriegsmüdigkeit mache sich überall bemerkbar.-

Hauptmann und Nachrichtenoffizier.

Verteilung :

Chef, O.Qu., I a/b, IIc, A 6,
Kofl. Koflak., Fl.N.S.St.,
Nachr.Abt., III b West, N.O.Heeresgr.N.O.Deutscher Kr.
N.O.1,2,4,Kr.N.St.Antw.
Sämtl.Gen.Kdos. & selbst.Div.

40 Squadron, CO Major Tillney (former CO Major Robert Lorraine, now Colonel and Wing Commander), three flights of F.E.8 single-seat biplanes. CO Major Tillney, only 21 years old, used to lead another squadron, was then a flight instructor at Salisbury Plain, and recently became a squadron CO again; an excellent officer aviator who still flies himself, e.g. to give one of his men a day off to take part in a football match; however, squadron COs are no longer allowed to fly over the front.

43 Squadron with Sopwith two-seaters belongs to the same Wing as 40 Squadron; prisoner claims not to know the other Wing squadrons.

Fienvillers and Candas (Candas is the train station) are two names for the same airfield. On the photograph presented to him, the prisoner identified the hangars for Army Aviation Depot Candas and 19 Squadron (which was established here for the first time).

3. Tactical Use of 40 Squadron

During the three weeks that the prisoner was with the squadron, only offensive and defensive patrol flights and escorting photography airplanes on combat patrols: La Bassée to north of Arras. On 6 March, the prisoner took part in a barrier flight near Roclincourt where several flights joined together into a formation of over 15 aircraft. When they returned, they were strongly reprimanded for not having attacked, despite their superior number, German combat pilots on the other side of the lines,

4. Orders on 9 March:

To fly an offensive patrol over the German lines in a flight group of nine airplanes. They were soon attacked by German Albatros single-seaters who had their way with them due to their superiority. Apparently, the prisoner's entire flight was annihilated. The F.E. single-seater is hopelessly inferior to the new Albatros single-seater that has suddenly appeared here.

5. Training:

The private flight schools in England have become less important than the military ones. There are 2 types of the latter.

2/Lt. G.J. Haseler

1. Personal:

2/Lt. G.J. Haseler, who joined the infantry when the war began, has been with 40 Squadron for a year now. The prisoner, who gives the impression of a very educated and thoughtful man, refused [to answer] other personal and military questions. His behavior while looking at photographs that were presented to him of Aire Airport seemed to confirm 40 Squadron's presence there.

2. Aviation Associations:

The German Air Arm's superiority since the appearance of the Albatros fighter was clearly emphasized by this prisoner. The English single-seaters were 9 against the Germans' 4 in his last combat, but the fight became hopeless from the onset.

2/Lt. W.B. Hills:

1. Personal:

2/Lt. W.B. Hills worked on a sugar plantation in South Africa (Zululand) before the war, took part in the Southwest Africa Campaign as a volunteer up to the end, was then in the 9th Hampshire Regiment; has been in the field as a pilot for 4 months.

2. Aviation Associations:

Aire Airport of 40 Squadron, part of 10 Wing, confirmed. The affiliation of 43 Squadron (Sopwith two-seater) to the same wing confirmed. 43 Squadron's task is photography and reconnaissance for which 40 Squadron will provide protection.

3. Orders on 9 March:

Two months ago, the squadron could have flown over German lines for hours without being attacked or disrupted. Since the appearance of the Albatros fighters, things have changed so much that on 6 March the pilots of 40 Squadron protested against the order to conduct an offensive patrol over German lines, feeling that their machines were inferior. But since the order had to be followed, things turned out as expected. The F.E. single-seater has completely inadequate climbing ability when compared to its opponent. One is lost once one is below the opponent.

Statement made by the English Aviation Officer Shot Down by Lt. Kurt Wolff, Jasta 11, on 31 March 1917.

Lt. A. Strange

Personal:

Lt. A. Strange, an engineer with the London Underground before the outbreak of the war, in an infantry regiment in 1915, went to England in August 1916 to train as a pilot at a military aviation school, and in mid-January 1917, after just two days with the St.Omer Aircraft Depot, was with his squadron as a pilot.

2. Aviation Associations:

11 Squadron, F.E. two-seaters, CO Major Atkinson.

Airbase: Izel lès Hameau, west airfield.

11 Squadron works with fighter squadron 60, infrequently with 29 Squadron.

Combat sector: Adinfer—Arras.

60 Squadron airbase: Hermaville.

14 Squadron was still at Auchel 14 days ago, CO Major Cherry. Also confirmed that 12 Squadron is at Avenes le Comte.

Use of 11 Squadron: photography, bombing.

A flight from 11 Squadron, accompanied by 3 Nieuports from 60 Squadron, flew to Douai on 25 March 1917 to photograph the airbase. The prisoner heard that they were surprised to find so many aircraft sheds, supposedly 37, in their photos. On the return trip, one of the accompanying Nieuports, manned by Lt. Gilbert, was shot down (whose aviation association has been determined). The formation from which Lt. Garnett's Nieuport was shot down on 30

Right & Folloing 3 Pages: The original interrogation report of Lt. T. Shepard, 2/Lt. G.J. Haseler und 2/Lt. W.B. Hills. (Bayerisches Hauptstaatsarchiv, Abt. 4, München)

hrichtenoffizier.
A. O. K. 6.
Nr. 8 7 1 0.

A. H. Qu., 12.3.1917.

Aussagen der am 9.3.17 von J.St. 11 abgeschossenen englischen Flieger-Offiziere.

Persönliches:

Oberleutnant T.Shepard, Architekt, Sohn eines aktiven engl.Majors; Vater und 5 Söhne im Felde, wovon 1 gefallen und 2 verwundet; gleich bei Beginn des Krieges bei der Jnfanterie eingetreten, August 15 zum ersten Male ins Feld nach Maricourt nördlich der Somme, später bei Roclincourt eingesetzt, im Juli 16 an der Sommeschlacht teilgenommen; August 16 ins R.F.C. eingetreten, Ausbildung in England bis Januar 17, dann für 3 Wochen zum Armee-Flugpark Candas zum Einfliegen von Maschinen kommandiert, seit 3 Wochen bei der 40.squadron an der Front.

Fliegerverband:

40.squadron, Kommandeur Major Tillney (früherer Kommandeur Major Robert Lorraine, jetzt Oberst und wing-commander), 3 flights von F.E. Einsitzer D.D. Der Kommandeur Major Tillney, erst 21 Jahre alt, führte früher eine andere squadron, war dann Fluglehrer in Salisbury Plain, seit kurzem wieder squadron commander; ein hervorragender Flieger, der noch selbst fliege, z.B. um einem seiner Herren einen freien Tag zur Teilnahme an einem Fussball Match zu verschaffen; allerdings sei den squadron commanders das Ueberfliegen der Front nicht mehr gestattet.

Zum selben wing wie die 40.squadron gehört noch die 43.squadron mit Sopwith Zweisitzern; die anderen squadrons des wing gibt Gef. nicht zu wissen an.

Fienvillers und Candas sind 2 Namen für denselben Flughafen (Candas ist die Bahnstation). Gef. bezeichnet auf dem ihm vorgelegten Bilde die Hallen des Armee-Flugparks und die der 19.squadron (die hiermit zum 1.Male dort sichergestellt wird.).

3.) Taktische Verwendung der 40.squadron:

Jn den 3 Wochen der Zugehörigkeit des Gefangenen zur squadron nur offensive und defensive Patrouillenflüge und Begleitung von Photographier-Flugzeugen. Gefechtsstreifen: La Bassée bis nördlich Arras. Der flight des Gefangenen war am 6.3. an einem Sperrflug bei Roclincourt beteiligt, wo sich nach und nach mehrere flights zu einem Geschwader von über 15 Flugzeugen zusammenballten. Sie wurden bei ihrer Rückkehr sehr getadelt, dass sie nicht trotz ihrer Ueberzahl die sich jenseits der feindl. Linien haltenden deutschen Kampfflieger angegriffen hatten.

4.) Auftrag am 9.3. :

Jn dem auf 9 Flzg. verstärkten flight-Verband über die deutschen Linien hinüber eine "offensive patrol" zu fliegen. Sie wären sehr bald von deutschen Albatros-Einsitzern angegriffen worden, die infolge ihrer Ueberlegenheit mit ihnen gemacht hätten, was sie wollte. Anscheinend sei das gesamte flight des Gef. aufgerieben worden. Gegen den neuen Albatros-Einsitzer, der auf einmal hier auftrete, sei der F.E.Einsitzer hoffnungslos unterlegen.

5.) Ausbildung :

Die privaten Flugschulen in England traten neben den militärischen sehr zurück. Von letzteren gäbe es 2 Arten, solche für Nicht-Angehörige der Armee, die erst in die Armee eintreten wollen, und solche für Angehörige der Armee, bes. für Offiziere. Gef. ist in einer Flugschule in Salisbury Plain unter seinem jetzigen Kommandeur Major Tillney, ausgebildet worden, und dann in den letzten Monaten selbst Fluglehrer gewesen. Die besten Flieger wähle man für Einsitzer aus, eine besondere Ausbildung aber genössen sie nicht, wie es denn eigentlich leichter sei, die leichten Kampf-Einsitzer als z.B. die schweren F.E. Zweisitzer zu fliegen.

6.) Ersatz :

Gef. ist direkt von der Flugschule Salisbury Plain zum Armee-Flugpark Candas versetzt worden. Jn Salisbury Plain gäbe es ausser der Flugschule noch eine reserve squadron, über deren Aufgaben Gef. nicht recht unterrichtet ist. Er meint, sie gäbe wohl ganze flights zur Front ab, bezweifelt aber, ob einzelne Maschinen oder Flieger von dort als Ersatz zu den Squadrons gesandt würden.

7.) Die Ueberlegenheit der deutschen Photographien wird unumwunden anerkannt. Das sei auch nicht zu verwundern, die Deutschen hätten eben ihre weit überlegenen Objektive. Wer in der engl. Armee ein Zeiss-Fernglas besitze, werde beneidet.

1.) Persönliches :

Unter-Leutnant G.J.Haseler, zu Beginn des Krieges bei der Jnfanterie eingetreten, seit 1 Jahre beim R.F.C., jetzt 40.Squadron. Sonstigen persönlichen wie militärischen Fragen weicht der, den Eindruck eines sehr gebildeten und überlegenden Mannes machende Gef. aus. Sein Verhalten beim Betrachten der vorgelegten Aufnahmen des Flughafens Aire scheint aber die Anwesenheit der 40.Squadron dort zu bestätigen.

2.) Die Ueberlegenheit der deutschen Fliegerwaffe seit dem Auftreten der Kampf-Albatrosse wird auch von diesem Gef. lebhaft betont. Bei seinem letzten Gefecht seien die engl. Kampf-Einsitzer 9 gegen 4 deutsche gewesen, aber der Kampf habe sich von vornherein für sie aussichtslos gestaltet.

3.) Den deutschen Rückzug an der Ancre erklärt sich der Gef. so, dass die Deutschen künftig immer einer Offensive durch Zurücknahme von Frontteilen ausweichen werden, um den Gegner zu Wochen und Monate dauerndem Neuaufmarsch zu nötigen.

4.) Ueber die Aussichten des U-Boot-Krieges gäben sich die Deutschen Täuschungen hin. Weder könnten die Deutschen in einem Monat so viel Schiffe versenken, als die Engländer neu bauten, noch könnten sie so viele U-Boote bauen, als die Engländer ausser Gefecht setzten. Wer insbesondere in den englischen Häfen die Anzahl von gefangenen, vollständig unbeschädigt erbeuteten U-Booten gesehen habe, könne nicht den geringsten Zweifel über den Ausgang haben. Jm Verhältnis

4.

zu dem riesigen Schiffsverkehr, der noch immer in den engl.Häfen vor sich gehe, seien die versenkten Schiffe belanglos.

III. 1.) Persönliches :

Unterleutnant W.B.Hills, vor dem Krieg in Südafrika(Zululand) in Zuckerplantagen tätig, hat als Freiwilliger den Feldzug in Südwestafrika bis zu seiner Beendigung mitgemacht, war dann im IX/Hampsh.R. und ist seit 4 Monaten als Pilot im Felde.

2.) Fliegerverband: 40.squadron, zum X.wing gehörig, Flughafen Aire bestätigt. Die Zugehörigkeit der 43.squadron(Sopwith Zweisitzer) zu dem gleichen wing bestätigt. Die Aufgabe dieser 43.squadron sei Photographieren und Erkundung, wozu 40.squadron Schutzflugzeuge stellt.

3.) Ausbildung: Die Ausbildung sei anfangs für alle Flieger gleich; während dieser Zeit erhielten auch alle Flugschüler denselben Unterricht in Materialkenntnis, Funken und Beobachten. Die besten Flieger würden dann weiter auf Kampfeinsitzer gesetzt. Besondere Flugschulen für Kampfeinsitzer scheint es nach den Ausführungen des Gefangenen nicht zu geben.

4.) Vor 2 Monaten hätte die squadron noch Stunden lang über den deutschen Linien fliegen können, ohne angegriffen und gestört zu werden. Seit dem Erscheinen der Kampf-Albatrosse habe sich das dermassen geändert, dass die Flieger der 40.squadron am 6.3. gegen den Auftrag, einen offensiven Patrouillenflug über die deutschen Linien zu machen im Gefühl der Unterlegenheit ihrer Maschinen, protestiert hätten. Da aber der Befehl befolgt werden musste, sei es gekommen, wie nicht anders zu erwarten war. Der F.E. Einsitzer besitze gegenüber diesem Gegner eine ganz ungenügende Steigfähigkeit. Sei man einmal unter diesem Gegner, dann sei man verloren.

5.) Bezüglich der U-Boote bestehen in England garkeine ernsten Befürchtungen. Man verstünde nicht, dass die Deutschen solche Hoffnungen darauf setzen könnten.

Verteilt nach

bes. Plan.

Kommandeur d. Flak. 6
Deutsche Feldpost Nr. 687
empf. 12.3.17

Hauptmann und Nachrichtenoffizier.

March 1917 was jointly provided by 11 and 60 Squadrons. Only one aircraft is equipped with a camera on such flights; the other F.E. and Nieuport aircraft are only for protection.
3. Orders for Last Flight:
To fly cover for one of the squadron's photography aircraft along with a third F.E.; Nieuports should have been there for cover too, but were not seen once again. The three aircraft obviously were flying low because of their orders. The Albatros' first shot seriously wounded the observer, Lt. W.G.T. Clifton. Since both of the aircraft's machine guns were mounted with the observer, the prisoner was therefore utterly defensless and had to land, particularly since his fuel tank was also shot through; his clothes were riddled with bullet holes.
4. Airplane:
The large F.E. is considered quite inferior. It is too slow and too cumbersome, flies 70 miles per hour. The Spad is considered the best and newest machine. Prisoner persistently inquired whether Lt. Baker's Spad machine had fallen into German hands undamaged.
A red Albatros is known among English aviators as particularly dangerous.
Some time ago, a Halberstadt that had gotten lost landed at an English airfield in the Savy region.
According to the interrogated, 29 Squadron will be equipped entirely with Nieuports; at present, maybe half [of their aircraft] are De Havilland scouts.
5. Very Lights:
They are named after their inventor, Very. They are red, white, and green flares that English aircraft use as signals.
6. German Flak:
Incomparably better than their English counterparts. Currently, there is an outstandingly good [German] flak gun in front of Arras. The prisoner's machine itself has been hit by it. Another airplane in the squadron was shot down by it. However, the occupants escaped with their lives. The prisoner was not able to discover exactly where the flak gun was located, but came under its fire just over Arras. There is also good flak near Adinfer.
7. The German railway system Cantin—Arleux—Marquion was handwritten on the map the prisoner carried.

Tattenburg [?]
Hauptmann and Intelligence Officer

...ichtenoffizier
A.O.K. 6.
Nr. 8994.

A. H. Qu., 4. 4. 17.

A u s s a g e n

des am 31.3.17 bei Gavrelle abgeschossenen englischen Fliegeroffiziers.

I. Persönliches :

Leutnant A. Strange, vor Kriegsausbruch Jngenieur bei der Londoner Untergrundbahn, Signalabteilung, im Jahre 1915 bei einem J.R. eingetreten, im August 16 zur Ausbildung als Flieger nach England in eine militärische Fliegerschule, seit Mitte Januar 1917, nach nur 2 tägigem Kommando zum Aircraft Depot St. Omer, bei seiner Squadron Pilot.

II. Fliegerverband :

11. Squadron. F.E. Zweisitzer. Führer : Major Atkinson.
Flughafen : Jzel le Hameau, westlicher Flugplatz.
Die 11. Squadron arbeitet mit der Jagd-Squadron 60, selten mit 29 zusammen. Gefechtsstreifen Adinfer – Arras.
Flughafen der 60. Squadron : Hermaville.
Die 25. Squadron lag vor 14 Tagen noch in Auchel, Führer war noch Major Cherry. Ebenso 12. Squadron in Avesnes le Comte bestätigt.
Verwendung der 11. Squadron : Photographieren, Bombenangriff.
Eine flight der 11. Squadron unternahm von 3 Nieuports der 60. Squadron begleitet, am 25.3.17 einen Flug nach Douai, um den Flughafen zu photographieren. Gef. hat gehört, dass man sehr erstaunt gewesen sei, auf der Aufnahme so viele Flugzeug-Schuppen, angeblich 37, zu finden. Auf dem Rück-weg wurde einer der begleitenden Nieuports mit Lt. Gilbert abgeschossen (dessen Fliegerverband damit erwiesen ist).
Auch das Geschwader, aus dem am 30.3.17 der Nieuport des Lts. Garnett abgeschossen wurde, ist von der 11. und 60. Squadron gemeinsam gestellt worden. Nur 1 Flugzeug ist bei diesen Flügen mit Kamera ausgerüstet, die anderen F.E. und die Nieuports dienen nur zum Schutz.

Left & Following 2 Pages: The original interrogation report of Lt. L. A. Strange. (Bayerisches Hauptstaatsarchiv, Abt. 4, München)

- 2 -

III. Auftrag beim letzten Flug :

Ein photographierendes Flugzeug der Squadron mit noch einem dritten F.E. zu decken. Auch Nieuports hätten zum Schutz da sein sollen, man habe sie aber wieder einmal nirgends gesehen. Die 3 Flugzeuge seien (sichtlich wegen des Auftrages) niedrig geflogen. Schon der erste Schuss des Albatros habe den Beobachter, Leutnant W.G.T.Clifton, schwer verwundet. Da beide M.G. des Flugzeuge beim Beobachter angebracht waren, sei Gef. somit ganz wehrlos gewesen und habe landen müssen, zumal auch sein Benzinbehälter durchschossen worden sei. Seine Kleider sind mehrfach von Kugeln durchlöchert.

IV. Flugzeug :

Der grosse F.E. gilt als sehr minderwertig (it is rotten) Er sei zu langsam und schwerfällig; fliegt 70 Meilen die Stunde. Als beste und neueste Maschine gilt der Spad. Gef. erkundigt sich angelegentlichst, ob die Spad-Maschine des Lts.Baker unversehrt in die Hände der Deutschen gefallen sei.
Ein roter Albatros ist bei den englischen Fliegern als besonders gefährlich bekannt.
Vor einiger Zeit sei ein Halberstädter, der sich verflogen hatte auf einem englischen Flugplatz in Gegend Savy gelandet.
Die 29.Squadron soll dem Vernehmen nach auch ganz mit Nieuports ausgerüstet werden; jetzt habe sie vielleicht zur Hälfte noch de Havilland Scouts.

V. Very Lights. Heissen so nach dem Erfinder Very. Es sind die roten, weissen, grünen Leuchtkugeln, die die englischen Flugzeuge als Signale benutzen.

VI. Deutsche Flaks. Unvergleichlich besser als die englisch Ein hervorragend guter Flak stehe jetzt noch vor Arras. Die Maschine des Gef. selbst sei von ihm schon getroffe worden. Ein anderes Flugzeug der Squadron sei von ihm abgeschossen worden. Die Jnsassen seien aber mit dem

\- 3 -

Leben davon gekommen. Wo der Flak genau steht, vermochte Gef. noch nicht zu entdecken, man gerate aber über Arras schon in sein Feuer. Ein sehr guter Flak habe auch bei Adinfer gestanden.

VII. Auf der mitgeführten Karte sind deutsche Bahnanlagen Cantin - Arleux - Marquion handschriftlich eingetragen.

Hauptmann und Nachrichtenoffizier.

eilung :

ef., Chef, O.Qu., Ia/b, M.S.O.,
I, Kofl.(40), Koflak., Fl.N.St., Art.N.St.,
u, Ara 6, Vernehm.Offz., Stoluft(8),
Kdos., Et.Jnsp., Kdtr.Lille,
b West, Nachr.Abt., Kogenluft,
Heeresgr.Kr.Rupprecht, N.O. 1,2,4,
.St.Antw., N.O.Berlin.

mandeur d. Flak. 6
6.4.17 Nr. 18835
Abt. I a I

1.) M.R. Flakgruppe 25. 10.4.17.

zu VI. ... 24.3.17. ...

Bibliography

Publication

Bodenschatz, Karl, Jagd in Flanderns Himmel, München 1935
Bowyer, Chaz, Royal Flying Corps Communiqués 1917-1918, 1998
Bronnenkant, Lance, Blue Max Airmen Vol. 5, Manfred von Richthofen, Aeronaut Books 2014
Bronnenkant, Lance, Blue Max Airmen Vol. 7, Bernert, Schäfer, Wolff, Aeronaut Books 2015
Bronnenkant, Lance, Blue Max Airmen Vol. 8, Gontermann und Lothar von Richthofen, Aeronaut Books 2016
Bronnenkant, Lance, Blue Max Airmen Vol. 9, Allmenröder, Brandenburg, Pechmann, Tutschek, Aeronaut Books 2017
Buddecke, Hans-Joachim, El Schahin – Der Jagdfalke, Berlin 1918
Chef des Generalstabes der Feldheeres, Weisungen über den Einsatz von Jagdstaffeln, Oktober 1917
Castan, Joachim, Der rote Baron, Deutschland 2016
Cole, Christopher, Royal Flying Corps 1915-1916, Chatham 1969
Cross & Cockade Great Britain Journal, Vol. 14, Leicester 1983.
Cross & Cockade, Vol. 1, Nb. 3, Society of World War I Areo Historians, USA
Ferko, A.E., Richthofen, Berkhamsted, 1995
Fiedler, Beate-Christine, Verfolgung und Ausgrenzung sogenannter „Mischlinge". Der Fall Friedrich Wilhelm Lübbert, in: Hamburger Schlüsseldokumente zur deutsch-jüdischen Geschichte, 22.09.2016.
Flieger-Ring-Nachrichtenblatt, Herausgeber Otto Fuchs 2. Jahrgang Nr. 8, 1. August 1922
Gibbons, Floyd, The Red Knight of Germany, New York 1927
Hoeppner von, Ernst, Deutschlands Krieg in der Luft 1914/1918, Berlin, Leipzig 1921
Imrie, Alex, German Fighter Units 1914 – May 1917
Imrie Alex, Pictorial History of the German Army Air Service, 1971
Knötel d.J., Herbert, Pietsch, Paul, Baron Collas, Das deutsche Heer, Friedensuniformen bei Ausbruch des Weltkrieges, Band I, II and III, Stuttgart 1982.
Marsa-Wona Taschenatlas vom westlichen Kriegsschauplatz, Königswartha, o.A.
Miller, James. F., Albatros D III, Oxford 2014
Miller, James. F., Inside the Victories of Manfred von Richthofen Vol. 1 Aeronaut Books 2016
Neumann, Georg, Die deutschen Luftstreitkräfte im Weltkrieg, Berlin 1920
Norman Franks, Hal Giblin, Nigel McCrery, Under the Gun of the Red Baron, London, 1998
Nowarra, Heinz, Eisernes Kreuz und Balkenkreuz, Mainz 1968.
Herris, Jack, Siemens-Schuckert Aircraft of World War I, Aeronaut Books 2014
Hoeppner von, Ernst, Deutschlands Krieg in der Luft, Leipzig 1921.
Henshaw, Trevor, The Sky Their Battlefield II (Second revised and extended edition Edition), Hertfordshire, 2014
Kölner Zeitung vom 22. April 1938
Paasche's Spezialkarten der Westfront, Blatt 2, Verlag Paasche und Lutz, Stuttgart, without year.
Pietsch Thorsten, Mein Lieber Hans - Feldbriefe einer Mutter 1914-1917, Germany 2006
Richthofen von, Manfred, Der rote Kampfflieger, Berlin – Wien 1917.
Richthofen, Ein Heldenleben, Berlin 1920
Richthofen von, Kunigunde, Mein Kriegstagebuch, Berlin 1937
Schäfer, Emil, Vom Jäger zum Flieger, Berlin 1918
Schmäling, Bruno, Leckscheid, Jörn, Jasta Color Volume 2, Aeronaut Books, USA 2023
Schmäling, Bruno, Bock, Winfried, Royal Prussian Jagdstaffel 30, Aeronaut Books, USA 2014
Schnitzler, E., Carl Allmenröder, der bergische Kampfflieger, Wald, 1927.
Stegemann, Hermann, Geschichte des Krieges, Band 4, Stuttgart/Berlin 1921
Schweckendieck, O., Der Kampfflieger Lothar Freiherr von Richthofen, Hamburg 1938.
Udet, Ernst, Kreuz wider Kokarde -„Jagdflüge des Leutnants Ernst Udet", S. 132, Berlin 1918.
VanWyngarden, Richthofens's Circus, Jagdgeschwader Nr. 1, Osprey Publishing 2011.
White, Andrew, Five-Step to Fokker Fodder, From the Trenches to the Red Baron, the First World War Diaries of William 'Jack' Lindsey, Fighting High, Ltd., 2019.
Zuerl, Walter, Pour le mérite-Flieger, München 1938
Die Woche 23, 9 June 1917
Die Woche Nr. 28, 14 July 1917
Die Woche Nr. 34, 25 August 1917

https://www.anciens-aerodromes.com
https://www.deutsche-biographie.de/pnd116777206.html,
https://www.pressreader.com/uk/iron-cross/20210331/281775631945724
https://de.wikipedia.org/wiki/Hans_Lübbert
https://cuxpedia.de/index.php?title=Lübbert_Hans_Julius

Archival Records

Bayerisches Hauptstaatsarchiv, Abteilung 4, München

A.O.K. der 6. Armee, Flieger-Nachrichten und Verfolgungsstelle Douai, Flugmeldebuch XV, XVI, XVII, Bestand AOK 6/318,
Wöchentlicher Tätigkeitsbericht des Stabsoffiziers der Flieger der 6. Armee, General der Pioniere 6 Band 58
Oberkommando der 6. Armee, Kraftfahr- und Fliegertruppe (WK) Bund 2150, Interrogation reports of captured allied crew members:
Capt. O. Greig
2/Lt. J.E. Mac Lennan
Lt. C.D. Bennet
Lt. T. Sheppard
2/Lt. W.B. Hills
2/Lt. G.J.Haseler
Lt. R.P. Backer
Lt. R.G. Gilbert
Flugzeuge Armeeflugpark 6, La Briquette, BHStA/KA ILUFT203,
Flugzeuge Armeeflugpark 6, La Briquette, 31.10.1916 / BHStA/KA München Akt Iluft 206
Flugzeuge Armeeflugpark 6, La Briquette, 31.12.1916 / BHStA/KA München Akt Iluft 205
Flugzeugbestand der Jagstaffel 11 in Douai 31.12.1916 / BHStA/KA München Akt Iluft 204
Flugzeuge Armeeflugpark 6, Pont a Chin, 28.2.1917 / BHStA/KA München Akt Iluft 205
Personal Records
Allmenröder, Carl, Kriegsstamm-Listen (KrStL.) 18031/173, Armee-Flug-Park 6
Baierlein, Josef Kriegsstamm-Rolle (KrStr.) 18153/17; 18039/626; 18076/71
Braumüller, Georg, KrStR_31421_BH18040-00029, Armee-Flug-Park 6,
Festner, Sebstian, KrStR. 18154/659; 17961/48
Howe Erwin, KrStR. 18038/454, Armee-Flug-Park 6
Kleinhenz, Karl, OP 25963
Krefft, Konstantin, OP 11364, Kr.StR. 17956/114
Lang, Rudolf, OP 15292
Lübbert, Hans-Georg, Eduard, KrStL. 32301/211, Armee-Flug-Park 6
Oesterreicher Albert, OP 1188
Prehn, Alfred, KrStR. 18035/402, and 18040/1184, Armee-Flug-Park 6
Richthofen von, Lothar, Kriegsstammliste 18031/267, Armee-Flug-Park 6,
Schäfer, Carl-Emil, KrStL. 18031/264, Armee-Flug-Park 6
Simon, Georg KrStL. 18031/208, Armee-Flug-Park 6
Wolff, Kurt KrStL. 18031/208, Armee-Flug-Park 6

Badisches Generallandesarchiv

OP Franz Anslinger, Badisches Generallandesarchiv, Karlsruhe

National Archives UK

AIR1 Files
Combat report 43. Sqn. RFC 06.03.1917
Combat report 16. Sqn. RFC 06.03.1917
Combat report 29. Sqn. RFC 09.03.1917
Combat report 25. Sqn. RFC 17.03.1917
Combat report 43. Sqn. RFC 17.03.1917
Combat report 25. Sqn. RFC 28.03.1917
The Secretary, Air Ministry: Repatriated or escaped Officer of war,

2/Lt. W B Hills, 40 Sqn.
2/Lt. G F Haseler, 40. Sqn
2/Lt. T A Shepard
Lt. Baker, 19. Sqn.
2Lt. C.G. Gilbert, 29. Sqn.
Lt. L. A. Stange, 11. Sqn.
Report on casualties to personal and machines when flying:
16. Sqn. RFC 06.03.1917
43. Sqn. RFC 21.03.1917
16. Sqn. RFC 21.03.1917

Stadtarchiv Solingen

Carl Allmenröder, postcard from 21.03.1917

Taktisches Luftwaffengeschwader 71 Richthofen

Niederhoff, Kurt, Jasta 11, photo album and personal notes
Oltersdorf, Käte, nurse at the Kriegslazarett 76 Kortrijk (Courtrai), photo albums, personal diary.
Wolff, Kurt, letters written to his sisters and relatives.

Privat Collections
Rainer Absmeier Collection

Various photos

Anonymous Owner

Wolff, Kurt, handwritten personal diary, Kampfstaffel 26, Kampfgeschwader VII, Jasta 11.

Winfried Bock Collection

Combat report 40. Sqn. RFC, summarized in the daily intelligence summary of the RFC "War dairy" from:
23. January 1917
4. March 1917
9. March 1917
30 March 1917
Combat report 2. Sqn. RFC:
4. March 1917
14. February 1917
Combat report 43. Sqn. RFC:
17 March 1917
Combat report 60. Sqn. RFC
30 March 1917
Identification of the opponents of Jagdstaffel 11, January 1917 – March 1917
Germany 6th Army, Compilation, and elaboration of the Kofl reports of the 6th Army, day by day.
Germany 4th Army, Compilation, and elaboration of the Kofl reports of the 6th Army, day by day.
Jagdstaffel 11, War diary excerpt by Erich Tornuß, revised and completed by Dr. Gustav Bock and Winfried Bock.
Jagdstaffel 11, tabular compilation of the air victories.
Jagdstaffel 12, War diary 12, by Erich Tornuß, revised and completed by Dr. Gustav Bock and Winfried Bock.
Jagdstaffel 4, War diary excerpt by Erich Tornuß, revised and completed by Dr. Gustav Bock and Winfried Bock.
Jagdstaffel 37, War diary excerpt by Erich Tornuß, revised and completed by Dr. Gustav Bock and Winfried Bock.

Lance Bronnenkant Collection

Various photos
Identification of British aircrafts shot down by Jasta 11
Information about the results of his research on Jasta 11
Identification of photos Jasta 11

Alex Imrie Collection

Allmenröder, Wilhelm, Jasta 11, correspondence in the 1960s
Bodenschatz, Carl, Adjutant Jagdgeschwader I, handwritten notes of the conversation in the 1960s
Groos, Gisebert-Wilhelm, Jasta 11, handwritten notes of the conversation in the 1960s
Heldmann, Alois, Jasta 10, handwritten notes of the conversation in the 1960s
Hemer, Franz, Jasta 6, handwritten notes of the conversation in the 1960s.
Osten von der, Hans-Georg, Jasta 11 und 4, handwritten notes of the conversation in the 1960s.
Simon, Georg, Jasta 11, handwritten notes of the conversation in the 1960s
Richthofen von, Manfred, English translation of his combat reports.
Richthofen von, Manfred, Geschwaderflüge, approx.1918

Reinhard Kastner Collection

Compilation of the biographic dates of Jasta 11 pilots
Compilation of the protocols of the interrogation officers of A.O.K. 6, AOK 6 / Bund 305 - 19/1,
Various photos

Thorsten Pietsch Collection http://www.frontflieger.de/

Hinsch, Hans, Photos,

Bruno Schmäling Collection

Allmenröder, Carl, Jasta 11, Photo album, Notes, Documents, Excerpts of letters to his fiancée Helene Kortenbach..
Allmenröder, Helene, Handwritten notes of the conversation 1975, 1976
Allmenröder, Wilhelm, Jasta 11, photo album, notes, documents.
Beaulieu-Marconnay, Heinz, Feldflieger-Abteilung. 2, Photo album, notes.
Bodenschatz, Carl, handwritten notes of the conversation 1978
Böning, Walter, letter to his parents 14. April 1918, handwritten notes of the conversation.
Deilmann, Carl, Letter to Dr. Heinz Penzlin, Newspaper „Die Welt",3. May 1978
Delang, Kurt, Jasta 10, photo album, letter to Bruno Schmäling, 7. April 1980, handwritten notes of the conversation, 1980
Döring, Kurt-Bertram, photos and notes, personal information from his widow Frau von Döring
Fuchs, Otto, Fliegerabteilung (A) 292, Jasta 11, handwritten notes of the conversation, 1975 – 1979
Groos, Gisbert-Wilhelm, handwritten notes of the conversation, 1976
Heldmann, Alois, Jasta 10, handwritten notes of the conversation, 1976
Hemer, Franz, Jasta 6, Handwritten notes of the telephone-conversation
Jacobs, Josef, Staffelführer Jagdstaffel 7, handwritten notes of the conversation and taped interviews 1975 – 1979
Jagdstaffel 11, War diary excerpt by Erich Tornuß, Copy of the original handwritten excerpt.
Leinemüller, Xaver, mechanic Jasta 10, photos, notes, handwritten notes of the conversation, 1976
Lohmeyer, Hermann, Mechanic Jasta 11, Photo album, Notes, Handwritten notes of the conversation 1982, Letter to Bruno Schmäling 17.02.1982
Lübbert, Hans-Georg, Eduard, war letters 1916 – 1917, Fight-log, photos, notes.
Lübbert, Friedrich-Wilhelm, photos, notes, memories passed on to his son Eddy Lübbert
Lübbert Eddy, Handwritten notes of the conversation 1982.
Müller, Oskar, Offz.z.b.V., Jasta 12, photo-album, notes, documents.
Osten von der, Hans-Georg, photos, handwritten notes of the conversation 1974 – 1975.
Rumpel, Theodor, Jasta 16 and 23, handwritten notes of the conversation.
Schoenebeck von, Carl-August, , photos, notes, Documents, handwritten notes of the conversation 1974 – 1979
Zacharias, Ludwig, correspondence 1974 - 1978

Michael Schmeelke Collection

Brauneck, Otto, wartime letters to his family, provided by his sister Frau Papendieck.
Brauneck, Otto, photos and handwritten notes
Müller, Oskar, Offz.z.b.V. Jagdstaffel 12, photo album, notes.
Brauneck, Otto, phots and handwritten notes.

Joachim Tinnerholm Collection
Report by Joachim von Alt-Stütterheim about Lothar von Richthofen
Report by Wolf-Manfred von Richthofen on the characterization of his father by his father's father-in-law, Robert Graf von Keyserlingk, his wife Margarete and his grandmother Kunigunde von Richthofen.
Report by Wolf-Manfred von Richthofen on the memories of former pilots of Jagdgeschwader I about his father.
Diary von Busso von Alvensleben, Jagdstaffelschule Valenciennes, Jasta 4 und Jasta 21.

Tobias Weber Collection http://www.buddecke.de/
Various photos

Dr. Hannes Täger Collection
Nachrichtenblatt der Luftstreitkräfte 1917 (Kopie)

Greg VanWyngarden Collection
Information about the results of his research on Jasta 11.
Identification of photos Jasta 11.
Identification of allied aircraft shot down by Jasta 11
Various British combat reports
Various photos

Ludwig Zacharias Collection
Allmenröder, Wilhelm, Handwritten notes of the conversation in the 1960is

Reinhard Zankl Collection
Compilation of the "Flugzeugbestandsnachweises Flugpark La Briquette und Pont a Chin" October 1916 – March 1917 form the documents in the Bayerisches Hauptstaatsarchiv, Abt. 4 München

Made in the USA
Las Vegas, NV
06 November 2026

57642759R00159